80

Selected Titles in This Series

200 **F. Dias, J.-M. Ghidaglia, and J.-C. Saut,** Mathematical problems in the theory of water waves, 1996

199 **G. Banaszak, W. Gajda, and P. Krasoń, Editors,** Algebraic K-theory, 1996

198 **Donald G. Saari and Zhihong Xia, Editors,** Hamiltonian dynamics and celestial mechanics, 1996

197 **J. E. Bonin, J. G. Oxley, and B. Servatius, Editors,** Matroid theory, 1996

196 **David Bao, Shiing-shen Chern, and Zhongmin Shen, Editors,** Finsler geometry, 1996

195 **Warren Dicks and Enric Ventura,** The group fixed by a family of injective endomorphisms of a free group, 1996

194 **Seok-Jin Kang, Myung-Hwan Kim, and Insok Lee, Editors,** Lie algebras and their representations, 1996

193 **Chongying Dong and Geoffrey Mason, Editors,** Moonshine, the Monster, and related topics, 1996

192 **Tomek Bartoszyński and Marion Scheepers, Editors,** Set theory, 1995

191 **Tuong Ton-That, Kenneth I. Gross, Donald St. P. Richards, and Paul J. Sally, Jr., Editors,** Representation theory and harmonic analysis, 1995

190 **Mourad E. H. Ismail, M. Zuhair Nashed, Ahmed I. Zayed, and Ahmed F. Ghaleb, Editors,** Mathematical analysis, wavelets, and signal processing, 1995

189 **S. A. M. Marcantognini, G. A. Mendoza, M. D. Morán, A. Octavio, and W. O. Urbina, Editors,** Harmonic analysis and operator theory, 1995

188 **Alejandro Adem, R. James Milgram, and Douglas C. Ravenel, Editors,** Homotopy theory and its applications, 1995

187 **G. W. Brumfiel and H. M. Hilden,** $SL(2)$ representations of finitely presented groups, 1995

186 **Shreeram S. Abhyankar, Walter Feit, Michael D. Fried, Yasutaka Ihara, and Helmut Voelklein, Editors,** Recent developments in the inverse Galois problem, 1995

185 **Raúl E. Curto, Ronald G. Douglas, Joel D. Pincus, and Norberto Salinas, Editors,** Multivariable operator theory, 1995

184 **L. A. Bokut′, A. I. Kostrikin, and S. S. Kutateladze, Editors,** Second International Conference on Algebra, 1995

183 **William C. Connett, Marc-Olivier Gebuhrer, and Alan L. Schwartz, Editors,** Applications of hypergroups and related measure algebras, 1995

182 **Selman Akbulut, Editor,** Real algebraic geometry and topology, 1995

181 **Mila Cenkl and Haynes Miller, Editors,** The Čech Centennial, 1995

180 **David E. Keyes and Jinchao Xu, Editors,** Domain decomposition methods in scientific and engineering computing, 1994

179 **Yoshiaki Maeda, Hideki Omoro, and Alan Weinstein, Editors,** Symplectic geometry and quantization, 1994

178 **Hélène Barcelo and Gil Kalai, Editors,** Jerusalem Combinatorics '93, 1994

177 **Simon Gindikin, Roe Goodman, Frederick P. Greenleaf, and Paul J. Sally, Jr., Editors,** Representation theory and analysis on homogeneous spaces, 1994

176 **David Ballard,** Foundational aspects of "non"standard mathematics, 1994

175 **Paul J. Sally, Jr., Moshe Flato, James Lepowsky, Nicolai Reshetikhin, and Gregg J. Zuckerman, Editors,** Mathematical aspects of conformal and topological field theories and quantum groups, 1994

174 **Nancy Childress and John W. Jones, Editors,** Arithmetic geometry, 1994

173 **Robert Brooks, Carolyn Gordon, and Peter Perry, Editors,** Geometry of the spectrum, 1994

172 **Peter E. Kloeden and Kenneth J. Palmer, Editors,** Chaotic numerics, 1994

(Continued in the back of this publication)

Mathematical Problems in the Theory of Water Waves

CONTEMPORARY MATHEMATICS

200

Mathematical Problems in the Theory of Water Waves

A Workshop on the Problems
in the Theory of Nonlinear Hydrodynamic Waves
May 15–19, 1995
Luminy, France

F. Dias
J.-M. Ghidaglia
J.-C. Saut
Editors

American Mathematical Society
Providence, Rhode Island

These proceedings issue from a meeting held at CIRM (International Center for Research in Mathematics) in Luminy, France, on May 15–20, 1995, to review recent developments in the mathematical theory of water waves.

1991 *Mathematics Subject Classification.* Primary 76B15, 35Q53, 35Q51, 76B25.

Library of Congress Cataloging-in-Publication Data

Mathematical problems in the theory of water waves : a workshop on the problems in the theory of nonlinear hydrodynamic waves, May 15–19, 1995, Luminy, France / F. Dias, J.-M. Ghidaglia, J.-C. Saut, editors.

p. cm. — (Contemporary mathematics, ISSN 0271-4132 ; 200)

Includes bibliographical references.

ISBN 0-8218-0510-X (alk. paper)

1. Wave-motion, Theory of. 2. Water waves—Congresses. I. Dias, F. (Frederic), 1962– . II. Ghidaglia, J.-M. (Jean-Michel) III. Saut, J.-C. (Jean-Claude) IV. Series: Contemporary mathematics (American Mathematical Society); v. 200.

QA927.M298 1996

532′.593′0151—dc20

96-22083

CIP

♾ The paper used in this book is acid-free and falls within the guidelines
established to ensure permanence and durability.

10 9 8 7 6 5 4 3 2 1   01 00 99 98 97 96

Dedicated to
the memory of
T. Brooke Benjamin

Contents

Preface

Stemming from a familiar physical problem, the theory of water waves leads to a variety of difficult mathematical issues, involving several domains of mathematics: partial differential equations, dynamical systems, hamiltonian systems, numerical analysis, etc.

These Proceedings are issued from a meeting held at the CIRM (International Center for Research in Mathematics) in Luminy, France (May 15–20, 1995), which gathered around 40 applied mathematicians and physicists. The contributions therein review some of the recent developments in the mathematical theory of water waves. In particular the following aspects are considered: modelling of various wave systems, mathematical and numerical analysis of the full water wave problem (the Euler equations with a free surface) and of asymptotic models (Korteweg–de Vries, Boussinesq, Benjamin–Ono, Davey–Stewartson, Kadomtsev–Petviashvili, etc.), existence and stability of solitary waves.

This volume is dedicated to the memory of T. B. Benjamin (1929–1995), whose contributions to the theory of water waves are deep, durable and connot be overestimated.

This meeting was made possible thanks to generous grants from the following French organisations: Ministère des Affaires Etrangères (Mission de l'Appui Scientifique et Technique), Ministère de l'Enseignement Supèrieur et de la Recherche (Mission Scientifique et Technique, DSPT1), Centre International de Rencontres Mathèmatiques (Luminy), Direction des Recherches et Etudes Techniques, Centre National de la Recherche Scientifique (Sciences Physiques pour l'Ingènieur).

An Appreciation of T. Brooke Benjamin (1929–1995)

On the eve of the commencement of this conference, Professor T. Brooke Benjamin was diagnosed with the cancer that claimed his life in August of 1995. It is appropriate for several reasons that this volume be dedicated to his memory.

First and foremost are Brooke's accomplishments in theoretical and experimental aspects of the study of fluid mechanics, including seminal work on water waves. Benjamin was among the very best scientists in the world in this active and important area, and held that place for most of the four decades his career spanned. Indeed, a reading of the corpus of his work puts him on a par with such earlier giants as G. I. Taylor and T. von Karman.

Secondly, though Brooke was not able to attend the conference in Luminy, I subsequently communicated the program to him. He read the abstracts from start to finish, and, in late July, we had a long conversation, which, in part, centered around the content of many of the lectures. As was his usual style, he was critically disposed toward my reporting of various of the ideas put forward in the meeting. Indeed, it was always a point in one's favour when Brooke took enough interest to comment. However, his overall opinion of the conference was that it must have been a huge success, especially as it brought together such a wide range of expertise and points of view.

Finally, Brooke's name appeared early and often in the lectures and discussions at the conference. His influence on the topics and the technical aspects of the Luminy conference was pervasive.

Benjamin hailed from Merseyside in the North of England. He read Electrical Engineering at Liverpool University not long after World War II. It is worth noting that he was apparently equally talented in his formal studies and in his musical pursuits. By the time he had reached the age of 20, he had written more than 100 complete pieces of music! A Rotary Club Fellowship took him to Yale University for a Master's degree, awarded in 1952. After his year's sojourn in Connecticut, during which he developed a life-long interest in American football, he matriculated in Cambridge University. He completed his Ph.D. there in 1955 under the direction of A. M. Binnie, and was immediately awarded a Fellowship in King's College. During this period, he wrote his first paper jointly with James Lighthill on periodic surface water waves and their application to understanding bore propagation. The paper has become a classic, and has been revisited many times by scientists all over the world, including Benjamin himself. In fact, one of his last published works settled a conjecture made some 40 years earlier in that paper. During the ensuing

15 years at Cambridge, Benjamin looked very deeply into important phenomena in fluid mechanics. His works during this period included novel experimental investigations of cavitation, especially its effect on solid boundaries, associated theoretical developments of stability of bubbles, his well-known theory of vortex breakdown, the famous Benjamin-Feir instability of periodic surface waves, the derivation and analysis of the so-called Benjamin-Ono equation, a comprehensive theory for flow in articulated pipes and, more generally, in containers with flexible boundaries, and his influential studies of flow of thin films down inclined planes. His work during this era of his life was recognized when, in 1966, he was elected a Fellow of the Royal Society of London, one of the youngest ever.

In the late 1960's, Brooke became enamoured of the powerful methods of nonlinear functional analysis being pioneered in France, Russia, and North America. He was especially intrigued by their potential for explication of general issues in fluid mechanics. The work of Krasovskii on periodic water waves, the work of Arnold on stability theory, the general theory of positive operators put forward by Krasnoselskii, the older work of Leray and Leray and Schauder, and a number of other works had a great influence on him during this period. He came to the view that fluid mechanics might benefit from a closer alliance of the classical theoretical and experimental studies with methods from modern mathematical analysis. This point of view accords very well with the philosophy shown by Dias, Ghidaglia, and Saut in organizing the present conference. Brooke noted this explicitly as he reviewed the conference abstracts with me in July 1995. Deciding that this point of view could best be developed away from a great center like Cambridge, Benjamin moved in 1970 to set up and become the director of the Fluid Mechanics Research Institute at the University of Essex.

Supported initially by a U.K. Science Research Council grant, Brooke's new center was already attracting world-wide attention in the early 1970's. As Brooke envisioned, it featured a first-class laboratory facility and a cadre of young experimentalists, theoreticians and mathematicians. In large measure because of Brooke's influence, this Institute saw an extraordinary range of collaboration among its members. In addition to putting together this new Institute, Brooke wrote several papers showing very clearly how well he had mastered some of the ideas from nonlinear analysis. His theory of stability of solitary waves was a seminal contribution, as was his work on conjugate flows in internal wave propagation. Later, he wrote a series of papers applying Leray-Schauder theory to problems of hydrodynamic stability. He followed these up with specific studies of steady flow problems for viscous fluids. While developing his skill with nonlinear analysis and its applications, he did not neglect the experimental side of the subject. While at Essex, he initiated a new laboratory study of the Taylor problem of flow between rotating, concentric cylinders and, in collaboration with Scott, he completed experiments on gravity-capillary waves in a narrow channel. He also supervised several students in experimental work at Essex. His choice of colleagues at Essex was good enough that some of the young folks Brooke had assembled were enticed elsewhere in due course. This should not have presented a problem since Brooke had an understanding with the Vice Chancellor and the Mathematics Department regarding continued support of his Institute; however, the support was not forthcoming. Eventually, the funding for the Institute fell below critical mass and Brooke moved on to Oxford.

In the fall of 1978, Brooke accepted the Sedleian Professorship of Natural Philosophy at Oxford, and a Fellowship at The Queen's College. He brought along Tom

Mullin at an early stage in his time at Oxford to help with setting up a laboratory facility, this time sited in the Physics Department. Benjamin and Mullin published an influential series of works on the Taylor flow, exposing aspects of the problem that had eluded earlier investigators. At a later stage, they also collaborated on work concerned with buckling instabilities of shear flows. At about the same time, Benjamin and a young post-doctoral fellow, Peter Olver, wrote a wonderful paper cataloging all the symmetries and associated conservation laws for the full, inviscid water-wave problem. He also brought an old fascination with soap films to a new level with a couple of very substantial papers with Cocker. Using pipe cleaners and washing-up liquid, Brooke on many occasions amused his friends and colleagues with his simple but effective demonstrations of the statics and dynamics of soap films. Building on his very early work, he also put forward a new Hamiltonian theory of bubble motion in an unbounded fluid. This latter paper was also partly an outgrowth of a long paper he wrote on one of his favorite themes, namely, the connection between impulse and flow force and the associated variational principles.

As he came up on the last decade of his career, Benjamin's range of inquiry stretched even further. For example, he worked on reflection of waves from ripply bottoms with Pritchard and Karakiewicz, and wrote a paper that was connected with work on wave-driven sediment motion and sand-bar formation, on a very general theory for the existence of solitary waves involving refined ideas from functional analysis with Bose and me, on the shape oscillations of bubbles, on gravity currents, more on Hamiltonian structure, and with Pritchard and Tavener on interesting low-Reynolds number free-surface flows. He revisited his theory of vortex breakdown, but with decidedly more sophisticated tools in his arsenal. Very recently, he put forward and partially analysed a new equation for solitary internal waves in situations where damping can be ignored, but surface tension cannot. He also wrote the aforementioned work on the Benjamin-Lighthill theory. During early 1995, he was hard at work on questions about damping of surface water waves in preparation for some experiments to be run at Penn State.

In 1991, Benjamin held forth in Orlando, Florida, at a National Science Foundation sponsored CBMS-lecture series. In a collection of ten lectures, he outlined his overall view of much of modern fluid mechanics. It was tough going for the audience, but there were many veins of gold revealed during this week-long conference. It is a great pity that Brooke never finished writing these lectures for publication.

Brooke Benjamin's work was formally celebrated many times during his productive career. As mentioned before, in 1966 he was elected a Fellow of the Royal Society of London and in the same year was presented with the L. H. Moody award by the American Society of Mechanical Engineers. In 1970, he received the Hopkins Prize from the Cambridge Philosophical Society. In 1989, a conference was convened at Penn State in honor of his 60th birthday. He was awarded honorary doctorates from the University of Bath, the University of Liverpool, and Brunel University. An accolade he very much appreciated was his election as a Foreign Member of the French Academy of Science in 1992. In 1994, Brooke was named Honorary Member of the American Society of Mechanical Engineers, their highest award.

Brooke supervised about 35 Ph.D. theses in his lifetime. Most of his students went on to substantial careers of their own. On the surface, he was tough and somewhat inaccessible to these students, but the real level of his concern for their development was always clear to his more senior colleagues. In addition, he acted very successfully as a mentor for many post-doctoral level scientists. A full accounting of those on whom he had direct influence, including students, young associates and collaborators, would surely number more than 100.

Throughout his career, Brooke was concerned in a very substantive way with science policy. This was already apparent during his Essex years, but in the 1990's, when British government policy turned away from the development of science, Brooke helped to mount the U.K. National Conference of University Professors as a countervailing force. He served as its first President, and was the author of several closely reasoned essays on why the then-current policy was going awry. His feelings about such matters ran very deep.

I conclude this appreciation of Brooke Benjamin with some personal notes. In addition to his love of music, which never flagged throughout his life, he was deeply interested in mechanical things, especially steam locomotives. During the last decade of his life, he was attached to Penn State as a permanent Adjunct Professor in addition to his Chair at Oxford. In this position, he visited Penn State each fall for periods from one to three months.

During his stays there, we often trekked around the countryside to find steam locomotives to view. To his great delight, we even found a few that were in working order. At Penn State, he also indulged his love of American football, becoming one of the football team's ardent supporters. He always expressed in many, many ways a deep love and commitment to his wife Natalia and his young daughter Victoria. Above all else, they were the light of his world. He will be missed for his brilliance and insight, and for his taste in choice of problems to confront. He will be missed also for his technical expertise, his sage advice, and for his rigorous criticism and holding himself and others to the very highest standards. At a more prosaic level, we will miss long, philosophical discussions with him that went on far into the night. Perhaps most, we will miss the first sight of his tall angular frame in the late morning, the broad smile that creased his face, and the outstretched hand when he would greet a good friend.

Jerry Bona
Austin, Texas
1 July 1996

List of Participants

Y. Agnon
Technion
Israel Institute of Technology
Haifa 32000
ISRAEL
cvragno@technion.technion.ac.il

T.R. Akylas
Dept of Mechanical Engineering
MIT
Cambridge, MA 02139-4307
USA
drblnc@mit.edu

J. Albert
Dept of Mathematics
University of Oklahoma
Norman, OK 73019-0001
USA
albert@nsfuvax.math.uoknor.edu

C. Bennett
Dept of Mathematics
University of South Carolina
Columbia, SC 29208
USA
bennett@math.scarolina.edu

J. Bona
Dept of Mathematics
Penn. State University
University Park, PA 16802
USA
bona@math.psu.edu

A. de Bouard
CNRS URA 760
Université Paris-Sud
91405 Orsay
FRANCE
bouard@lanors.matups.fr

T. Bridges
Dept of Math. and Comput. Sci.
University of Surrey
Guildford, GU2 5XH
UK
t.bridges@mcs.surrey.ac.uk

M. Chen
Dept of Mathematics
Penn. State University
University Park, PA 16802
USA
chen_m@math.psu.edu

P. Chossat
Inst. Non-Linéaire de Nice
UMR 129 CNRS - UNSA
06560 Valbonne
FRANCE
chossat@ecu.unice.fr

T. Colin
CEREMAB
CNRS URA 226 Univ. Bordeaux I
33405 Talence Cedex
FRANCE
colin@ceremab.u-bordeaux.fr

W. Craig
Mathematics Dept
Brown University
Providence, RI 02912
USA
craigw@math.brown.edu

F. Dias
Inst. Non-Linéaire de Nice
UMR 129 CNRS - UNSA
06560 Valbonne
FRANCE
dias@ecu.unice.fr

V. A. Dougalis
Mathematics Dept
National Technical University
15780 Zographou, Athens
GREECE
dougalis@theseas.ntua.gr

T. Fokas
Dept Mathematical Sciences
Loughborough Univ. of Technology
Loughborough, LE11 3TU
UK
a.s.fokas@lut.ac.uk

J.-M. Ghidaglia
CMLA
CNRS URA 1611 - ENS Cachan
94235 Cachan Cedex
FRANCE
jmg@cmla.ens-cachan.fr

R. Grimshaw
Dept of Mathematics
Monash University
Clayton, Victoria 3168
AUSTRALIA
rhjg@wave.maths.monash.edu.au

M. Hărăguş
Mathematisches Institut A
Universität Stuttgart
D-70569 Stuttgart
GERMANY
haragus@mathematik.uni-stuttgart.de

K.R. Helfrich
Woods Hole Oceanographic Inst.

Woods Hole, MA 02543
USA
khelfrich@whoi.edu

D. Henderson
Dept of Mathematics
Penn. State University
University Park, PA 16802
USA
dmh@math.psu.edu

A. Il'ichev
Mathematical Steklov Institute
Vavilova str. 42
117966 Moscow
RUSSIA
ilichev@mian.su

G. Iooss
INLN
UMR 129 CNRS - UNSA
06560 Valbonne
FRANCE
iooss@ecu.unice.fr

C. Kharif
IRPHE
163 Avenue de Luminy
13288 Marseille cedex 9
FRANCE
kharif@castor.univ-mrs.fr

K. Kirchgässner
Mathematisches Institut A
Universität Stuttgart
D-70569 Stuttgart
GERMANY
kirchg@mathematik.uni-stuttgart.de

E. Lombardi
Inst. Non-Linéaire de Nice
UMR 129 CNRS - UNSA
06560 Valbonne
FRANCE
lombardi@ecu.unice.fr

Y. Matsuno
Dept Physics, Faculty of Liberal Arts
Yamaguchi University
Yamaguchi 753
JAPAN
matsuno@ccyi.ccy.yamaguchi-u.ac.jp

C. C. Mei
Dept Civil and Environ. Eng.
MIT
Cambridge, MA 02139
USA
ccmei@mit.edu

D. Menasce
Inst. Non-Linéaire de Nice
UMR 129 CNRS - UNSA
06560 Valbonne
FRANCE
david@ecu.unice.fr

A. Mielke
Inst. für Angewandte Mathematik
Universität Hannover
Welfengarten 1, D–30167 Hannover
GERMANY
mielke@server.ifam.uni-hannover.de

T. Miloh
School of Engineering
Tel-Aviv University
Ramat, 69978 Tel Aviv
ISRAEL
miloh@eng.tau.ac.il

S. Pierini
Istituto di Meteo. e Oceanografia
Istituto Universitario Navale
Corso Umberto I, 174, 80138 Napoli
ITALY
pierini@naval.uninav.it

A. Ramamonjiarisoa
IRPHE
163 Avenue de Luminy
13288 Marseille cedex 9
FRANCE
fred@castor.univ-mrs.fr

J.-C. Saut
Mathématiques
Université Paris-Sud
91405 Orsay
FRANCE
saut@lanors.matups.fr

H. Segur
Program in Appl. Mathematics
University of Colorado
Boulder, CO 80309-0526
USA
segur@newton.colorado.edu

V. Shrira
Shirshov Inst. of Oceanology
Acad. Sci., 117218
Moscow, Krasikova 23
RUSSIA
shrira@glas.apc.org

E. Soewono
Dept of Mathematics
Institut Teknologi Bandung
Ganesha IO, Bandung 40132
INDONESIA
soewono@ma.itb.ac.id

M. Stiassnie
Dept of Civil Eng.
Technion
Haifa 32000
ISRAEL
cvrvvmm@technion.technion.ac.il

S.-M. Sun
Dept of Mathematics
Virginia Polytechnic Inst.
Blacksburg, VA 24061-0123
USA
sun@math.vt.edu

Li-y. Sung
Dept of Mathematics
University of South Carolina
Columbia, SC 29208
USA
sung@math.scarolina.edu

M. Tanaka
Dept of Applied Mathematics
Gifu University
1-1 Yanagido, Gifu 501-11
JAPAN
tanaka@cc.gifu-u.ac.jp

J. Toland
School of Mathematical Sciences
University of Bath
Claverton Down, Bath BA2 7AY
UK
j.f.toland@maths.bath.ac.uk

M. Tom
Dept of Mathematics
Louisiana State University
Baton Rouge, LA 70803-0001
USA
tom@marais.math.lsu.edu

B. van Groesen
Faculty of Applied Mathematics
University of Twente
7500 AE Enschede
THE NETHERLANDS
groesen@math.utwente.nl

J.-M. Vanden-Broeck
Mathematics Dept
University of Wisconsin-Madison
Madison, WI 53706
USA
vandenbr@sunset.huji.ac.il

M. Weinstein
Dept of Mathematics
University of Michigan
Ann Arbor, MI 48109
USA
miw@math.lsa.umich.edu

List of Lectures

Y. Agnon, M. Glozman : *Standing Stokes waves - Are they stable?*

T. Akylas : *Nonlinear coupling of long waves with short-scale oscillations*

J. Albert, J. Bona, M. Rose, J.-C. Saut : *Model equations for waves in stratified fluids*

J. Bona : *Two-way propagation of nonlinear, dispersive waves*

A. de Bouard, J.-C. Saut : *Solitary waves of generalized Kadomtsev-Petviashvili equations*

T. Bridges : *Reappraisal and validity of the Whitham modulation theory*

P. Chossat : *1:2 mode interaction with O(2) symmetry and water waves*

T. Colin, F. Dias, J.-M. Ghidaglia : *On rotational effects in the modulations of weakly nonlinear water waves over finite depth*

W. Craig : *Normal forms for water waves*

V. Dougalis : *Numerical approximation of blow-up of solutions of some nonlinear, dispersive wave equations*

A. Fokas : *The KdV equation on the quarter plane*

R. Grimshaw : *Solitary waves with oscillatory tails*

E. van Groesen : *Variational description of soliton-splitting during run-up*

M. Hărăguş, K. Kirchgässner : *Breaking the dimension of steady waves*

K. Helfrich, J. Pedlosky : *Weakly nonlinear and finite-amplitude isolated anomalies in baroclinic zonal flows*

D. Henderson, C. Chow, H. Segur : *A generalization of Hasselmann's stability result for resonant triad interactions*

A. Il'ichev : *Nonlinear wave-guides for resonant triads in the water with surface effects*

G. Iooss, P. Kirrmann : *Gravity–capillary solitary and cnoidal waves for an inviscid fluid of infinite depth*

C. Kharif, C. Skandrani, J. Poitevin : *Nonlinear evolution of water surface waves: the frequency down-shift phenomenon*

Y. Matsuno : *Forced Benjamin-Ono equation in a two-layer fluid system*

C. Mei, J.-J. Lee : *Stationary waves on an inclined sheet of viscous fluid at high Reynolds and moderate Weber numbers*

A. Mielke : *Mathematical treatment of sideband instabilities*

T. Miloh, A. Minzoni : *On the number of solitons for the ILW equation*

S. Pierini : *Coupled nonlinear wave models: applications to internal tides and tsunamis*

A. Ramamonjiarisoa, M. Joelson : *Random fields of water surface waves: a preliminary investigation*

H. Segur, J. Curry, B. Dubrovin, J. Hammack, D. MacCallister, N. Sheffner : *Waves in shallow water*

V. Shrira, S. Badulin, C. Kharif : *Model of water wave 'horse-shoe' patterns*

E. Soewono, S. van Gils : *Mode dynamics of perturbed Korteweg-de Vries equation*

M. Stiassnie : *On the equilibrium spectrum of gravity–capillary waves*

S.-M. Sun : *Solitary waves in stratified fluids of great depth*

L.-y. Sung : *Solving Davey-Stewartson equations by inverse scattering*

M. Tanaka : *On the crest instability of steep gravity waves*

J. Toland : *A plethora of solitary gravity–capillary water waves with nearly critical Bond and Froude numbers*

M. Tom : *On a generalized Kadomtsev-Petviashvili equation*

J.-M. Vanden-Broeck : *Numerical studies of solitary waves with vorticity and surface tension*

M. Weinstein : *Stability and instability in infinite dimensional conservative systems*

Contemporary Mathematics
Volume **200**, 1996

Radiation and Modulational Instability Described by the Fifth-Order Korteweg - De Vries Equation

Igor Bakholdin and Andrej Il'ichev

ABSTRACT. Some aspects of modulational instability for the fifth-order Korteveg - de Vries equation are considered. We discuss the character of instability of radiating wave tail propagating outwards the Korteweg - de Vries solitary wave core and also some features of interaction of two envelope solitary waves of small amplitude. The analysis of the interaction is performed within the framework of the Gorshkov - Ostrovskii - Papko asymptotic theory.

Introduction

To understand the basic features of complex wave processes in dispersive media it is natural to use the so-called model equations obtained as a result of derivation via some asymptotic procedure. For example, the celebrated Korteweg - de Vries equation, obtained as a small amplitude - long wave limit in a great number of different wave theories, supports the existence of solitons (solitary waves) - a very popular and attractive phenomenon in the nonlinear wave theory. In many cases (especially in water wave theories) the full system of equations also supports the existence of solitary waves, which shape in the corresponding limit is given by that one of the Korteweg - de Vries solitons (see, for example, [1],[2]). Solitary waves in this case appear as a result of bifurcation from a wavenumber zero of the linear wave spectrum. In other words, there exists a branch of the linear dispersion equation which curve lies at one side of the line tangent to it at the zero point for wavenumbers $k > 0$. Moreover, this line does not intersect any other dispersion curve. Consequently, dispersive systems which dispersion relation does not satisfy the mentioned property can not be described in any limit by the Korteweg - de Vries equation, and we should expect the appearance of new types of solitary waves, and this solitary waves do arise. For example, the existence of new types of solitary waves is given by the solutions of the fifth order generalization of the Korteweg - de Vries equation

$$\frac{\partial u}{\partial t} + u\frac{\partial u}{\partial x} + \frac{\partial^3 u}{\partial x^3} + s\frac{\partial^5 u}{\partial x^5} = 0, \qquad s = \pm 1 \tag{0.1}$$

1991 *Mathematics Subject Classification.* Primary 35Q12; Secondary 76B25.

The both authors are supported by Russian Fond of Fundamental Investigations Grant No. 93-013-17355; the second author is also supported by the International Science Foundation Grant No. MDM300.

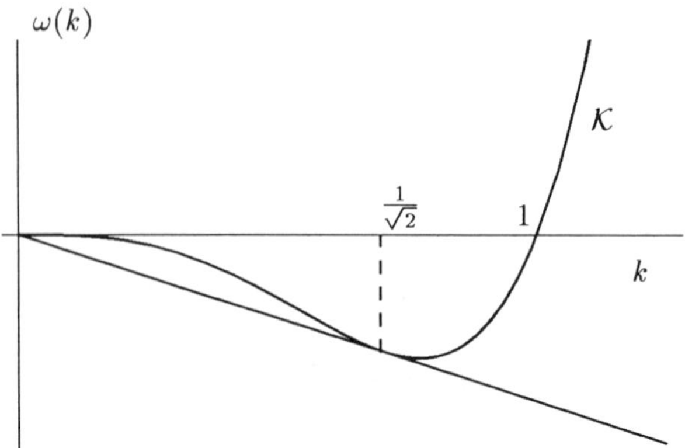

Fig. 1 The dispersion curve $\mathcal{K}$ given by (0.2) for s=1. The bifurcation from $k = 1/\sqrt{2}$ corresponds to the appearance of envelope solitary waves having the velocity $V = -1/4 + \mu$, with $\mu < 0$. The generalized solitary wave arises as a result of nonlinear resonance of a solitary wave with the velocity $V = \mu$, $\mu > 0$ and a periodic wave with $k > 1$.

which appears to be the model equation for a wide class of wave processes in dispersive media (see, for example, [3],[4]). The equation (0.1) is written here in the reference frame moving with the phase velocity of linear long waves and by the use of special scaled variables. The equation (0.1) besides the classical solitary wave of the Korteweg - de Vries type ($s = -1$) also posseses two different types of solitary waves for $s = 1$. All three types of solitary waves are common for nonlinear wave processes in dispersive media. Therefore, the effects in question, described by the equation (0.1) have the universal character.

The second type of a solitary wave (we further call it generalized solitary wave) arises as a result of nonlinear resonance of a solitary wave with a high wavenumber periodic wave. The possible appearance of a generalized solitary wave may be detected by the special property of a dispersion relation, namely, when there exists a branch of the dispersion relation and a line, tangent to the curve $\mathcal{K}$ of this branch at the zero point, which intersects only once for positive wavenumbers $k > 0$ the curve $\mathcal{K}$, or some other dispersion curve. When we have only one brunch of the dispersion relation, the dispesion curve must necessarily have the turning point to support the existence of the generalized solitary wave.

The third type of a solitary wave - the envelope solitary wave, appears as a result of the so called $1:1$ resonance [5], namely, in the case of the equality of group and phase velocities of a wave, corresponding to a nonzero wavenumber. When the dispersion relation has only one branch, the generalized and envelope solitary waves coexist (for different velocities). Moreover, the presence of modulational instability, which causes the appearance of envelope solitary waves may influence the radiation processes taking place in such systems.

The dispersion relation (see Fig.1) for the equation (0.1) is given by

$$\omega^* = -k^3 + sk^5 \tag{0.2}$$

As it was mentioned, the classical solitary waves, bifurcating from the quiescent state, exist for $s = -1$. Their global existence was proved in [6],[7], and nonlinear stability in [8]. In this case the unique dispersion curve $\mathcal{K}$ for $k > 0$ lies below the k - axis tangent to $\mathcal{K}$ at $k = 0$. The amplitude of solitary wave is positive, and the bifurcation, giving rise to the appearance of the solitary wave corresponds to the increase of the angle of inclination of the line, initially tangent to $\mathcal{K}$. For $s = 1$ the dispersion curve $\mathcal{K}$ have an intersection with the k - axis at $k = 1$ (see Fig.1). In this case the classical solitary wave does not exist, but for the speed $V = \mu > 0$, and μ small enough there exists a generalized solitary wave, bearing non-decreasing periodic tails at infinity. The dynamical system, describing travelling wave solutions of the equation (0.1) have the form in this case:

$$\begin{aligned} \dot{u} &= u_1 & \dot{u}_1 &= u_2 \\ \dot{u}_2 &= u_3 & \dot{u}_3 &= -u_2 - \frac{u^2}{2} + \mu u \end{aligned} \tag{0.3}$$

where μ - is the velocity of the travelling wave (considered to be small), the dot in (0.3) denotes the differentiation with respect to $\xi = x - \mu t$. The system of equations (0.3) may be written in the form

$$\dot{\mathbf{w}} = \mathcal{A}\mathbf{w} + F(\mu, \mathbf{w}) \tag{0.4}$$

where $\mathbf{w} = \{u, u_1, u_2, u_3\}^t$, $\mathcal{A}$ is the constant matrix

$$\mathcal{A} = \begin{pmatrix} 0 & 1 & 0 & 0 \\ 0 & 0 & 1 & 0 \\ 0 & 0 & 0 & 1 \\ 0 & 0 & -1 & 0 \end{pmatrix}$$

The eigenvalues of $\mathcal{A}$ - double zero and $\pm i$ lie on the imaginary axis. Make the following change of the unknown functions

$$\mathbf{w} = a_0(\xi)\varphi_0 + a_1(\xi)\varphi_1 + a_+(\xi)\varphi_+ + a_-(\xi)\varphi_-$$

where $\mathcal{A}\varphi_0 = 0$, $\mathcal{A}\varphi_1 = \varphi_0$, $\mathcal{A}\varphi_+ = i\varphi_+$, $\varphi_- = \overline{\varphi}_+$, $a_+ = \overline{a}_-$. The generalized eigenvectors of the matrix $\mathcal{A}$ are given by $\varphi_0 = (1,0,0,0)^t$, $\varphi_1 = (0,1,0,0)^t$, $\varphi_+ = (1, i, -1, -i)^t$, and the system of equations (0.3) takes the form in new variables

$$\begin{aligned} \dot{a}_0 &= a_1 \\ \dot{a}_1 &= \mu a_0 - \frac{a_0^2}{2} - a_0(a_+ + a_-) - \frac{(a_+ + a_-)^2}{2} + \mu(a_+ + a_-) \\ \dot{a}_+ &= ia_+ + \frac{i}{2}[-\mu a_0 - \mu(a_+ + a_-) + \frac{1}{2}a_0^2 + a_0(a_+ + a_-) + \frac{(a_+ + a_-)^2}{2}] \\ \dot{a}_- &= -ia_- - \frac{i}{2}[-\mu a_0 - \mu(a_+ + a_-) + \frac{1}{2}a_0^2 + a_0(a_+ + a_-) + \frac{(a_+ + a_-)^2}{2}] \end{aligned} \tag{0.5}$$

$u = a_0 + a_+ + a_-$. The advantage of the mentioned change of variables consists in the decomposition of the unknown function u into the long-wave part a_0 of order μ and the high wavenumber part a_+, having the order μ^2. Further make the scaling transformations

$$a_0 = \mu\alpha_0(\zeta), \quad a_1 = \mu^{3/2}\alpha_1(\zeta), \quad a_+(\zeta) = \mu^2 z(\xi), \quad a_- = \mu^2\overline{z}(\xi), \quad \zeta = \sqrt{\mu}\xi$$

The system of equations (0.5) is equivalent to the system (prime denotes the differentiation with respect to ζ):

$$\begin{aligned} \alpha_0' &= \alpha_1 \\ \alpha_1' &= \alpha_0 - \frac{\alpha_0^2}{2} + O(\mu) \\ \frac{dz}{d\xi} &= iz + \frac{i}{2}\left(\alpha_0 - \frac{\alpha_0^2}{2}\right) + O(\mu) \\ \frac{d\overline{z}}{d\xi} &= -i\overline{z} - \frac{i}{2}\left(\alpha_0 - \frac{\alpha_0^2}{2}\right) + O(\mu) \end{aligned} \tag{0.6}$$

It follows immediately from (0.6) that at the lowest order in μ, the first pair of equations (0.5) have the solution $a_0^* = 3\mu\ \cosh^{-2}(\sqrt{\mu}\xi/2)$. To compute a_+ at the lowest order introduce $\alpha = a_+ + a_-$, $\beta = a_+ - a_-$. Using the smallness of the high wavenumber part of the generalized solitary wave in comparison with the long-wave part we obtain from (0.5) up to the terms of order μ^3

$$\ddot{\alpha} + \alpha - a_0^*\alpha = \mu a_0^* - \frac{1}{2}(a_0^*)^2 \tag{0.7}$$

The homogeneous equation for $\alpha = a_+ + a_-$ corresponding to (0.7) may be solved by the WKBJ method up to the terms of arbitrary order in μ. Make the ansatz

$$\alpha = exp\left(\int \varphi d\zeta\right)$$

where

$$\varphi = \varphi_0 + \sqrt{\mu}\varphi_1 + \mu\varphi_2 + \dots \tag{0.8}$$

Consdering (0.8) up to the terms of order $\sqrt{\mu}$ we get from (0.7) $\varphi_0 = \pm i$, $\varphi_1 = 0$,

$$\alpha = 2(C\sin\xi + C_1\cos\xi)$$

where C and C_1 are real arbitrary constants (if considering the terms of order μ in (0.8) we obtain the phase shift: $\alpha = 2\{C\sin[\xi - 3\sqrt{\mu}\tanh(\sqrt{\mu}\xi/2)] + C_1 cos[\xi - 3\sqrt{\mu}\tanh(\sqrt{\mu}\xi/2)]\}$). The function β is determined from the equation $\dot{\alpha} = i\beta$ and is given by (at the lowest order in μ)

$$\beta = -2i(C\cos\xi - C_1\sin\xi)$$

Substi tuting $a_+ = 0.5(\alpha + \beta)$ into the third equation (0.5) and solving it by the method of variation of the arbitrary constant we get the expression for the odd function C

$$C = \frac{1}{2}\int_{-\xi}^{\xi}\left[\left(\frac{a_0^{*2}}{4} - \frac{\mu a_0^*}{2}\right)\cos\xi\right]d\xi \tag{0.9}$$

and for the even function C_1

$$C_1 = \int_{-\infty}^{\xi}\left[\left(\frac{a_0^{*2}}{4} - \frac{\mu a_0^*}{2}\right)\sin\xi\right]d\xi \tag{0.10}$$

It can be easily seen from the expressions (0.9), (0.10), that $C_1(\infty) = 0$,

$$2C(\infty) = \int_{-\infty}^{\infty} \left[\left(\frac{a_0^{*2}}{4} - \frac{\mu a_0^*}{2} \right) \cos\xi \right] d\xi \neq 0$$

and it is proportional to $\exp[-\pi/\sqrt{\mu}]$ at the lowest order in μ (the last integral can be computed using the contour integration in the low half plane). The precise numerical value of the constant of proportionality D at the zeroth order in μ can be computed taking into account the terms of all orders in μ in the second and third equations in (0.5) (we suppress this analysis here). The approximate value of the constant D ($\approx 119.82\pi$) was computed for the first time in [9] using the method of the matched asymptotic expansions. Hence, the asymptotics of the symmetric travelling wave - the generalized solitary wave is given by

$$u = D \exp\left(-\frac{\pi}{\sqrt{\mu}}\right) \sin x + O\left\{\sqrt{\mu}\exp\left(-\frac{\pi}{\sqrt{\mu}}\right)\right\}, \quad as \;\; x \to \infty$$

The constant D was computed also in [10] by the asymptotic analysis in the spectral k - plane. Boyd [11] confirmed the results of Pomeau *et al* [9] by using the precise numerical technique. It was also pointed out by Benilov *et al* [12] that the existence of nonsteady assymetric generalized solitary waves is impossible. This conclusion means that any localized solution must radiate energy in the form of oscillating periodic wave tails. For small μ the amplitude in the lowest order equals to two amplitudes of a steady periodic component of the symmetric generalized solitary wave [12]. Consequently, the time of the decay of a solitary wave of small amplitude is exponentially lardge. When the amplitude of solitary wave is moderate, the amplitude of the radiating tale is sufficiently lardge, as was shown by the numerical experiments in [12], and , therefore, the process of the decay of the initial solitary wave may be effected by the modulational instability which is present in the system.

The dispersion curve for $s = 1$ in (0.1) also has $1 : 1$ resonance point (at $k = 1/\sqrt{2}$), i.e. there exists a point on the dispersion curve where the phase speed is equal to the group speed (see Fig. 1). This resonance corresponds to the appearance of envelope solitary waves with velocities less than the group velocity $V_0 = -1/4$. In [13], [14] it was shown that steady envelope solitary wave solutions of (0.1) exist as a result of modulational instability which takes place for wavenumbers k in the vicinity of $k = 1/\sqrt{2}$. To explain this make a decomposition

$$u = (\epsilon\Psi(T,X)\exp[i\phi] + c.c) + (\epsilon^2\Psi_1(T,X)\exp[2i\phi] + c.c) + \epsilon^2\Psi_0 + ...,$$
$$\phi = -\omega t + kx, \quad T = \epsilon t, \quad X = \epsilon x \tag{0.11}$$

where Ψ_0 is a real function representing the mean flow. It was shown in [14] that for waves weakly depending on T in the reference frame moving with the group velocity ω', at the lowest order in ϵ the Schrödinger equation holds:

$$i\left(\frac{\partial\Psi}{\partial T} + \omega'(k)\frac{\partial\Psi}{\partial X}\right) + \epsilon\frac{\omega''(k)}{2}\frac{\partial^2\Psi}{\partial X^2} - \epsilon\Delta|\Psi|^2\Psi = 0 \tag{0.12}$$

where $\omega(k) = \omega^*(k)$ is given by (0.2) for $s = 1$ and

$$\Delta = \Delta^* = \frac{25k^2 - 3}{6k(-3+5k^2)(1-5k^2)}$$

The prime in (0.12) denotes the differentiation with respect to the wavenumber k. In [13] it was shown for the first time, that the non-steady wave solution given by (0.11) for $\Psi = \Psi(X - VT)$, $\Psi \to 0$ for $X \to \pm\infty$, obeying (0.12) become steady for $k = 1/\sqrt{2}$, where phase and group velocities of a monochromatic wave propagating under the envelope are equal. The form of the envelope solitary wave solution of (0.1) is described by the solitary wave solution of (0.12) for $k = 1/\sqrt{2}$ at the lowest order in ϵ, as it was shown in [14]. The result of the mentioned paper may be easily extended to the general case. Assume, that for the general reversible dispersive media with dispersion relation having at least one branch, where there exists a point $k = q$ where the group velocity equals to the phase one we have the equation (0.12) for the envelope amplitude of the lowest mode. Then the steady solitary wave solution of (0.12) has the form

$$\Psi = \Phi(X - VT)\exp(i\epsilon\rho qT), \quad V = \omega^{'}(q) + \rho\epsilon^2, \quad \rho = -sgn(\omega^{''}(q))$$

The amplitude Φ at the lowest order in ϵ satisfies the equation

$$\ddot{\Phi} = -\frac{2\rho q}{\omega''(q)}\Phi + \frac{2\Delta(q)}{\omega''(q)}\Phi^3 \tag{0.13}$$

where dot denotes the differentiation with respect to X. The well known solitary wave solution of (0.13) takes place in the presence of modulational instability: $\Delta(q)\omega^{''}(q) < 0$. We consider here the dispersion media where the dispersion relation have the additional property: the line passing through $k = 0$ and $k = q$ does not intersect any dispersion curve elsewhere. For a wide class of nonlinearities, differential equations, describing the steady one dimensional wave propagation in such media, may be approximated by the normal form system for the case of 1:1 resonance [15], [5] (dot denotes the differentiation with respect to $\xi = x - Vt$, V being a travelling wave velocity):

$$\begin{aligned} \dot{A} &= iqA + B + iAP[\mu, |A|^2, K] \\ \dot{B} &= iqB + iBP[\mu, |A|^2, K] + AQ[\mu, |A|^2, K] \end{aligned} \tag{0.14}$$

where P and Q are real polynomials:

$$\begin{aligned} P &= p_1\mu + p_2|A|^2 + \ldots \\ Q &= q_1\mu + q_2|A|^2 + \ldots \end{aligned}$$

and $K = i/2(A\overline{B} - \overline{A}B)$ is the first integral of (0.14) and $\mu = \omega^{'}(q) - V$. Introduce the new variables

$$A = r_0 \exp i(q\xi + \theta_0), \quad B = r_1 \exp i(q\xi + \theta_1) \tag{0.15}$$

As it was indicated in [5], the homoclinic orbits of (0.14) take place for $q_2 < 0$ and $K = 0$. The last equality implies $\theta_0 = \theta_1$ and the system (0.14) takes the form

$$\begin{aligned} \dot{\theta}_0 &= P(\mu, r_0^2, 0) \\ \dot{r}_0 &= r_1 \\ \dot{r}_1 &= r_0 Q(\mu, r_0^2, 0) = \mu q_1 r_0 + q_2 r_0^3 + \ldots \end{aligned} \tag{0.16}$$

The solution of the system (0.16) at the lovest order in μ is given by

$$
\begin{aligned}
r_0 &= \pm\sqrt{2|\mu q_1 q_2|^{-1}}\cosh^{-1}(\sqrt{|q_1\mu|}\xi)\\
r_1 &= \mp q_1\mu\sqrt{2|q_2|^{-1}}\sinh(\sqrt{|q_1\mu|}\xi)\cosh^{-2}(\sqrt{|q_1\mu|}\xi)\\
\theta_0 &= p_1\mu\xi + 2\frac{p_2}{q_2}\sqrt{|q_1\mu|}\tanh(\sqrt{|q_1\mu|}\xi)
\end{aligned}
\tag{0.17}
$$

The comparison of the last two equations in (0.16) with (0.13) yields

$$
\mu = \rho\epsilon^2, \quad \Phi = \epsilon r_0, \quad q_1 = -2\frac{q}{\omega''(q)}, \quad q_2 = 2\frac{\Delta(q)}{\omega''(q)}
\tag{0.18}
$$

In [5] it is shown that homoclinic orbits of (0.14) given by (0.15), (0.17) persist for the full system. This fact and the expression for q_2 in (0.18) mean that steady envelope solitary waves in the case of 1:1 resonance always appear as a product of the modulational instability.

The other interesting problem of the modulational instability for (0.1) is the interaction of the products of such an instability - envelope solitary waves. It is known that solitary waves, having an oscillatry structure of fronts (with damped oscillations) may form the so-called coupled state (breather) of two solitary waves, oscillating around the common center. For example, as follows from the asymptotic analysis given in [16],[17], two solitary waves of the equation

$$
\frac{\partial u}{\partial t} + u\frac{\partial u}{\partial x} - \frac{\partial^5 u}{\partial x^5} = 0
$$

of arbitrary (even small) and approximately equal amplitudes do form a breather when interacting. This theoretical conclusion was justified by the authors of [16] in their experiments with electric circuits. Moreover, when the distance between two solitary waves of equal amplitudes are precisely equal to the critical value, the period of oscillations tends to zero and a new solitary wave with two humps appears (Boyd [17] call it "bion"). The possibility of the formation of new steady solitary waves leads to the conclusion about the nonuniquess of solitary wave, i. e. the existence of a finite (or infinite) number of solitary waves having the same velocity. It is known [6] that for envelope solitary wave solutions of the equation (0.1) of moderate amplitude we have an infinite number of such waves with the same velocity. Yet, there exist only two solitary waves (of elevation and depression) corressponding to small amplitudes and velocities. In the present paper we demonstrate that in spite of the mentioned features of interaction of solitary waves with oscillatory fronts, the envelope solitary waves of the equation (0.1) of small amplitude interact without the formation of breathers.

The present paper is organized as follows. In section 1 we give the analysis of the modulational instability of radiating waves, outcoming from the solitary wave core, and prove the convective character of this instability in the reference frame moving with solitary wave core, which means that the process of radiation is always quasi-stationary. We also present some numerical results, describing the radiation process. Section 2 is devoted to the analysis of the interaction of envelope solitary waves via the Gorshkov-Ostrovskii-Papko asymptotic theory.

1. Character of the instability of radiating oscilatory waves

For the equation (0.1) ($\omega(k) = \omega^*(k)$ and $\Delta(k) = \Delta^*(k)$) the modulational instability takes place for $k \in (0, \sqrt{3}/5) \cup (1/\sqrt{5}, \sqrt{3/5})$. It means that a periodic wave, corresponding to $k > 1$ which is radiated by a solitary wave core is stable. In the general case of the media, which dispersion relation behaves itself like that one of (0.1), the equation (0.12) supports the instability for k in a sufficiently small vicinity of k_0 when $\omega''(k_0)\Delta(k_0) < 0$. Here k_0 denotes the point of the unique intersection of the line tangent to a certain dispersion curve $\mathcal{K}$ at $k = 0$ with $\mathcal{K}$ or any other dispersion curve. This means that radiated periodic waves must be unstable.

First make the transformation $\Psi \to \Psi/\epsilon$ in (0.12) to get

$$i\left(\frac{\partial\Psi}{\partial t} + \omega'(k)\frac{\partial\Psi}{\partial x}\right) + \frac{\omega''(k)}{2}\frac{\partial^2\Psi}{\partial x^2} - \Delta|\Psi|^2\Psi = 0 \tag{1.1}$$

The solutions of (1.1) have the amplitude of order ϵ. We investigate the question about the character of the instability of the homogeneous periodic solution of (1.1)

$$\Psi = a_0 \exp\left[-i\Delta a_0^2 t\right]$$

where a_0 is a constant of order ϵ. Introducing the new real variables given by $\Psi = a\exp(i\varphi)$ one obtains [18]

$$\begin{aligned} &\frac{\partial\varphi}{\partial t} + \omega'\frac{\partial\varphi}{\partial x} + \frac{\omega''}{2}\frac{\partial^2\varphi}{\partial x^2} - \frac{\omega''}{2a}\frac{\partial^2 a}{\partial x^2} + \Delta a^2 = 0 \\ &\frac{\partial a^2}{\partial t} + \omega'\frac{\partial a^2}{\partial x} + \omega''\frac{\partial}{\partial x}\left(a^2\frac{\partial\varphi}{\partial x}\right) = 0 \end{aligned} \tag{1.2}$$

Substituting

$$a = a_0 + \delta a, \quad \varphi = -\Delta a_0^2 t + \delta\varphi$$

into (1.2), neglecting the second order terms and eliminating $\delta\varphi$ we obtain the linear equation on perturbations δa

$$\left(\frac{\partial}{\partial t} + \omega'\frac{\partial}{\partial x}\right)^2\delta a + \frac{\omega''^2}{4}\frac{\partial^4\delta a}{\partial x^4} - \omega'' a_0^2\Delta\frac{\partial^2\delta a}{\partial x^2} = 0 \tag{1.3}$$

Substitute now $\delta a = exp(i\kappa x - i\Omega t)$ in (1.3) to obtain the following dispersion relation (in the reference frame moving with a velocity V)

$$\Omega(\kappa) = (\omega'(k) - V)\kappa \pm \sqrt{\omega''(k)a_0^2\Delta\kappa^2 + \frac{\omega''^2(k)}{4}\kappa^4} \tag{1.4}$$

The quantity $\Delta(k)$ for k in a vicinity of k_0 is negative, therefore from (1.4) it follows that the radiating wave outcoming from a solitary wave core is unstable. The solution of the initial value problem for (1.3) have the form

$$\delta a(x,t) = \frac{1}{2\pi}\int_{-\infty}^{\infty}\widehat{\delta a}(\kappa)\exp[i(\kappa x - \Omega(\kappa)t)]d\kappa \tag{1.5}$$

where $\widehat{\delta a}(\kappa)$ is a Fourier transform of the initial perturbation

$$\widehat{\delta a}(\kappa) = \int_{-\infty}^{\infty} \delta a(x,0) \exp[-i\kappa x] dx$$

We are interested in the asymptotics of the integral in the right hand side of (1.5) for lardge t. According to the method of stationary phase [19] this asymptotics is determined by the values of the integrand in saddle points of $\Omega(\kappa)$ $(d\Omega(\kappa_0)/d\kappa = 0)$. Further we assume that $d^2\Omega(\kappa_0)/d\kappa^2 \neq 0$. The instability is convective when $Im(\Omega(\kappa_0) \leq 0$ and it is absolute when $Im(\Omega(\kappa_0) > 0$. From (1.4) we have that

$$\kappa_0^2 = \frac{1}{2}\left\{-\frac{4c-b^2}{\omega''^2(k)} \pm \frac{|b|}{\omega''^2(k)}\sqrt{b^2+8c}\right\}, \tag{1.6}$$

$$b = \omega'(k) - V, \quad c = \omega''(k) a_0^2 \Delta$$

It follows then from the expression (1.6) that κ_0^2 is always positive, and therefore all $\kappa_{0,i}$, $i = 1,2,3,4$ are real. To write the asymptotic expression for the rootes $\kappa_{0,i}$ recall, that the constant c has the order $\delta = \epsilon^2$:

$$\kappa_{0,k}^2 = \frac{b^2}{\omega''^2(k)} + O(\delta^2), \quad \kappa_{0,j}^2 = -\frac{4c}{\omega''^2(k)} + \frac{4c^2}{\omega''^2 b^2} + O(\delta^3), \tag{1.7}$$

$$k = 1,2, \quad j = 3,4$$

for all $b \neq 0$. When $b = 0$ i.e. $\omega'(k) = V$:

$$\kappa_{0,i}^2 = -\frac{2c}{\omega''^2(k)}, \quad i = 1,2,3,4$$

After the substitution of this of value into (1.4) we obtain

$$\Omega(\kappa_0) = \pm\sqrt{-\frac{c^2}{\omega''^2(k)}}$$

i.e. the imaginary part of $\Omega(\kappa_0)$ is either positive or negative. It means that perturbations in a reference frame moving with a group speed of the radiating wave are growing with time and we have the absolute instability in this reference frame. It is evident, that the instability has also the absolute character in reference frames moving with speeds sufficiently close to $\omega'(k)$. The solitary wave core moves with a speed which value is in the vicinity of the phase speed $V = \omega(k)/k$ of radiating wave. The contant b being the difference between the phase and group speeds for k in the vicinity of k_0 does not depend on ε. Consequently, we have that for finite b, $\Omega(\kappa_0)$ corresponding to first two rootes $\kappa_{0,k}$ in (1.7) is real and also

$$\Omega(\kappa_{0,j}) = b\kappa_{0,j} \pm \sqrt{-\frac{4c^3}{\omega''^2(k)b^2}}(1 + O(\delta)), \quad j = 3,4$$

Therefore the rate of the decay of perturbations in the reference frame moving with the speed of the solitary wave core is algebraic: the main contribution to the integral (1.5) is given by the linear combination of the following temporal terms

$$\frac{\exp[-i\Omega(\kappa_{0,i})t]}{\sqrt{t}}, \quad i = 1, 2, 3, 4$$

As it was mentioned in the Introduction of the present paper the amplitude of the radiating wave in (0.1) is proportional to $2D(\mu)exp(-\pi/\sqrt{\mu})$, (the initial amplitude of the decaying soliton equals to 3μ), and the quantity $D(\mu)$ depend on μ and tends to D as μ tends to zero (see [20],[11]). To perform the numerical investigation of radiation process we use the finite-difference method of the crest type. The application of such a method for the case of the modified Schrödinger equation is described in [21]. There is no essential difference in apllication of this method for the numerical solution of the equation (0.1) from that of the paper [21]. The numerical computations were performed on the closed ring of a large length. This permits not to pose the boundary conditions, which inevitably leads to the radiation of the waves of small amplitude. Therefore, this method makes it possible to make the numerical computations with exponentially small amplitude, radiating by the soliton and for the same values of the parameter μ as in [12], where for this purpose the special spectral method was developed.

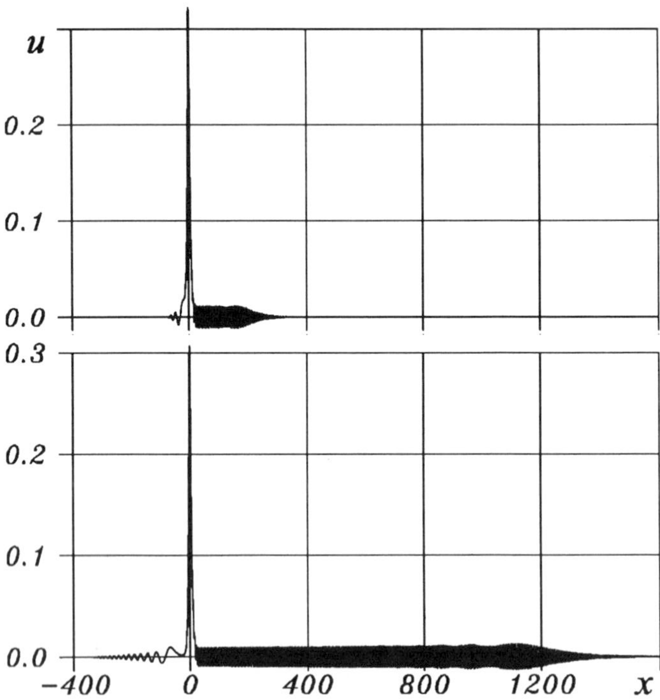

Fig.2. The evolution of a soliton with the amplitude 3μ, $\mu = 0.1$ at t=110 and t=560.

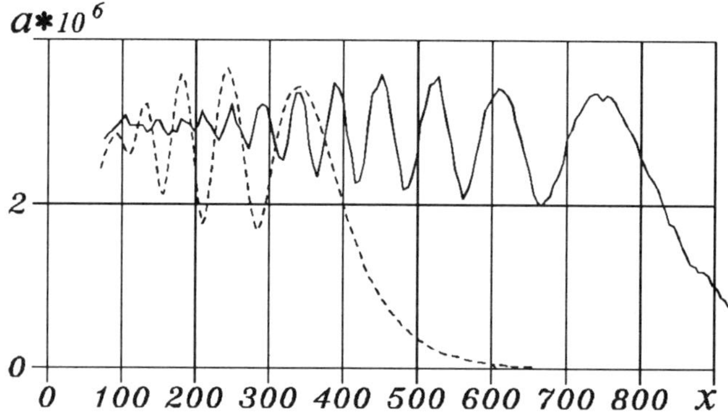

Fig.3. The shape of an envelope of the radiating wave at $t = 235$ (dashed line) and $t = 460$ (solid line).

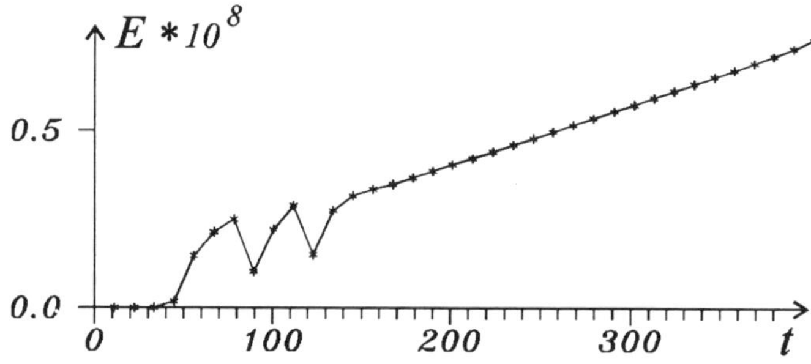

Fig.4. The radiating energy for $\mu = 0.03$.

In the present contribution we investigate mainly the stabilty of the radiating wave. In Fig.2 the shape of the solution u is shown (in the reference frame moving with the velocity μ) at different values of the time t: $t = 110$ and $t = 560$ for $\mu = 0.1$, i.e for comparatively large amplitudes of the radiating soliton. In Fig.3 the envelope of the radiating wave is given for $\mu = 0.03$ at $t = 460$ (solid line) and at $t = 235$ (dashed line). Fig.4 shows the time dependence of the energy of the radiating wave for the same value of the parameter μ. It can be seen from the last figure, that the process of the propagation of radiating wave tail is a quasi-stationary one.

2.Interaction of envelope solitary waves of small amplitude

Envelope solitary waves of (0.1) were found in [14]. Their asymptotic form is given by

$$u = -\frac{2}{9}(A^2 + c.c) - 4|A|^2 - \frac{64}{27\sqrt{2}}i(AB - \overline{AB}) + O(\mu^2) \tag{2.1}$$

where A and B are given by (0.15), (0.17), $q_1 = -1/2$, $q_2 = -19/9$, $p_1 = -\sqrt{2}/4$, $p_2 = -383/108\sqrt{2}$. The typical form of such a wave is given by Fig.5.

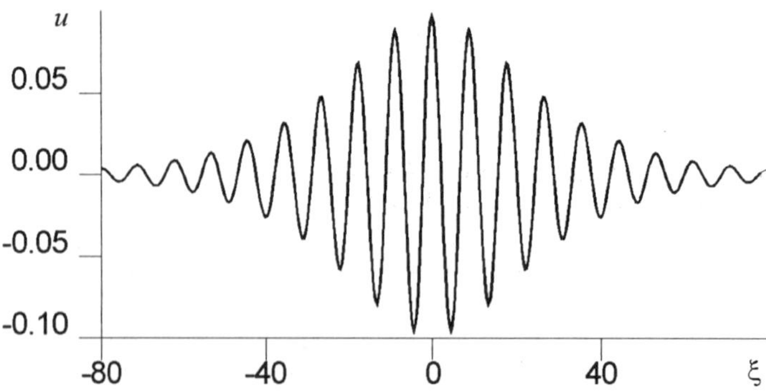

Fig.5 Crest-form of the envelope solitary wave of elevation for $\mu = -0.005$.

Solutions with the same envelopes but arbitrary phase shift λ (in the formula (0.15) θ_0 is replaced by $\theta_0+\lambda$) were obtained by the numerical asymptotic method in [22], namely, the sequence of solitary waves with arbitrary values of λ was obtained. According to classification proposed in [22] generalized solitary waves and envelope solitary waves in question are special cases of two-wave resonances. Other types of resonances and corresponding envelope solitary waves with rational ratio S between wavelengths of two interacting waves were obtained also. This solitary waves are limiting solutions for $S = 1/n$ and $S = n/(n+1)$ resonance series, where the integer n tends to infinity.

Consider the interaction of two envelope solitary waves which velocities V_1 and V_2 differs slightly: $V_2 - V_1 << 1$. We look for the solution of the equation (0.1) in the asymptotic form [17]

(2.2) $u(x,t) = U_1(X_1) + U_2(X_2) + \varepsilon[\Delta_1(X_1) + \Delta_2(X_2) + \alpha(x,t)] + O(\varepsilon^2)$

where U_1, U_2 are the envelope solitary waves, $X_1 = x - \int^t V_1(\tau)d\tau$, $X_2 = x - \int^t V_2(\tau)d\tau$, $\tau = \varepsilon t$, ε is a small parameter, characterizing the interaction, and $\alpha(x,t)$ is a radiating wave tail during interaction, we furter suppress its contribution. Substituting (2.2) into the equation (0.1) we get at the zeroth order for the $i-th$ wave ($i = 1,2$)

$$\frac{\partial U_i}{\partial t} + U_i\frac{\partial U_i}{\partial X_i} + \frac{\partial^3 U_i}{\partial X_i^3} + \frac{\partial^5 U_i}{\partial X_i^5} = 0, \quad i = 1,2 \tag{2.3}$$

The equation (2.3) is satisfied in the vicinity of each solitary wave. For the perturbation Δ_1 (at the first order in ε) we have

$$-V_1\frac{\partial \Delta_1}{\partial X_1} + \frac{\partial (U_1\Delta_1)}{\partial X_1} + \frac{\partial^3 \Delta_1}{\partial X_1^3} + \frac{\partial^5 \Delta_1}{\partial X_1^5} = -\left\{\frac{\partial U_1}{\partial \tau} + \varepsilon^{-1}\frac{\partial\left[U_1(X_1)U_2(X_2)\right]}{\partial X_1}\right\} \tag{2.4}$$

Next specify the mutual position of the interacting waves. Assume that the wave 2 having the greater speed, and hence the smaller amplitude, has a position in front of the wave 1, i.e. $X_2 = X_1 - l$, where l is the distance between the crests of the waves. The product $U_1(X_1)U_2(X_2) = U_1(X_1)U_1(X_1 - l)$ with the required accuracy. Taking this into account we obtain from (2.4)

$$L\Delta_1 = G, \quad G = -\frac{\partial U}{\partial \tau} - \frac{\partial [U(X)U(X-l)]}{\partial X} \tag{2.5}$$

where L is the operator of the left hand side in (2.5), (here we suppress the subscript 1 at U_1 and X_1). The necessary condition for the compartibility of the equation (2.5) is given by the equality $(G, \phi) = 0$, where ϕ is the eigenfunction of the operator L^* ajoint to L, corresponding to the zero eigenvalue and $(\cdot, \cdot)$ denotes the scalar product in $L^2(\mathbf{R})$. It can be easily seen that $\phi = U$, and the condition of the compartibility has the form

$$\frac{d}{dt}\int_{-\infty}^{\infty} U^2 dX = \frac{d}{dl}\int_{-\infty}^{\infty} U^2(X)U(X-l)dX \tag{2.6}$$

To obtain the equation on $\mu_1 = V_1 - V_0$ we use (2.1):

$$U = A + \overline{A} - \frac{2}{9}(A^2 + \overline{A}^2) - 4|A|^2 + O(|\mu|^{3/2}) \tag{2.7}$$

where A is given by (0.15), (0.17). Note, that after the integration of the terms in (2.6), containing the factors $\exp(inqx)$, n is an integer, we get the terms, which are proportional to $\exp(-2n/\sqrt{\mu_1})$, of the smallness beyond all orders in μ_1. Therefore, we neglect their contribution.

As a result of the substitution of U from (2.7) and use of (0.15), (0.17) we get

$$\frac{d\mu_1}{dt} = -8\mu_1^2 q_1 |q_2|^{-1} \frac{d}{ds}\int_{-\infty}^{\infty} \cosh^{-2}(x)\ \cosh^{-2}(x - \sqrt{\mu_1 q_1} l)dx + O(\mu_1^3) \tag{2.8}$$

Making the same considerations for the wave 2 we obtain

$$\frac{d\mu_2}{dt} = 8\mu_2^2 q_1 |q_2|^{-1} \frac{d}{ds}\int_{-\infty}^{\infty} \cosh^{-2}(x)\ \cosh^{-2}(x + \sqrt{\mu_2 q_1} l)dx + O(\mu_2^3) \tag{2.9}$$

At the right hand sides of (2.8), (2.9) we may take $\mu_1 = \mu_2 = \mu$, with the required accuracy, where μ is a half of the sum of the velocities of solitary waves situated at a large distance, minus V_0. Substracting next (2.8) from (2.9) we get

$$\frac{d^2 l}{dt^2} = -\frac{dV(l)}{dl}$$

where

$$V(l) = \frac{152}{19}\mu^2 \int_{-\infty}^{\infty} \cosh^{-2}(x)\ \cosh^{-2}(x - \sqrt{\frac{|\mu|}{2}} l)dx + O(\mu^3)$$

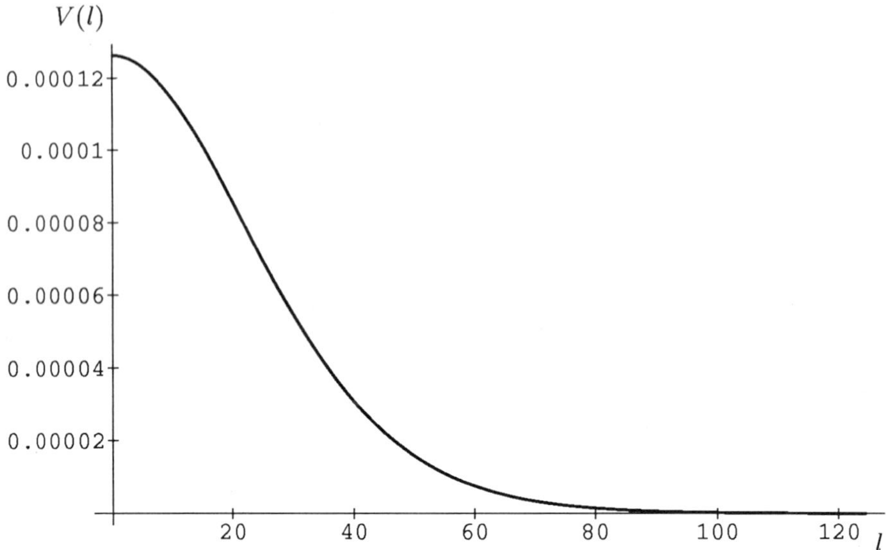

Fig.6. Potential energy $V(l)$ for $\mu = -0.005$.

The picture of a typical form of the potential energy $V(l)$ is given by the Fig 6. It could be seen from this figure that the envelope solitary waves are drifting apart and no breathers are formed (unlike interaction of solitary wave solutions of the equations discussed in [17]). The reason of the absence of the breathers, resulting the interaction of two envelope solitary waves of small amplitude, is the exponential smallness of the contributions to the potential energy from the terms which may influence the appearance of the potential pit.

Acknowledgement

The authors are grateful to Professor Andrei Kulikovskii for usefull discussions.

References

1. J. T. Beale *The existence of solitary water waves.* Comm. Pure. Appl. Math. **30** (1977), 373-389.
2. J. L. Bona, R. L. Sachs *The existence of internal solitary waves in a two fluid system near the KdV limit.* Geophys. Astrophys. Fluid Dynamics **47** (1989), 25-51.
3. J. Hunter, J. Sheurle *Existence of perturbed solitary wave solutions to a model equation for water waves.* Phys. D **32** (1988), 253-268.
4. A. V. Marchenko *Long waves in a shallow liquid beneath an ice sheet.* Appl. Math. Mech. **52** (1988), 180-186.
5. G. Iooss, M. C. Peroueme *Perturbed homoclinic solutions in reversible 1:1 resonance vector fields.* Journ. Diff. Eqns. **102** (1193), 62-88.

6. B. Buffoni, A. R. Champneys and J. F. Toland *Bifurcation and coalescense of a plethorna of homoclinic orbits for a Hamiltonian system.* Preprint, University of Bath, 1994.

7. A. T. Il'ichev *Existence of a family of soliton-like solutions for the Kawahara equation.* Math. Notes **52** (1992), 662-668.

8. A. T. Il'ichev, A. Yu. Semenov *Stability of solitary waves in dispersive media described by a fifth-order evolution equation.* Theoret. Comput.Fluid Dynamics **3** (1992), 307-326.

9. Y. Pomeau, A. Ramani and B. Grammaticos *Structural stability of the KdV solitons under a singular perturbation.* Phys. D **31** (1988), 127-134.

10. T. R. Akylas, T.-S. Yang *On short-scale oscillatory tails of long wave disturbances.* Stud. Appl. Math. **94** (1995), 1-20.

11. J. P. Boyd *Weakly non-local solitons for capillary-gravity waves: fifth degree Korteweg-de Vries equation.* Phys. D **48** (1991), 129-146.

12. E. S. Benilov, R. Grimshaw and E. P. Kuznetsova *The generation of radiating waves in a singularly-perturbed Korteweg-de Vries equation.* Phys. D **69** (1993), 270-278.

13. A. T. Il'ichev, A. V. Marchenko *Propagation of long nonlinear waves in a ponderable fluid beneath an ice sheet.* Fluid Dynamics **24** (1989), 73-79.

14. R. Grimshaw, B. Malomed, E. Benilov *Solitary waves with damped oscillatory tails: an analysis of the fifth order Korteweg-de Vries equation.* Phys. D **77** (1994), 473-485.

15. G. Iooss, M. Adelmeyer *Topics in Bifurcation Theory and Applications,* Advanced Series in Nonlinear Dynamics vol. 3, World Scientific, Singapore, 1992.

16. K. A. Gorshkov, L. A., Ostrovskii and V. V. Papko *Interactions and coupled states of solitans as classical particles.* Phys JETP **44** (1976), 306-315.

17. J. P. Boyd *Solitons from sine waves: analytical and numerical methods for non-integrable solitary and cnoidal waves.* Phys. D **21** (1986), 227-246.

18. V. I. Karpman *Linear and Nonlinear Waves in Dispersive Media.* Moscow: Nauka, 1973 (in Russian).

19. E. T. Copson *Asymptotic Expansions.* Cambridge at the University Press, 1965.

20. R. Grimshaw, N. Joshi *Weakly nonlocal solitary waves in a singularly perturbed Korteweg-de Vries equation.* SIAM J. Appl. Math. **55** (1995), 124-135.

21. I. B. Bakholdin *Wave jumps in media described by a modified Schrödinger equation.* Fluid Dynamics **29** (1994), 528-538.

22. I. B. Bakholdin *Nonlinear resonances in models described by the modified Korteweg-de Vries and Schrödinger equations* Bull. Russ. Acad. Sci. Phys./Suppl.: Phys. Vib. **59** (1995), 208-221.

Russian Academy of Sciences, Keldysh Institute of Applied Mathematics , Moscow, Russia, 125047 , Miusscaya sq. 4

Russian Academy of Sciences, Mathematical Steklov Institute , Moscow, Russia, 117966 , Vavilova 42

E-mail address:ilichev@mian.su

Current address: Mathematisches Institut A, Stuttgart Universität, Pfaffenwaldring 57, 70569 Stuttgart, Germany

Contemporary Mathematics
Volume **200**, 1996

Numerical Simulation of Singular Solutions of the Generalized Korteweg-de Vries Equation

J. L. Bona, V. A. Dougalis,
O. A. Karakashian, & W. R. McKinney

ABSTRACT. Presented here are some detailed results from a computer-assisted study of singularity formation in solutions of generalized Korteweg-de Vries equations $u_t + u^p u_x + u_{xxx} = 0$. This report supplements an earlier paper by the same authors. Of special interest here will be the path of the peak of the solution as it nears blow-up. Also, more detailed information is provided than heretofore about the structure of potentially singular solutions in the critical exponent case $p = 4$.

1. Introduction. This paper is concerned with the initial-value problem for the generalized Korteweg-de Vries equation (GKdV equation henceforth)

$$u_t + u^p u_x + \epsilon u_{xxx} = 0. \tag{1.1a}$$

In (1.1a), $u = u(x,t)$ is a real-valued function of the spatial variable $x \in [0,1]$ and the temporal variable $t \geq 0$, p is a positive integer, ϵ is a positive constant and subscripts connote partial differentiation. The partial differential equation (1.1a) is considered in conjunction with initial data

$$u(x,0) = u_0(x) \quad \text{for} \quad 0 \leq x \leq 1, \tag{1.1b}$$

where u_0 is a given smooth, real-valued function of x that is periodic of period 1. Solutions u of (1.1) are sought that are likewise 1-periodic in the spatial variable.

Equation (1.1a) arises in many instances in modeling wave propagation in weakly nonlinear, weakly dispersive regimes where dissipative effects can safely be ignored. Because of their considerable role in describing a variety of physical systems, and because of their very interesting mathematical properties, these equations have come in for substantial scrutiny in the last three decades.

It is our purpose here to study the initial- and periodic-boundary-value problem (1.1) for values of $p \geq 4$. We will use high-order accurate numerical methods for approximating solutions of (1.1). These tools will cast light on some fairly delicate aspects of the evolution of relatively large initial data to be discussed presently.

1991 *Mathematics Subject Classification.* Primary 35Q53, Secondary 65M50.

This article fits into the development by many authors of theory and intuition about (1.1). Making no claim whatever to completeness, we mention as a sample the papers of Kato [**7**], Bona, Strauss & Souganidis [**4**], Bona, Dougalis & Karakashian [**1**], Pego & Weinstein [**11**], [**12**], some of which will inform the further discussion. The present contribution particularly supplements two earlier studies by the same group of authors (Bona *et al.* [**2**], [**3**]).

One of the primary focuses of Bona *et al.* [**2**] was to better understand the instability of the special traveling-wave solutions of (1.1a) called solitary waves that was predicted theoretically for $p \geq 4$ by Bona *et al.* [**1**] (see also Pego & Weinstein [**12**]). Using a specially designed, adaptive version of our high-order numerical scheme, it was observed that small perturbations of solitary-wave initial profiles form similarity structures under the evolution (1.1a), and that these similarity profiles lead to the formation of singularities in the solution at a finite time. Indeed, the numerical simulations point to the conjecture that there is a point (x^*, t^*) such that $u(x,t) \to +\infty$ as $(x,t) \to (x^*, t^*)$. A detailed analysis of computed blow-up rates of various spatial norms of u as t approaches t^* suggest that u has the form

$$u(x,t) = \frac{1}{(t^* - t)^{2/3p}} \chi\left(\frac{x^* - x}{(t^* - t)^{1/3}}\right) + \text{bounded term}, \tag{1.2}$$

where χ is a smooth, bounded function. The existence of such similarity solutions χ has recently been settled by Bona & Weissler [**5**], but this does not at once imply the validity of (1.2) for a broad class of initial data, nor even for appropriate perturbations of solitary waves.

In our companion paper Bona *et al.* [**3**], consideration was given to the effect of dissipation on the singularity formation imputed to exist from the paper [**2**]. In the main, we studied the initial-value problem for the GKdV-Burgers equation

$$\begin{aligned} &u_t + u^p u_x - \delta u_{xx} + \epsilon u_{xxx} = 0, \quad 0 \leq x \leq 1, \ \ t \geq 0, \\ &u(x,0) = u_0(x), \quad 0 \leq x \leq 1, \end{aligned} \tag{1.3}$$

where δ is a positive constant, p is an integer greater than or equal to 4, and u_0 is as in (1.1b). Among other things, the following interesting result came to light. Let u_0 be given initial data and suppose the solution u emanating from u_0 blows up in finite time in case $\delta = 0$ in (1.3) (i.e. for (1.1)). Our theory and simulations indicate that in this circumstance, there is a $\delta_c > 0$ so that if $\delta > \delta_c$, then the associated solution $u = u_\delta$ is uniformly bounded for $(x,t) \in [0,1] \times [0,\infty)$. However, if $\delta < \delta_c$, the solutions u_δ appear to form singularities in finite time with similarity-form blow-up just as displayed in (1.2).

In the present paper, two issues are addressed that were not elaborated previously, but which are related to the results just described. The first is concerned with the putative blow-up due to instability, of perturbed solitary waves. Very careful observations indicate that the peak of a blowing-up solution, which is the spatial point $x = X(t)$ at which the solution at time t makes its maximum positive excursion, propagates as it becomes infinite. In Section 3, we report computational evidence that $X(t)$ moves according to the law $X(t) \sim x^* - c(t^* - t)^{1/3}$ as t approaches t^*, where c is a positive constant. Another point taken up here which has been largely neglected in the earlier work is the structure of solutions in the critical case $p = 4$. In this case it was noted before that the computations relating to the instability of solitary waves yielded much less definite results than for supercritical values $p > 4$. A very careful set of numerical experiments is reported in Section

4 that suggests strongly there is singularity formation as a result of the instability, but that the structure of the blow-up may not follow so closely the paradigm illustrated in (1.2).

The description of the numerical simulations in Sections 3 and 4 are preceded in Section 2 by a summary of the algorithms used and a report of the analytical facts pertaining to these algorithms. The paper concludes with a short section in which open questions suggested by and related to the contents of Sections 3 and 4 are brought out.

2. The Computational Procedures. The computer-generated data presented later in this report was obtained by a fully discrete Galerkin-finite-element scheme that utilizes splines for the spatial discretization and high-order, conservative, implicit Runge-Kutta methods of the Gauss-Legendre class for the time stepping. The stability limitations are kept minimal by using an implicit scheme, but there results as a consequence a nonlinear system of algebraic equations to solve at each time step. These systems are solved approximately by a few iterations of Newton-type, starting with an appropriate extrapolation of the solution based on previous time levels.

The methods just outlined are developed in detail in Bona *et al.* [**2**]. They comprise what we will refer to as the *base scheme*, and they feature uniform spatial and temporal grids. The base scheme as realized by a Fortran code has been extensively checked for accuracy and convergence as reported in Section 3 of [**2**]. It is worth mentioning the recent, very satisfactory theoretical results pertaining to the base scheme detailed in Karakashian & McKinney [**6**]. Let U^n connote the fully discrete approximation to the exact solution u at the time $t_n = n\,k$, where k is the constant time step, generated by applying the base scheme with splines of order r and a constant spatial grid of size h and a q-stage Gauss-Legendre time-stepping. (Referring to the detailed aspects of the determination of U^n set forth in [**2**], this involves one "outer" and two "inner" Newton iterations.) The theorem that applies here is that there is a constant $c = c(u,t)$ depending on a low-order Sobolev-norm of the solution $u(\cdot,t)$ for $0 \le t \le T$ such that

$$\max_{0\le n\le T/k} \|U^n - u(\cdot,t_n)\|_{L_2} \le c(h^r + k^{2q}),$$

uniformly for small h and k. The exponents r and $2q$ appearing above are the optimal rates achievable by these specific approximation techniques.

Since most of the solutions we aim to approximate feature large variations and very steep gradients, it will be necessary to equip the base scheme with adaptive capabilities in both space and time. The decision to refine spatially is based on the inverse inequality

$$\|v\|_{L_\infty} \le \frac{c_*}{h^{1/2}}\|v\|_{L_2}$$

which is valid for members of the finite-dimensional subspace S_h^r consisting of splines of order r on a mesh of size h. The temporal adaptivity employed is based upon limiting the variation of a discrete version of an integral invariant of (1.1a) (a Hamiltonian functional in fact). We refer to [**2**] for more details.

3. Trajectory of the Center of the Peak Near Blow Up. The solitary-wave solution of the GKdV-equation has the form $u(x,t) = \varphi(x - ct)$ where

$$\varphi(x) = A\text{sech}^{2/p}(K(x - x_0)) \tag{3.1}$$

with $A > 0$. Here K, A and c are related by the formulas

$$K = p\left[\frac{A^p}{\epsilon(p+1)(p+2)}\right]^{1/2} \quad \text{and} \quad c = \frac{2kA^p}{(p+1)(p+2)}.$$

It was found in our earlier study (Bona *et al.* [**2**]) that if initial data u_0 in (1.1) is taken as

$$u_0(x) = \lambda\varphi(x), \tag{3.2}$$

say, where $\lambda > 1$, and $p > 4$, then the associated solution u apparently evolves into a similarity structure of the form depicted in (1.2). Thus the solution forms a single peak that becomes infinite in finite time. Detailed aspects of the formation of the singularity are presented in [**2**] in terms of the rate of blow-up of various norms of the solution. In addition to evolving into a similarity structure and forming a single peak, the peak itself propagates in the process of blowing up. It is this propagation we wish to investigate in the present section. By means of numerical experiments, the details of the propagation of the peak for times t near the blow-up time t^* will be investigated both for initial-value problems for the GKdV-equation (1.1) and for the GKdV-Burgers equation (1.3). A couple of different types of initial profiles will be considered.

In the first groups of numerical experiments, the GKdV-equation (1.1a) was integrated numerically with initial data u_0 as indicated in (3.1)–(3.2), a perturbed solitary wave. We took $\lambda = 1.01$, $A = 2$ and $\epsilon = 5 \times 10^{-4}$ and computed the solution for values of $p = 5$, 6 and 7. The relevant parameters appearing in our numerical scheme were taken to be the same as those appearing in Tables 11, 15 and 16 for $p = 5$, 6, and 7, respectively, in [**2**].

For solutions u that develop a dominant peak like those considered here, let $X(t)$ connote the location of the peak, which is to say

$$u(X(t), t) = \max_{0 \le x \le 1} u(x, t).$$

Let (x^*, t^*) denote the point of blow-up. As an Ansatz, suppose that as t approaches t^*, $X(t)$ has the form

$$X(t) = x^* + C(t^* - t)^\gamma, \tag{3.3}$$

where C and γ are constants to be determined. As a consequence of the presumption (3.3), for t near t^* the speed of the peak of the solution would be

$$\dot{X}(t) = -\gamma C(t^* - t)^{\gamma - 1}.$$

The evaluation of the parameters C and γ and the estimation of how closely the power law (3.3) describes what actually transpires is effected by making use of technical aspects of the adaptive computer code to do with cutting the temporal step, refining locally the spatial mesh and translating the peak into the region of finest spatial mesh. These procedures are explained in Bona *et al.* [**2**]. The refinement of the spatial structure takes place statically, with new grid points being inserted in smaller and smaller, nested neighborhoods of $x = 1/2$. This refined mesh is effective in following a blowing-up solution because the code systematically takes advantage of the translation-invariance of the differential equations to place the peak of the solution at or near $x = 1/2$. These spatial shifts occur at times τ_i, $1 \le i \le f$, where τ_i is the time when the ith spatial refinement takes place. The τ_i

get very close to t^* as i increases to f and $X(\tau_i)$ can be estimated by the quantities $x_i + 0.5$, where x_i is the accumulated amount of translation of the peak at the time of the ith refinement. Approximating x^* and t^* by $x_f + 0.5$ and τ_f, respectively, it is reasonable to expect that γ should be approximated well, for i large, but not very close to $i = f$, by the quantities

$$\gamma_i = \frac{\log((x_i - x_f)/(x_{i+1} - x_f))}{\log((\tau_i - \tau_f)/(\tau_{i+1} - \tau_f))} \,. \tag{3.4}$$

On the right-hand side of (3.4), there appear differences of terms that get extremely close to each other as i increases. Care must be exercised therefore in their computation in order to avoid loss of accuracy due to cancelation. For example, $\tau_f - \tau_i$ is actually computed as the sum of the differences $\Delta\tau_j = \tau_{j+1} - \tau_j$ formed by accumulating small, but positive quantities during the refinement process.

In Table 1 the values of γ_i computed with (3.4) are recorded for the GKdV equation with $p = 5$ and 6, for $i = 5, 10, \ldots, 30$. It was found in Bona *et al.* [**2**], that the $p = 5$ solitary wave with the aforementioned parameters evolved into a similarity profile whose evolution was followed numerically up to its apparent blow-up at $x^* \cong x_f = 0.61333$, $t^* \cong \tau_f = 0.022543$. The adaptive code performed $f = 42$ refinements achieving a maximum amplitude $U_{\max} = 224,766$. For $p = 6$ the analogous parameters were $x^* \cong x_f = .54732$, $t^* \cong \tau_f = 0.51541\times10^{-2}$, $f = 41$, $U_{\max} = 26,099$.

i	$p = 5$		$p = 6$	
	γ_i	C_i	γ_i	C_i
5	.3388	−.4128	.3482	−.3219
10	.3306	−.4082	.3152	−.3111
15	.3393	−.4168	.3052	−.3054
20	.3374	−.4148	.3362	−.3175
25	.3336	−.4181	.3409	−.3153
30	.3348	−.4086	.3101	−.3104

TABLE 1. Values of γ_i and C_i at the time τ_i of the ith spatial refinement for GKdV. Solitary-wave initial profile (3.2).

The data in Table 1 is consistent with the hypothesis that γ is a constant independent of p, quite probably equal to 1/3; the same value appears from the analogous data in the case $p = 7$ not shown here.

The parameter C postulated as a constant in (3.3) may be approximated by the numbers $(x_i - x_f)/(\tau_f - \tau_i)^{\gamma_i}$. This computation was found to be very sensitive to small variations in the computed values of the γ_i, since $\tau_f - \tau_i$ can become as small as 10^{-40} in our computations. Instead, taking into account the robust value $\gamma = 1/3$ that emerged by use of (3.4), the computation of the approximations C_i of C at τ_i was effected by the formula

$$C_i = (x_i - x_f)/(\tau_f - \tau_i)^{1/3}.$$

The entries in Table 1 for C_i stabilize and suggest that C is indeed a constant for each p. Its values are about $-.41$ for $p = 5$, $-.31$ for $p = 6$, and $-.28$ for the $p = 7$

data not shown here. Hence, for the family of solutions evolving from perturbed solitary-wave initial profiles of the form (3.1) with $A = 2$ and $\epsilon = 5 \times 10^{-4}$, C seems to be well approximated by $-2/p$.

When consideration is given to the Korteweg-de Vries-Burgers model equation (1.3) for small positive values of δ so that solutions emanating from, say, initial profiles of the type (3.1) still blow up in finite time as the computations in Bona *et al.* (1996) suggest, practically the same values of γ and C emerge. This is consistent with the findings of the last-quoted paper in that it is again verified that values of δ in the range of blow-up of GKdV-B do not affect the leading-order asymptotics of the blow-up structure.

Analogous computations were performed using as initial data the Gaussian profile

$$u_0(x) = \exp\left(-100\left(x - \frac{1}{2}\right)^2\right) \tag{3.5}$$

not specifically tied to a traveling-wave solution of this class of equations. For example, using (3.5) as initial data for the dissipative equation in (1.3) with $p = 5$, $\epsilon = 2 \times 10^{-4}$ and $\delta = 10^{-4}$, it was found that the solution apparently blew up in finite time at the point $(x^*, t^*) = (.72886, .37376)$ where the numerical solution achieved a value $U_{\max} = 17,148$. The mechanism of blow-up was identical to the example discussed in Bona *et al.* **[2]** (cf. Figure 9 and Table 18 of that reference): the initial Gaussian profile quickly resolved into a solitary-wave type pulse traveling to the right, followed by a hump. Subsequently, the solitary-wave portion became unstable, evolving into a similarity solution apparently of the form displayed in (1.2). The computed values of γ_i and C_i associated with the trajectory of this leading pulse as it blew up are given in Table 2.

i	5	10	15	20	25	30
γ_i	.3384	.3320	.3315	.3301	.3380	.3288
C_i	−.3049	−.3053	−.3019	−.3021	−.3067	−.3028

TABLE 2. Values of γ_i and C_i at the time τ_i of the ith spatial refinement for GKdV with $p = 5$. Gaussian initial profile as in (3.5).

The emerging value of γ is again approximately 1/3, a number that seems therefore to be stable under perturbations of ϵ, p, δ and u_0, provided of course that there is a well-defined, dominating peak that blows up. The value of the negative constant C seems to depend in general on the particular initial profile u_0 and on ϵ and p as well.

4. On the Critical Exponent Case $\mathbf{p = 4}$.

It was remarked in Bona *et al.* **[2]** that the outcome of the computations to determine whether the instability of solitary waves for (1.1a) in the critical exponent case $p = 4$ leads to blow-up in finite time was not as convincing as for $p \geq 5$. In fact, perturbations of solitary-wave initial profiles of the form (3.2) with perturbation factors $\lambda > 1$ did not apparently lead to blow-up. In the above-mentioned reference (cf. in particular formula (5.8)) we constructed a special initial condition u_0 obtained from the $p = 4$ solitary wave (corresponding to $A = 2$,

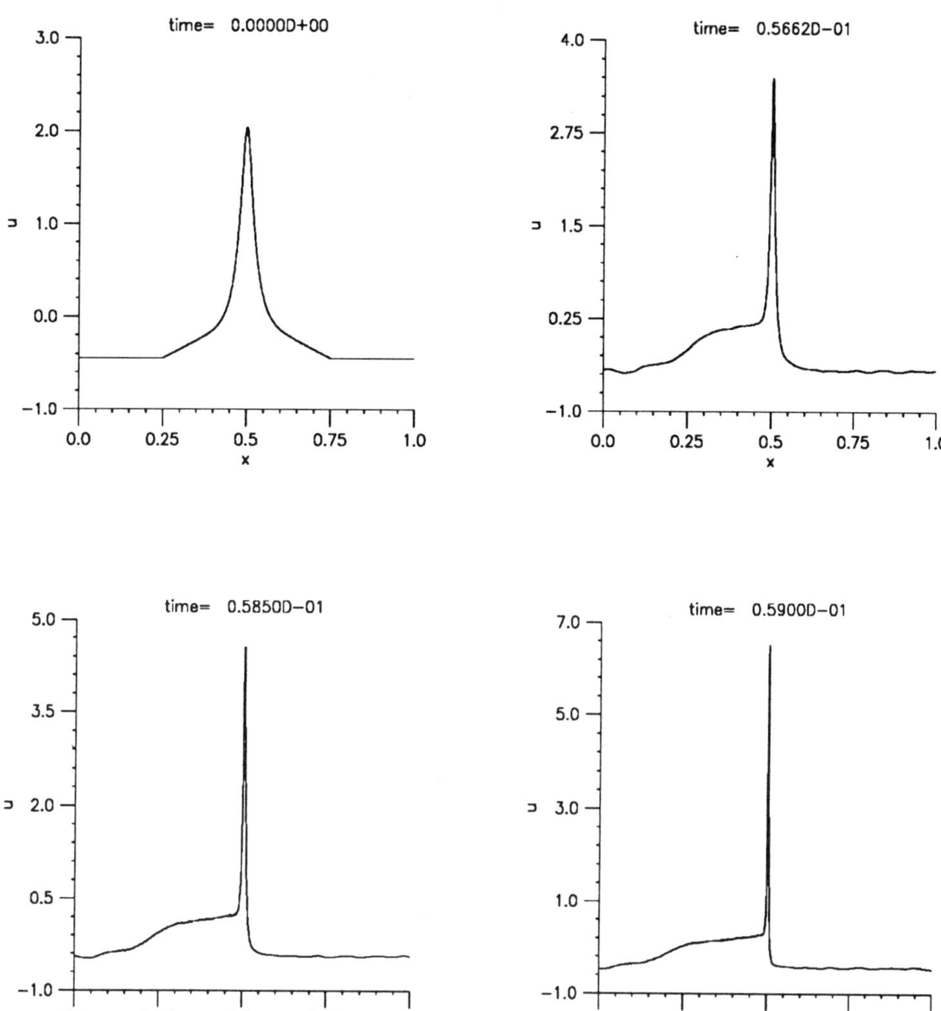

FIGURE 1. Presumed blow-up of a perturbed solitary-wave for $p = 4$. (Parameters as in Table 3.) (a): $u_0(x)$, (b)–(d): computed approximations at the times of the first three spatial grid refinements, respectively.

$\epsilon = 5\times 10^{-1}$) by first adding a constant which results in non-zero asymptotic values of about $-.446$ at $x = 0$ and $x = 1$. The resulting profile was then perturbed in a special direction (motivated by the theory in Bona *et al.* [**4**]) and then truncated for periodicity. The result is the initial datum shown in (a) of Figure 1. This was integrated numerically with the adaptive code starting with $h_0 = 0.1 \times 10^{-2}$, $k_0 = 0.25\times 10^{-3}$ and using tolerance levels (cf. [**2**]) TOL1 $= 0.1\times 10^{-5}$, TOL2 $= 0.1$. These parameters allowed the code to perform $f = 13$ spatial refinements while

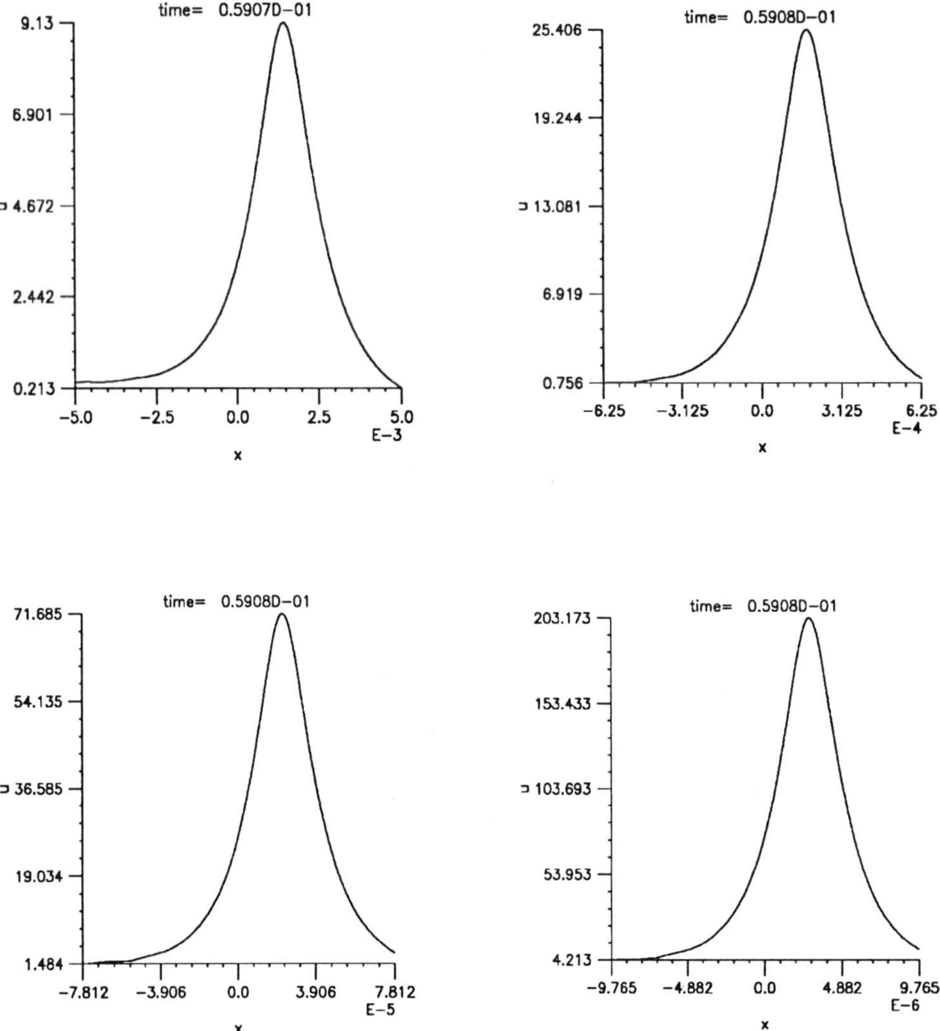

FIGURE 2. Presumed blow-up of a perturbed solitary-wave for $p = 4$. (Continuation of Figure 1). (a), (b), (c), (d): solution at the 4th, 7th, 10th, 13th spatial grid refinements, respectively. The x-axis comprises the finest and the next finest grid region.

sustaining good accuracy. The resulting solution traveled to the right and formed a thin spike which is shown (translated back to the region of the finest grid which is always a small interval symmetric about $x_0 = 1/2$ and called Ω^* in **[2]**, at the instances of the first three spatial refinements in frames (b), (c) and (d) of Figure 1. The spike grew in height achieving a maximum amplitude $U_{\max} = 203$ at $t^* \cong \tau_{13} = 0.059086$ and $x^* = 0.73244$, as may be seen in Figures 2 and 3, where the x- and the u-axes have been rescaled. In particular, in Figure 2 the profile of the

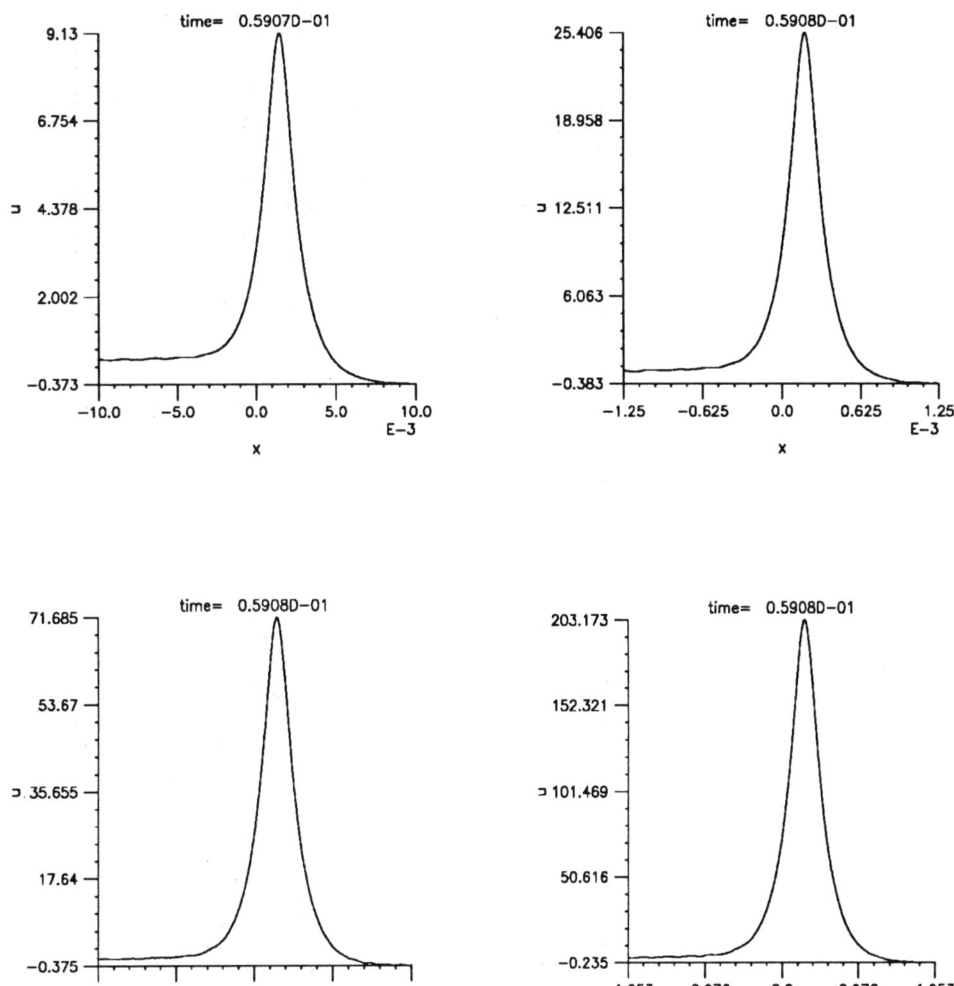

FIGURE 3. Data of Figure 2 with the x-axis scaled to contain the finest grid and the next two grid regions.

peak is shown on an interval about 1/2 (translated to a symmetric interval about 0 in the pictures) that contains the current interval Ω^* of finest grid size together with one more adjacent region of the next-to-finest mesh intervals at the instances of the 4th, 7th, 10th and 13th refinement ((a), (b), (c), and (d), respectively). In Figure 3 the x-interval has been enlarged to contain the interval Ω^* and the next two regions of successively coarser mesh enclosing Ω^*, with the purpose of exhibiting a detailed view of the peak profile and of the trailing, almost horizontal "shelf".

The maximum amplitude $U_{\max} = 203$ reached in this computation was an improvement by almost an order of magnitude on the analogous value that was achieved in the computations reported in Bona, *et al.* [**2**]. In addition, the monitoring of other quantities of interest in the refinement process (like the number of spatial refinements, the size of the time step etc.) adds confidence to the conjecture that the solution indeed blows up. Because of the better accuracy evinced in the present calculations, one expects more robust rates of blow-up of the various spatial norms of the solution and its derivative as $t \to t^*$. By a rate or exponent of blow-up for some spatial norm $\|u(\cdot,t)\|$ of the solution u, we mean a value $\rho > 0$ for which $\|u(\cdot,t)\| \sim c(t^* - t)^{-\rho}$ as $t \to t^*$. This is indeed what is observed; in Table 3 we record the computed blow-up rates of the norms used with the same notation in the analogous tables of [**2**] at the instances of spatial grid refinement τ_i, for $1 \le i \le 7$. For $i \ge 8$ the computed rates were not as stable, losing accuracy as i approached 13, the total number of refinements. (In Tables 3 and 4, the columns labeled $L_{2,D}$ and $L_{\infty,D}$ denote the computed rates of blow-up of the L_2-norm and L_∞-norm of u_x, respectively.)

i	L_4	L_5	L_6	L_∞	$L_{2,D}$	$L_{\infty,D}$
1	0.9124 (−1)	0.1111	0.1236	0.1854	0.3693	0.5568
2	0.9193 (−1)	0.1113	0.1238	0.1858	0.3712	0.5571
3	0.9228 (−1)	0.1112	0.1236	0.1854	0.3706	0.5570
4	0.9244 (−1)	0.1112	0.1235	0.1851	0.3700	0.5553
5	0.9391 (−1)	0.1128	0.1254	0.1882	0.3761	0.5671
6	0.9160 (−1)	0.1099	0.1221	0.1831	0.3654	0.5500
7	0.9289 (−1)	0.1115	0.1238	0.1853	0.3714	0.5541

TABLE 3. Blow-up rates. Perturbed Solitary wave, $p = 4$, $\epsilon = 5 \times 10^{-4}$, $\tau_f = 0.059086$, $f = 13$, $x^* = 0.73244$, $U_{\max} = 203.17$, $k_{\min} = 0.36 \times 10^{-14}$, $\Delta\tau_f = 0.40 \times 10^{-11}$, $\mathrm{TOL1} = 0.1 \times 10^{-5}$, $\mathrm{TOL2} = 0.1$.

In Bona *et al.* [**2**], assuming that the singular solution that develops from an unstable solitary wave of the GKdV equation is of the form

$$u = \frac{1}{(t^* - t)^\alpha}\, \chi\left(\frac{x^* - x}{(t^* - t)^\beta}\right) + \text{bounded term},$$

we computed the rates of blow-up of the L_q-norms of $u(\cdot,t)$ and $u_x(\cdot,t)$ as $t \to t^*$ as functions of α, β and q. Then using the fact that the third invariant of (1.1), namely the quantity

$$I_3 = \int_0^1 (u^{p+2} - \frac{1}{2}(p+1)(p+2)\epsilon u_x^2)dx,$$

is constant in time (implying that $\|u_x\|_{L_2}^2$ and $\|u\|_{L_{p+2}}^{p+2}$ have the same rates of blow-up) one may deduce that blow-up can occur only for $p \ge 4$ and that α and β must be related by the formula

$$\alpha = 2\beta/p.$$

As a consequence of the foregoing considerations, for $q > p/2$ the rates of blow-up ρ of the L_q-norms of u and u_x were predicted to be

$$\rho = \rho(\|u\|_{L_q}) = \alpha(1 - p/2q) \tag{4.1}$$

and

$$\rho = \rho(\|u_x\|_{L_q}) = \alpha(1 + p(q-1)/2q). \tag{4.2}$$

The evidence accumulated in Bona *et al.* [**2**] strongly suggested that for $p \geq 5$, $\rho(\|u\|_{L_\infty}) = 2/3p$. This determines a conjectured value of $\alpha = 2/3p$ (implying $\beta = 1/3$). Moreover, this conjecture for the value of α was strongly supported for $p \geq 5$ by the good agreement of the other blow-up rates determined from the numerical experiments with the predicted values displayed in (4.1) and (4.2).

In the critical exponent case $p = 4$ at hand, the predicted value of $\alpha = 1/6$ does not agree with the outcome of the numerical experiment. For example, examination of the column with the blow-up rate entries for the L_∞-norm in Table 3 shows that $\rho(\|u\|_{L_\infty})$ might well be .185, which is some 11% larger than the value 1/6 predicted by descent from higher values of p. Similarly, the blow-up rates suggested by the data of Table 3 for the other norms were also about 11% higher than the values suggested by (4.1) and (4.2) for $p = 4$ and $\alpha = 1/6$.

The anomaly for the rates of blow-up in the critical case $p = 4$ was further investigated by attempting several other expressions for the blow-up, including

$$\|u(\cdot,t)\| \sim c\left(\frac{t^* - t}{|\log(t^* - t)|}\right)^{-\rho} \quad \text{as} \quad t \to t^*. \tag{4.3}$$

The idea of having a logarithmic term appear in the critical exponent for blow-up in nonlinear. dispersive evolution equations is well-known in the context of the nonlinear Schrödinger equation (cf. McLaughlin *et al.* [**10**], Landman *et al.* [**9**] and Kosmatov *et al.* [**8**]). The computed values of ρ using (4.3) as the form of the blow-up are provided in Table 4 and are seen to be closer to the theoretical rates than the values of ρ in Table 3.

i	L_4	L_5	L_6	L_∞	$L_{2,D}$	$L_{\infty,D}$
1	0.7939 (−1)	0.9665 (−1)	0.1076	0.1613	0.3213	0.4844
2	0.8212 (−1)	0.9941 (−1)	0.1106	0.1660	0.3316	0.4976
3	0.8408 (−1)	0.1013	0.1126	0.1689	0.3377	0.5075
4	0.8538 (−1)	0.1027	0.1141	0.1710	0.3418	0.5128
5	0.8762 (−1)	0.1053	0.1170	0.1756	0.3510	0.5291
6	0.8613 (−1)	0.1034	0.1148	0.1722	0.3436	0.5172
7	0.8789 (−1)	0.1055	0.1172	0.1753	0.3515	0.5243

TABLE 4. Blow-up rates. Perturbed Solitary wave, same data as in Table 3, rates computed with the log formula in (4.3).

Since the adaptive refinement procedure used in the numerical schemes is based (among other things) upon controlling the time step by checking that properly weighted values of I_3 remain close to each other from time step to time step, it is reasonable to expect that the computed values of $\|u\|_{L_6}^6$ become infinite like $\|u_x\|_{L_2}^2$, something that is confirmed experimentally by examining the columns labelled L_6 and $L_{2,D}$ in Table 3. On the other hand, assuming self-similarity of the blow-up profile and using the constancy of I_3 (for $p = 4$), one is led to the conjecture

$$\rho = \rho(\|u\|_{L_q}) = \alpha(1 - 2/q), \quad \rho = \rho(\|u_x\|_{L_q}) = \alpha(3 - 2/q),$$

which seems to be consistent with the data of Table 3 if $\alpha \cong .185$ and Table 4 if $\alpha \cong .172$ and $\beta = 2\alpha$. Of course the blow-up is not necessarily self-similar, though the graphs in Figure 2 make a good case for such a hypothesis.

Another point of interest is that the exponent γ in the assumed form of the trajectory of the center of the peak, (see 2.3), in this experiment was found to be about .26 and not 1/3 as in the cases where $p \geq 5$.

5. Conclusions. The present study has added some detailed information connected with the apparent singularity formation in solutions of the GKdV equation and the GKdV-Burgers equation. The results in Section 3 indicating the peak of a blowing-up solution propagates at speed $(t^* - t)^{1/3}$ have been incorporated into the theoretical study of similarity solutions of the GKdV equation in Bona & Weissler [**5**].

The computations connected with the critical case $p = 4$ are improved in various ways over those reported previously. Especially the better agreement with predicted values of blow-up rate one obtains by incorporating a logarithm into the Ansatz about the singularity is intriguing.

It must be candidly remarked, however, that the computations connected with $p = 4$ seem to stretch the capabilities of our computer code in its present form. One possible reason for this may be seen already in Figures 2 and 3. The peak in the profile of the solution has wider "support" than the analogous peaks for the cases where $p \geq 5$. Moreover, the support does not shrink as rapidly as occurs for $p \geq 5$. It may be that the refinement process of our code (or indeed as embodied in the choices of parameters) is so restrictive that it causes the grid in the vicinity of the peak to become unacceptably structured after 10 to 15 refinements. As it took on the order of 40 refinements to produce a peak of height about 10^5 for $p \geq 5$, we are seriously handicapped if we make no more than a dozen refinements. Together with associated theoretical work, improving our simulations of solutions when $p = 4$ is presently under study.

References

[1] J. L. Bona, V. A. Dougalis and O. A. Karakashian, *Fully discrete Galerkin methods for the Korteweg-de Vries equation*, Comp. & Maths. with Applics. **12A** (1986), 859–884.

[2] J. L. Bona, V. A. Dougalis, O. A. Karakashian and W. R. McKinney, *Conservative, high-order numerical schemes for the generalized Korteweg-de Vries equation*, Phil. Trans. Royal Soc. London A **351** (1995), 107–164.

[3] J. L. Bona, V. A. Dougalis, O. A. Karakashian and W. R. McKinney, *The effect of dissipation on solutions of the generalized Korteweg-de Vries equation*, to appear (1996).

[4] J. L. Bona, P. E. Souganidis and W. A. Strauss, *Stability and instability of solitary waves of KdV type*, Proc. Royal Soc. London A **411** (1987), 395–412.

[5] J. L. Bona and F. B. Weissler, *Similarity solutions of the generalized Korteweg-de Vries equation*, to appear (1996).

[6] O. A. Karakashian and W. R. McKinney *On the approximation of solutions of the generalized Korteweg-de Vries-Burgers equation*, Math. and Comp. in Simulation **37** (1994), 405–416.

[7] T. Kato, *On the Cauchy problem for the (generalized) Korteweg-de Vries equation*, Studies in Appl. Math., Advances in Mathematics Supplementary Studies, vol 8, Academic Press, 1983, pp. 93–130.

[8] N. E. Kosmatov, V. F. Shvets and V. E. Zakharov, *Computer simulation of wave collapses in the nonlinear Schrödinger equation*, Physica D **52** (1991), 16–35.

[9] M. J. Landman, G. C. Papanicolaou, C. Sulem and P. L. Sulem, *Rate of blowup for solutions of the nonlinear Schrödinger equation at critical dimension*, Phys. Rev. A **38** (1988), 3837–3843.

[10] D. W. McLaughlin, G. C. Papanicolaou, C. Sulem and P. L. Sulem, *Focusing singularity of the cubic Schrödinger equation*, Phys. Rev. A **34** (1986), 1200–1210.

[11] R. L. Pego and M. I. Weinstein, *Eigenvalues, and instabilities of solitary waves*, Phil. Trans. Royal Soc. London A **340** (1992), 47–94.

[12] R. L. Pego and M. I. Weinstein, *Asymptotic stability of solitary waves*, Comm. Math. Phys. **164** (1994), 305–349.

DEPARTMENT OF MATHEMATICS AND TEXAS INSTITUTE FOR COMPUTATIONAL AND APPLIED MATHEMATICS, UNIVERSITY OF TEXAS, AUSTIN, TX 78712 USA; AND CENTRE DE MATHÉMATIQUES ET DE LEURS APPLICATIONS, ECOLE NORMALE SUPÉRIEURE DE CACHAN, 94235 CACHAN CEDEX, FRANCE

E-mail address: bona@ticam.utexas.edu

DEPARTMENT OF MATHEMATICS, UNIVERSITY OF ATHENS, PANEPISTIMIOPOLIS, 15784 ZOGRAPHOU, GREECE; AND INSTITUTE OF APPLIED & COMPUTATIONAL MATHEMATICS, FORTH, P.O. BOX 1527, 71110 HERAKLION, GREECE

E-mail address: dougalis@softlab.ece.ntua.gr

DEPARTMENT OF MATHEMATICS, UNIVERSITY OF TENNESSEE, KNOXVILLE, TN 37996–13000, USA

E-mail address: ohannes@math.utk.edu

DEPARTMENT OF MATHEMATICS, NORTH CAROLINA STATE UNIVERSITY, RALEIGH, NC 27695–8205, USA

E-mail address: mckinney@math.ncsu.edu

Contemporary Mathematics
Volume **200**, 1996

Spatially quasi-periodic capillary–gravity waves

T. J. Bridges and F. Dias

Abstract. For linearized capillary-gravity waves there exists values of the Weber number and the Froude number at which two periodic waves travel at the same speed but have differing wavelengths whose ratio is, generically, irrational. In this paper, using the Hamiltonian structure of the steady water–wave problem and a variational principle of the form due to Percival for invariant tori, we construct weakly nonlinear approximations of the quasi-periodic travelling waves for the two extremes of deep water and shallow water. The analysis of these nonlinear quasi-periodic waves is new and moreover a new degeneracy of the water–wave problem which corresponds to the vanishing of the twist coefficient is revealed. We interpret the dynamical consequences of them by analogy with the dynamics of invariant tori in Hamiltonian systems.

1. Introduction

The dispersion relation for linearized capillary–gravity waves travelling on the surface of a fluid layer of thickness h can be written as

$$c^2 = \left(gk^{-1} + \sigma k\right) \tanh kh\,, \tag{1.1}$$

where k is the wavenumber, c the phase velocity, g the acceleration due to gravity and σ the coefficient of surface tension divided by the fluid density. We define the Froude number F and the Weber number W as

$$F = \frac{c}{\sqrt{gh}}\,, \qquad W = \frac{\sigma}{gh^2}\,.$$

As is well-known, the dispersion curve F versus kh exhibits a minimum, denoted by F^*, if the Weber number is smaller than $1/3$. This minimum lies between 0 and 1. If F is fixed such that $F^* < F < 1$, there are two distinct positive wavenumbers satisfying the dispersion relation, say k_1 and k_2 (see Figure 1). As a solution to the linearized water–wave problem, one can therefore take a superposition of two travelling waves, one with wavenumber k_1 and the other with wavenumber k_2. Since both waves travel at the same speed, such a solution is steady in a frame of reference moving with speed c. If the wavenumbers are rationally related, the motion is spatially periodic and corresponds to the well-known $M : N$ resonance (Wilton's ripples). If the wavenumbers are irrationally related, the motion is quasi-periodic in space. The goal of this paper is to provide a weakly nonlinear analysis of such quasi-periodic waves. Resonant waves have been studied extensively. However, as far as we are aware, little research has been devoted to spatially quasi-periodic

1991 *Mathematics Subject Classification.* Primary 76B15.

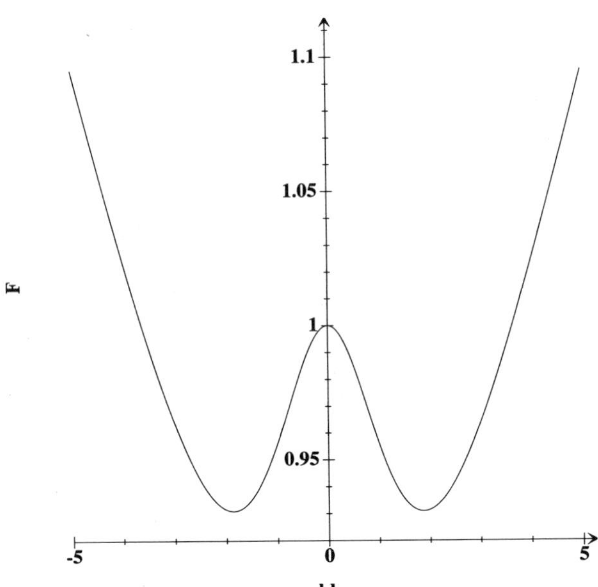

FIGURE 1. Dispersion relation. Finite depth. The plot corresponds to $W = 0.2$

water waves, except for the case when F is close to 1 which has been treated by Iooss & Kirchgässner [1992]. We will restrict the analysis to two special cases: deep water and shallow water. In a subsequent paper, we will consider the general case of finite depth and the stability of quasi-periodic waves.

As suggested above, a spatial dynamics approach will be used to study the nonlinear water–wave problem. By spatial approach, we mean that we work in a frame of reference moving with a wave of given wave speed c and we let the wavenumbers k_1 and k_2 vary with amplitude. Moreover, we will take advantage of the Hamiltonian structure of the steady wave problem and characterize the quasi-periodic waves using a variational principle that is analogous to Percival's variational principle for invariant tori.

2. Spatial Hamiltonian formulation

Let x (resp. y) denote the horizontal (resp. vertical) coordinate. Let $\phi(x,y)$ denote the velocity potential and $\eta(x)$ denote the shape of the free surface. The water–wave problem can be formulated as follows in a frame of reference moving from right to left with the velocity c:

$$\nabla^2\phi \equiv \phi_{xx} + \phi_{yy} = 0 \quad \text{for } -h < y < \eta(x)\,, \tag{2.1}$$

$$(c + \varphi_{(x)})\eta_x - \varphi_{(y)} = 0\,, \tag{2.2}$$

$$\tfrac{1}{2}\left(\varphi_{(x)} + c\right)^2 + \tfrac{1}{2}\varphi_{(y)}^2 + g\eta = \sigma\left(\frac{\eta_x}{\sqrt{1+\eta_x^2}}\right)_x + \tfrac{1}{2}c^2\,, \tag{2.3}$$

$$\phi_y = 0 \quad \text{on} \quad y = -h\,. \tag{2.4}$$

The following notation has been used:

$$\varphi(x) = \phi(x,\eta(x))\,, \qquad \varphi_{(*)}(x) = \phi_*(x,\eta(x))\,,$$

where the star stands for x or y. Consequently, φ_x and $\varphi_{(x)}$ have different meanings. They are however related since

$$\varphi_x = \varphi_{(x)} + \varphi_{(y)}\eta_x \,.$$

The steady equations can be characterized as an infinite-dimensional Hamiltonian system. Extend the set of dependent variables as follows:

$$Z = \begin{pmatrix} \varphi(x) \\ \eta(x) \\ w(x) \\ \phi(x,y) \\ u(x,y) \end{pmatrix} \quad \text{where} \quad u = \phi_x + c \ \text{ and } \ w = \frac{\eta_x}{\sqrt{1+\eta_x^2}} \,.$$

In terms of the above set of coordinates, the equations (2.1)–(2.4) correspond to an infinite-dimensional Hamiltonian system (M, ω_η, S) where M is a manifold, ω_η is a symplectic form and S a Hamiltonian functional (cf. Bridges [1992]). The explicit characterization of the manifold M is technical. A rigorous proof that M is a manifold, for the case of gravity waves travelling in finite depth, is given in Bridges & Mielke [1995]. For capillary-gravity waves travelling in infinite depth the manifold can be parametrized using the Baesens–MacKay chart (cf. Baesens & MacKay [1992]) leading to a formal definition of M (see Appendix A).

Define the one form α_η on M by

$$\alpha_\eta = \sigma\, w\, \mathbf{d}\eta + \int_{-h}^{\eta} u\mathbf{d}\phi\, dy \,, \tag{2.5}$$

where infinite depth corresponds to the limit $h \to +\infty$. Assuming α_η is suitably differentiable, $\mathbf{d}\alpha_\eta$, where $\mathbf{d}$ is the exterior derivative, results in the following closed two-form on M:

$$\omega_\eta = \mathbf{d}\alpha_\eta = \sigma\, \mathbf{d}w \wedge \mathbf{d}\eta + \int_{-h}^{\eta} \mathbf{d}u \wedge \mathbf{d}\phi\, dy + u_S\, \mathbf{d}\eta \wedge \mathbf{d}\varphi \,,$$

where $u_S = u|_{y=\eta}$. The Hamiltonian function, which physically can be identified with the flow force, is

$$S = \frac{1}{2}\int_{-h}^{\eta} [(u-c)^2 - \phi_y^2]\, dy - \frac{1}{2} g\eta^2 + \sigma\left(1 - \sqrt{1-w^2}\right) .$$

The Hamiltonian vectorfield, $Z_x = X_S$, is then defined by

$$\omega_\eta(X_S, V) = \langle \nabla S(Z), V\rangle_\eta \,, \tag{2.6}$$

for all tangent vectors V where

$$\langle U, V\rangle_\eta \stackrel{\text{def}}{=} U_1V_1 + U_2V_2 + U_3V_3 + \int_{-h}^{\eta} (U_4V_4 + U_5V_5)dy \,.$$

Note that the inner product is Z-dependent through η in the integral. Formally writing out the expression (2.6) results in

$$-u_S\eta_x = -\varphi_{(y)} \,, \tag{2.7}$$

$$u_S\varphi_x - \sigma w_x = \tfrac{1}{2}c^2 - g\eta + \tfrac{1}{2}(u_S^2 + \varphi_{(y)}^2) - cu_S \,, \tag{2.8}$$

$$\sigma\eta_x = \sigma w/\sqrt{1-w^2} \,, \tag{2.9}$$

$$-u_x = \phi_{yy} \,, \tag{2.10}$$

$$\phi_x = u - c \,, \tag{2.11}$$

or

$$\mathbf{K}(Z)Z_x = \nabla S(Z)\,, \tag{2.12}$$

where $\mathbf{K}(Z)$ is defined by $\omega_\eta(U,V) = \langle \mathbf{K}(Z)U, V\rangle_\eta$. It is important to note that the equations (2.12) make sense as a Hamiltonian system only in the form (2.6) because of the Z-dependence of the inner-product and symplectic form which in turn is due to the fact that M is a manifold and not a linear space (see Appendix A).

The action for the system is obtained by parametrizing the one-form α_η in (2.5) or

$$F(Z) = \alpha_\eta^* = \sigma w \eta_x + \int_{-h}^{\eta} u\phi_x\,dy\,.$$

The formulation (2.1)–(2.4) is then recovered formally via the least-action principle

$$\nabla \mathcal{F}(Z) = 0\,, \tag{2.13}$$

where

$$\mathcal{F}(Z) = \int_{x_i}^{x_f} [F(Z) - S(Z)]\,dx\,.$$

In this paper we consider spatially quasi-periodic travelling waves. The advantage of the above Hamiltonian formulation is that such waves can be characterized by a variational principle which is analogous to Percival's ([1974]) variational principle for invariant tori of finite-dimensional Hamiltonian systems.

Let us now introduce the variables $x_1 = k_1 x$ and $x_2 = k_2 x$. Then

$$\frac{\partial}{\partial x} = k_1\frac{\partial}{\partial x_1} + k_2\frac{\partial}{\partial x_2}\,.$$

Spatially quasi-periodic waves are 2π-periodic solutions of (2.13) in x_1 and in x_2. Restricting the functional F to doubly 2π-periodic functions and defining

$$\begin{aligned} F_1 &= \int_{-h}^{\eta} \phi_{x_1} u\,\mathrm{d}y + \sigma w \eta_{x_1}\,, \\ F_2 &= \int_{-h}^{\eta} \phi_{x_2} u\,\mathrm{d}y + \sigma w \eta_{x_2}\,, \end{aligned}$$

we can characterize quasi-periodic waves as critical points of $\overline{S}$ on level sets of $\overline{F}_1$ and of $\overline{F}_2$, or, with k_1 and k_2 as Lagrange multipliers, critical points of the unconstrained functional

$$\mathcal{F} = k_1\overline{F}_1 + k_2\overline{F}_2 - \overline{S}\,,$$

where the quantities with bars denote averages over x_1 and x_2. For example,

$$\overline{S} = \int_{\Pi^2} S\,\mathrm{d}x_1\,\mathrm{d}x_2 = \frac{1}{4\pi^2}\int_0^{2\pi}\int_0^{2\pi} S(x_1,x_2)\,\mathrm{d}x_1\,\mathrm{d}x_2\,.$$

In the next section, we will construct an approximation for $\mathcal{F}$ in deep water. Note that $\overline{S} = S$ since S is an absolute spatial invariant of the water–wave problem.

We now turn to the shallow water limit. In the region F close to 1 and W close to $\frac{1}{3}$, the capillary–gravity water–wave problem can be reduced to the Kawahara equation ([1972]), which can be integrated once to give

$$4(1-F)\eta + 3\eta^2 + 2\left(\frac{1}{3} - W\right)\eta_{xx} + \frac{2}{45}\eta_{xxxx} = 0\,, \tag{2.14}$$

where the free-surface elevation η and the horizontal coordinate x have been normalized by h.

This equation also has a Hamiltonian structure, as shown by Zufiria [1987]. First note that (2.14) is the Euler-Lagrange equation corresponding to the Lagrangian

$$L(\eta,\eta_x,\eta_{xx}) = \frac{1}{45}\eta_{xx}^2 - \left(\frac{1}{3} - W\right)\eta_x^2 + 2(1-F)\eta^2 + \eta^3 .$$

The Hamiltonian function then follows via Legendre transform resulting in the following Hamiltonian system: (R^4,ω,H) with variables $(q,p)\in R^4$ defined by

$$q_1 = \eta\,, \quad q_2 = \eta_x\,, \quad p_1 = -2\left(\frac{1}{3} - W\right)\eta_x - \frac{2}{45}\eta_{xxx} \;\text{ and }\; p_2 = \frac{2}{45}\eta_{xx}\,.$$

The symplectic form is the canonical one on R^4: $\omega = \mathbf{d}p_1\wedge\mathbf{d}q_1 + \mathbf{d}p_2\wedge\mathbf{d}q_2$ and

$$H(q,p) = \frac{45}{4}p_2^2 + p_1q_2 + (\frac{1}{3} - W)q_2^2 - 2(1-F)q_1^2 - q_1^3\,. \tag{2.15}$$

The Hamiltonian equations, dual to (2.14), are then

$$\frac{d}{dx}\begin{pmatrix} q \\ p \end{pmatrix} = \begin{pmatrix} \mathbf{0} & \mathbf{I}_2 \\ -\mathbf{I}_2 & \mathbf{0} \end{pmatrix}\begin{pmatrix} \delta H/\delta q \\ \delta H/\delta p \end{pmatrix}.$$

Introducing the pair of action functionals

$$\overline{I}_j = \int_{\Pi^2} p\cdot q_{x_j}\,\mathrm{d}x_1\,\mathrm{d}x_2\,, \quad j=1,2\,,$$

we can again characterize the spatially quasi-periodic waves using, in this case exactly, the variational principle of Percival [1974] since the system is finite-dimensional. The spatially quasi-periodic travelling waves correspond to critical points of H, averaged over T^2, on level sets of $\overline{I}_1$ and $\overline{I}_2$, with k_1 and k_2 treated as Lagrange multipliers.

3. Finite-dimensional approximation of spatially quasi-periodic waves in deep water

In infinite depth, the functional $\mathcal{F}$ defined in the previous section is

$$\mathcal{F} = \mathcal{F}_1 + \mathcal{F}_2\,, \tag{3.1}$$

where

$$\mathcal{F}_1 = \int_{\Pi^2}\int_{-\infty}^{\eta} \tfrac{1}{2}(2c\phi_x + \phi_x^2 + \phi_y^2)\,\mathrm{d}y\,\mathrm{d}x_1\,\mathrm{d}x_2 \tag{3.2}$$

and

$$\mathcal{F}_2 = \int_{\Pi^2} \left(\tfrac{1}{2}g\eta^2 + \tfrac{1}{2}\sigma\eta_x^2 - \tfrac{1}{8}\sigma\eta_x^4 + \cdots\right)\mathrm{d}x_1\,\mathrm{d}x_2\,. \tag{3.3}$$

We use the expansions

(3.4)

$$\eta = 2\,\mathrm{Re}\left[A_1\mathrm{e}^{ix_1} + A_2\mathrm{e}^{ix_2} + A_3\mathrm{e}^{2ix_1} + A_4\mathrm{e}^{2ix_2} + A_5\mathrm{e}^{i(x_1+x_2)} + A_6\mathrm{e}^{i(x_2-x_1)}\right] + \cdots,$$

and

$$\begin{aligned}\phi \;=\; & -2\,\mathrm{Im}\left[B_1\mathrm{e}^{ix_1}\mathrm{e}^{k_1y} + B_2\mathrm{e}^{ix_2}\mathrm{e}^{k_2y} + B_3\mathrm{e}^{2ix_1}\mathrm{e}^{2k_1y} + B_4\mathrm{e}^{2ix_2}\mathrm{e}^{2k_2y}\right. \\ & \left.+B_5\mathrm{e}^{i(x_1+x_2)}\mathrm{e}^{(k_1+k_2)y} + B_6\mathrm{e}^{i(x_2-x_1)}\mathrm{e}^{(k_2-k_1)y}\right] + \cdots,\end{aligned} \tag{3.5}$$

where the dots denote higher order terms. Without loss of generality we take $k_1 < k_2$. The sum of wavenumbers $k_1 + k_2$ is denoted by k_+ while the difference of wavenumbers $k_2 - k_1$ is denoted by k_-.

The expression for $\mathcal{F}_2$ is

$$\begin{aligned}\mathcal{F}_2 &= g(|A_1|^2 + |A_2|^2 + |A_3|^2 + |A_4|^2 + |A_5|^2 + |A_6|^2)\\ &+\sigma\left(k_1^2|A_1|^2 + k_2^2|A_2|^2 + 4k_1^2|A_3|^2 + 4k_2^2|A_4|^2 + (k_1+k_2)^2|A_5|^2\right.\\ &\left.+(k_2-k_1)^2|A_6|^2\right) - \tfrac{3}{4}\sigma\left(k_1^4|A_1|^4 + k_2^4|A_2|^4 + 4k_1^2k_2^2|A_1|^2|A_2|^2\right).\end{aligned}$$

The expression for $\mathcal{F}_1$ is

$$\begin{aligned}\mathcal{F}_1 &= \int_{\Pi^2}\int_{-\infty}^{\eta} \tfrac{1}{2}(2c\phi_x + \phi_x^2 + \phi_y^2)\,\mathrm{d}y\,\mathrm{d}x_1\,\mathrm{d}x_2\\ &= c\int_{\Pi^2}\int_{-\infty}^{\eta} \phi_x\,\mathrm{d}y\,\mathrm{d}x_1\,\mathrm{d}x_2 + \tfrac{1}{2}\int_{\Pi^2}\varphi(\varphi_{(y)} - \eta_x\varphi_{(x)})\,\mathrm{d}x_1\,\mathrm{d}x_2\\ &= -2c\,\mathrm{Re}\{1\} + \mathrm{Re}\{2\},\end{aligned}$$

where

$$\begin{aligned}\{1\} &= k_1A_1\overline{B_1} + k_2A_2\overline{B_2} + 2(k_1A_3\overline{B_3} + k_2A_4\overline{B_4}) + k_+A_5\overline{B_5} + k_-A_6\overline{B_6}\\ &+\tfrac{1}{2}(k_1^3A_1|A_1|^2\overline{B_1} + k_2^3A_2|A_2|^2\overline{B_2})\\ &+k_1^3A_1|A_2|^2\overline{B_1} + k_2^3A_2|A_1|^2\overline{B_2} + k_1^2\overline{A_1}A_3\overline{B_1} + k_2^2\overline{A_2}A_4\overline{B_2}\\ &+2k_1^2A_1^2\overline{B_3} + 2k_2^2A_2^2\overline{B_4} + k_+^2A_1A_2\overline{B_5} + k_-^2A_1\overline{A_2}B_6\\ &+k_1^2(\overline{A_2}A_5 + A_2\overline{A_6})\overline{B_1} + k_2^2(\overline{A_1}A_5 + A_1A_6)\overline{B_2},\end{aligned}$$

and

$$\begin{aligned}\{2\} &= k_1|B_1|^2 + k_2|B_2|^2 + 4(|A_1|^2 + |A_2|^2)(k_1^3|B_1|^2 + k_2^3|B_2|^2)\\ &+2(k_1|B_3|^2 + k_2|B_4|^2) + k_+|B_5|^2 + k_-|B_6|^2\\ &+4k_1k_2k_+A_1\overline{A_2B_1}B_2 + 4k_1k_2\overline{A_6B_1}B_2\\ &+8k_1^2A_1B_1\overline{B_3} + 8k_2^2A_2B_2\overline{B_4} + 4k_2k_-A_1\overline{B_2}B_6 + 4k_2k_+A_1B_2\overline{B_5}\\ &+4k_1k_+A_2B_1\overline{B_5}.\end{aligned}$$

Next we eliminate the coefficients B_i with $i = 1, \cdots, 6$ by setting the derivatives of $\mathcal{F}$ with respect to $\overline{B_i}$ to zero. The result is

$$\begin{aligned}B_1 &= c\left[A_1 - 3k_1\overline{A_1}A_3 - (k_1+2k_2)\overline{A_2}A_5 + (k_1-2k_2)A_2A_6\right.\\ &\left.+\tfrac{1}{2}k_1^2A_1|A_1|^2 - k_1(k_1-2k_2)A_1|A_2|^2\right],\\ B_2 &= c\left[A_2 - 3k_2\overline{A_2}A_4 - (2k_1+k_2)\overline{A_1}A_5 - k_2A_1\overline{A_6}\right.\\ &\left.+\tfrac{1}{2}k_2^2A_2|A_2|^2 - (2k_1^2 - k_2^2 - 2k_1k_2)A_2|A_1|^2\right],\\ B_3 &= c\left[A_3 - k_1A_1^2\right],\\ B_4 &= c\left[A_4 - k_2A_2^2\right],\\ B_5 &= c\left[A_5 - (k_1+k_2)A_1A_2\right],\\ B_6 &= c\left[A_6 - (k_1+k_2)\overline{A_1}A_2\right].\end{aligned}$$

Next we express $\mathcal{F}$ purely in terms of the coefficients $A_1, \cdots, A_6$. The result is

$$\begin{aligned}
\mathcal{F} \;=\; & |A_1|^2\left(g+\sigma k_1^2-k_1c^2\right)+|A_2|^2\left(g+\sigma k_2^2-k_2c^2\right)\\
&+c^2k_1^2(A_1^2\overline{A_3}+\overline{A_1}^2A_3)+c^2k_2^2(A_2^2\overline{A_4}+\overline{A_2}^2A_4)\\
&+2k_1k_2c^2(A_1A_2\overline{A_5}+\overline{A_1A_2}A_5)\\
&-2k_1(k_1-k_2)c^2(A_1\overline{A_2}A_6+\overline{A_1}A_2\overline{A_6})\\
&-k_1^2|A_1|^4\left(\tfrac{3}{4}\sigma k_1^2-k_1c^2\right)-k_2^2|A_2|^4\left(\tfrac{3}{4}\sigma k_2^2-k_2c^2\right)\\
&-k_1^2|A_1|^2|A_2|^2\left[3\sigma k_2^2-2(k_1+k_2)c^2\right]\\
&+|A_3|^2\left(g+4\sigma k_1^2-2k_1c^2\right)+|A_4|^2\left(g+4\sigma k_2^2-2k_2c^2\right)\\
&+|A_5|^2\left(g+\sigma k_+^2-k_+c^2\right)+|A_6|^2\left(g+\sigma k_-^2-k_-c^2\right).
\end{aligned}$$

Next we eliminate the coefficients A_3, A_4, A_5, A_6, resulting in

$$A_3\left(g+4\sigma k_1^2-2k_1c^2\right)=-c^2k_1^2A_1^2\,,$$

$$A_4\left(g+4\sigma k_2^2-2k_2c^2\right)=-c^2k_2^2A_2^2\,,$$

$$A_5\left(g+\sigma k_+^2-k_+c^2\right)=-2c^2k_1k_2A_1A_2\,,$$

$$A_6\left(g+\sigma k_-^2-k_-c^2\right)=-2c^2k_1(k_2-k_1)\overline{A_1}A_2\,.$$

The final expression for $\mathcal{F}$ is

$$\mathcal{F}=\alpha_1|A_1|^2+\alpha_2|A_2|^2+\tfrac{1}{2}\alpha_3|A_1|^4+\alpha_4|A_1|^2|A_2|^2+\tfrac{1}{2}\alpha_5|A_2|^4+\cdots$$

where

$$\alpha_1=\sigma(K_1-K_2)(k_1-K_1)\,,$$

$$\alpha_2=-\sigma(K_1-K_2)(k_2-K_2)\,,$$

$$\frac{\alpha_3}{2gK_1^2}=\frac{8K_2^2+K_1K_2+2K_1^2}{4K_2(K_2-2K_1)}\,,$$

$$\frac{\alpha_4}{gK_1K_2}=\frac{2K_1^2+10K_1K_2-K_2^2}{K_2(K_2-2K_1)}\,,$$

$$\frac{\alpha_5}{2gK_2^2}=\frac{8K_1^2+K_1K_2+2K_2^2}{4K_1(K_1-2K_2)}\,.$$

The wavenumbers K_1 and K_2 are the linear wavenumbers which satisfy the linear dispersion relation $g+\sigma K^2-Kc^2=0$. The parameter $r=K_1/K_2$ is the rotation number for the linear quasi-periodic waves. In the coefficients α_3, α_4 and α_5, the wavenumbers k_1 and k_2 have been replaced by their value from the linear dispersion relation. The coefficients α_3 and α_5 are not new and correspond to the coefficients for the nonlinear change in wavenumber for the two families of periodic travelling waves. The coefficient α_4, however, seems to be new in the water–wave literature.

The amplitude equations are

$$\begin{aligned}
\alpha_1A_1+\alpha_3A_1|A_1|^2+\alpha_4A_1|A_2|^2 &= 0\,,\\
\alpha_2A_2+\alpha_4A_2|A_1|^2+\alpha_5A_2|A_2|^2 &= 0\,.
\end{aligned}$$

If $A_1A_2\neq 0$, they become

$$\begin{aligned}
\alpha_1+\alpha_3|A_1|^2+\alpha_4|A_2|^2 &= 0\,,\\
\alpha_2+\alpha_4|A_1|^2+\alpha_5|A_2|^2 &= 0\,,
\end{aligned}$$

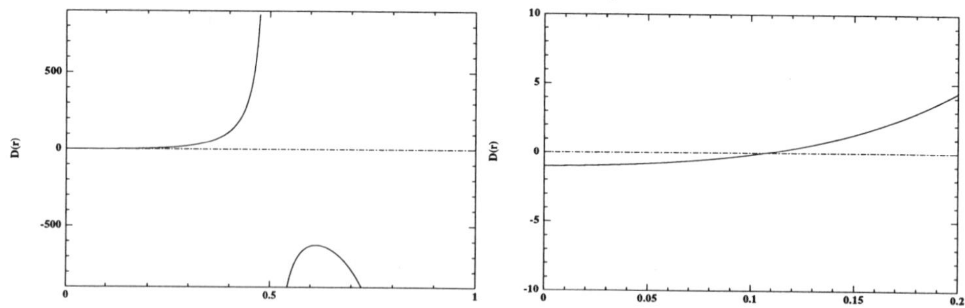

FIGURE 2. Twist coefficient $D(r)$ versus r. Infinite depth. The twist coefficient is negative in the interval $0 < r < 0.1072$ as shown in the blow up. The 1:2 resonance occurs when $r = \frac{1}{2}$.

and provide the variations of the wavenumbers k_1 and k_2 in terms of the amplitudes $|A_1|$ and $|A_2|$. It is useful to normalize the amplitude equations (see appendix B). The normalized normal form reads

$$k_1 - K_1 = d_{11}\overline{F}_1 - d_{12}\overline{F}_2 \tag{3.6}$$

$$k_2 - K_2 = -d_{12}\overline{F}_1 + d_{22}\overline{F}_2\,, \tag{3.7}$$

where the coefficients d are defined as follows:

$$d_{11} = -\frac{\alpha_3}{\sigma^2(K_1-K_2)^2}\,,$$

$$d_{12} = -\frac{\alpha_4}{\sigma^2(K_1-K_2)^2}\,,$$

$$d_{22} = -\frac{\alpha_5}{\sigma^2(K_1-K_2)^2}\,.$$

$\overline{F}_1$ and $\overline{F}_2$, which are the quadratic parts of the actions, are given by

$$\overline{F}_1 = \sigma(K_1-K_2)|A_1|^2\,,$$

$$\overline{F}_2 = \sigma(K_2-K_1)|A_2|^2\,.$$

There are several degeneracies associated with the amplitude equations: Wilton ripples, which correspond to $K_2 = 2K_1$ or $r = \frac{1}{2}$, and lead to the vanishing of the denominators of α_3 and α_4; vanishing of $\alpha_3\alpha_5 - \alpha_4^2$, which does not occur in infinite depth; vanishing of the twist coefficient defined in appendix B as $-2K_1K_2d_{12} - K_1^2d_{22} - K_2^2d_{11}$, which occurs when $K_2 = 9.3255K_1$ or $r = 0.1072$. Explicitly the twist coefficient, denoted $D(r)$, is given by

$$D(r) = \frac{g^2}{\sigma^3}\frac{10r^4 + 5r^3 - 72r^2 - 11r + 2}{(1-r)^2(2r-1)(2-r)}\,,$$

and is shown in Figure 2. It is interesting to compare Figure 2 with the twist coefficient for the restricted three body problem (cf. Deprit & Deprit-Bartholomé [1967, Figure 2]).

The twist coefficient is a fundamental property of the flow on an invariant torus (cf. Moser [1968], Moeckel [1990], Kaufmann [1986]). For example, for an autonomous Hamiltonian system on $\mathbb{R}^4$, with an invariant toral manifold, restriction to an energy surface and a Poincaré section normal to the main centerline results in a symplectic map for which the invariant torus reduces to an invariant circle (this construction is given explicitly in Moser [1968, §4]). The twist coefficient then gives information about the direction and magnitude of the twist. In the context of steady waves the quasi-periodic waves can be divided into two sets depending on the sign of the twist and it would be interesting to determine whether there are any physical consequences of the twist type.

4. Shallow water

In the neighborhood of $F = 1$, the capillary–gravity water–wave problem can be reduced to the generalized Korteweg–de Vries or Kawahara equation, which can be integrated once to give

$$4(1-F)\eta + 3\eta^2 + 2\left(\tfrac{1}{3} - W\right)\eta_{xx} + \tfrac{2}{45}\eta_{xxxx} = 0\,. \tag{4.1}$$

The linear dispersion relation for this model equation is

$$2(1-F) = \left(\tfrac{1}{3} - W\right)K^2 - \tfrac{1}{45}K^4\,. \tag{4.2}$$

The two positive roots K_1 and K_2 are related through

$$K_1^2 + K_2^2 = 45(\tfrac{1}{3} - W) \quad \text{and} \quad K_1^2 K_2^2 = 90(1-F)\,,$$

and explicit expressions for K_1 and K_2 are easily obtained. A further remark is that (4.2) corresponds to the Taylor expansion of (1.1) for small values of kh. The 1:2 resonance occurs when

$$\frac{5}{18}(1-F) = \left(\tfrac{1}{3} - W\right)^2\,,$$

leading to

$$K_1^2 = 9\left(\tfrac{1}{3} - W\right)\,, \quad K_2^2 = 36\left(\tfrac{1}{3} - W\right)\,.$$

Let us consider an expansion of the solution of the nonlinear problem for quasi-periodic waves in the form

$$\eta = A_0 + 2\,\mathrm{Re}\Big[A_1 e^{ix_1} + A_2 e^{ix_2} + A_3 e^{2ix_1} + A_4 e^{2ix_2} + A_5 e^{i(x_1+x_2)} + A_6 e^{i(x_2-x_1)}\Big] + \cdots \tag{4.3}$$

By performing the same analysis as in the deep water case, one finds

$$A_0 = -\frac{3}{2(1-F)}\left(|A_1|^2 + |A_2|^2\right)$$

$$A_3 = -\frac{3A_1^2}{4(1-F) - 4k_1^2[2(\frac{1}{3} - W) - \frac{8}{45}k_1^2]}$$

$$A_4 = -\frac{3A_2^2}{4(1-F) - 4k_2^2[2(\frac{1}{3} - W) - \frac{8}{45}k_2^2]}$$

$$A_5 = -\frac{6A_1A_2}{4(1-F) - (k_1+k_2)^2[2(\frac{1}{3} - W) - \frac{2}{45}(k_1+k_2)^2]}$$

$$A_6 = -\frac{6\overline{A_1}A_2}{4(1-F) - (k_2-k_1)^2[2(\frac{1}{3} - W) - \frac{2}{45}(k_2-k_1)^2]}\,.$$

The amplitude equations are

$$\begin{aligned} \alpha_1 A_1 + \alpha_3 A_1|A_1|^2 + \alpha_4 A_1|A_2|^2 &= 0\,, \\ \alpha_2 A_2 + \alpha_4 A_2|A_1|^2 + \alpha_5 A_2|A_2|^2 &= 0\,, \end{aligned}$$

where

$$\alpha_1 = \tfrac{2}{45}(k_1^2 - K_1^2)(K_2^2 - K_1^2)\,,$$

$$\alpha_2 = \tfrac{2}{45}(k_2^2 - K_2^2)(K_1^2 - K_2^2)\,,$$

$$\alpha_3 = \frac{135(24K_1^2 - 5K_2^2)}{K_1^2K_2^2(4K_1^2 - K_2^2)}\,,$$

$$\alpha_4 = \frac{810(27K_1^2K_2^2 - 4K_1^4 - 4K_2^4)}{K_1^2K_2^2(4K_1^2 - K_2^2)(4K_2^2 - K_1^2)}\,,$$

$$\alpha_5 = \frac{135(24K_2^2 - 5K_1^2)}{K_1^2K_2^2(4K_2^2 - K_1^2)}\,.$$

Again, the normalized amplitude equations are

$$k_1 - K_1 = d_{11}\overline{I}_1 - d_{12}\overline{I}_2 \tag{4.4}$$

$$k_2 - K_2 = -d_{12}\overline{I}_1 + d_{22}\overline{I}_2\,, \tag{4.5}$$

where the coefficients d are defined as follows:

$$\begin{aligned} d_{11} &= \left(\frac{45}{4}\right)^2 \frac{\alpha_3}{K_1^2(K_1^2 - K_2^2)^2}\,, \\ d_{12} &= \left(\frac{45}{4}\right)^2 \frac{\alpha_4}{K_1K_2(K_1^2 - K_2^2)^2}\,, \\ d_{22} &= \left(\frac{45}{4}\right)^2 \frac{\alpha_5}{K_2^2(K_1^2 - K_2^2)^2}\,. \end{aligned}$$

$\overline{I}_1$ and $\overline{I}_2$, which are the quadratic parts of the actions, are given by

$$\begin{aligned} \overline{I}_1 &= \tfrac{4}{45}K_1(K_2^2 - K_1^2)|A_1|^2\,, \\ \overline{I}_2 &= \tfrac{4}{45}K_2(K_1^2 - K_2^2)|A_2|^2\,. \end{aligned}$$

Let us define the rotation-like number $r = K_1^2/K_2^2$ and $R = r + 1/r$. There are several degeneracies associated with the amplitude equations: Wilton ripples, already mentioned above, which occur when $r = 1/4$; vanishing of $\alpha_3\alpha_5 - \alpha_4^2$, which occurs when $r = 0.0346$ and $r = 0.1789$; vanishing of the twist coefficient, which occurs when $r = 0.1877$ in shallow water. Explicitly the twist coefficient $D(r)$ is given by

$$D(r) = \frac{3}{2(1-F)K_2^4}\left(\frac{45}{4}\right)^2 \frac{r(20R^2 - 53R - 316)}{(1-r)^2(4r-1)(4-r)}\,,$$

and is shown in Figure 3. Note the qualitative similarity of the twist coefficient in Figure 3 with that of the Hénon map (cf. Moeckel [1990, Figure 1]).

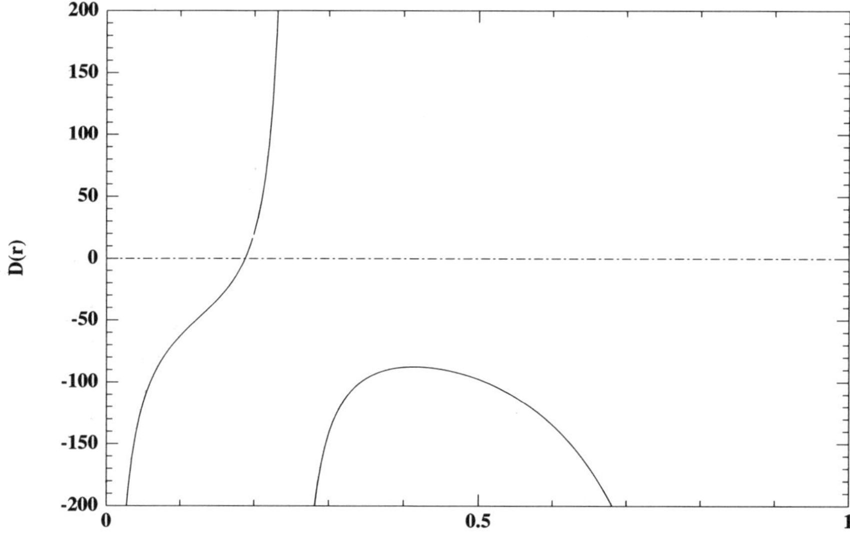

FIGURE 3. Twist coefficient $D(r)$ versus r. Shallow water. The 1:2 resonance occurs when $r = \frac{1}{4}$.

5. Concluding remarks

Formal expressions for weakly nonlinear spatially quasi-periodic travelling waves for the extreme cases of infinite depth and shallow depth, using the Hamiltonian structure and a variational principle, have been presented. A rigorous analysis which in turn leads to a dynamical systems formulation for quasi-periodic travelling waves is also possible, in finite depth, using the symplectic center manifold reduction of Mielke [1991]. For example, application of the symplectic center manifold theory to (2.12) leads to a reduced $\mathcal{C}^k$, for any finite k, differentiable canonical Hamiltonian system on $\mathbb{R}^4$. The reduced system can then be analyzed using standard techniques for invariant tori near equilibria of 2-DOF Hamiltonian systems (cf. Moser [1968]). In fact, Kaufmann [1986] gives explicit formulae for the KAM determinant and the twist coefficient for any 2-DOF system when the linearization is of the form studied here. Moreover such systems have weak homoclinic chaos between the invariant tori and so there should be uniformly travelling chaotic waves near the equilibrium for at least a Cantor set of values of the celerity. For gravity waves the possibility of such chaos is predicted in Baesens & MacKay [1992, §6] for large-amplitude gravity waves. For capillary-gravity waves it appears that such weak chaos should be pervasive even at very low amplitudes.

However in order to rigorously justify the existence of even quasi-periodic travelling waves it is necessary to prove persistence. For example Iooss & Los [1990] have proved rigorously the persistence of spatially quasi-periodic states for the Navier-Stokes equations when the quasi-periodic states are reversible. In the present case, with the addition of a Hamiltonian structure, the persistence question should be simpler since one can appeal to the $\mathcal{C}^k$-KAM theory.

Finally we note that the physical ramifications of these quasi-periodic travelling waves remain an interesting and open question.

Appendix A. Chart of the manifold

The Hamiltonian structure of the steady wave problem (2.1)-(2.4) is defined on the manifold M. There are many ways to chart this manifold, with the choice dependent on the particular application. For the case of gravity waves travelling in finite depth, Bridges & Mielke [1995] chart the manifold in such a way that it is also compatible with the time-dependent problem, which is necessary for a stability analysis, and as a by-product prove that M is a Hilbert manifold. To illustrate the parametrization of the manifold M we will use the Baesens-MacKay chart for the system when the surface tension is non-zero and the depth infinite: the case considered in §2.

Since there is an infinite number of ways to chart the manifold M it is important to note the distinction between charting the manifold and transforming the governing equations first and then characterizing the Hamiltonian structure. The latter approach implies – incorrectly – that there are many Hamiltonian structures for steady waves when there is only known to be one with many diffeomorphic representations. In other words there is an important distinction between a system with more that one symplectic structure (which is rare) and a system with a single symplectic structure with various (an infinity of) symplectic equivalents obtained by introducing particular morphisms (charts).

The Baesens-MacKay chart is constructed as follows. Transform the coordinates $(Z; x, y)$ using the diffeomorphism $(\widetilde{Z}; x, Y) = \mathcal{E} \circ (Z; x, y)$ defined by

$$Y = y - \eta(x) \quad \text{and} \quad \widetilde{Z} = \begin{pmatrix} G \\ \eta \\ \widetilde{w} \\ \Phi \\ U \end{pmatrix},$$

with

$$\Phi(x, Y) = \phi(x, y)|_{y=Y+\eta}\,, \ U(x, Y) = u(x, y)|_{y=Y+\eta}\,, \ \ G(x) = \Phi(x, 0)$$

and

$$\widetilde{w} = \sigma w - \int_{-\infty}^{\eta} u\phi_y \, dy\,.$$

Denote the image Hamiltonian system, under the transformation $\mathcal{E}$, by $(\widehat{M}, \widetilde{\omega}, \widetilde{S})$. Then,

$$\widetilde{S} \stackrel{\text{def}}{=} S \circ \mathcal{E} = \frac{1}{2} \int_{-\infty}^{0} [(U-c)^2 - \Phi_Y^2] \, dY - \frac{1}{2} g\eta^2 + \sigma \left(1 - \sqrt{1 - w^2}\right)$$

and

$$\begin{aligned} \widetilde{\alpha} \stackrel{\text{def}}{=} \mathcal{E}^* \alpha_\eta &= \mathcal{E}^* \left(\sigma w \mathbf{d}\eta + \int_{-\infty}^{\eta} u \mathbf{d}\phi \, dy \right) \\ &= \left(\widetilde{w} + \int_{-\infty}^{0} U \Phi_Y dY \right) \mathbf{d}\eta + \int_{-\infty}^{0} U(\mathbf{d}\Phi - \Phi_Y \mathbf{d}\eta) dY \\ &= \widetilde{w} \mathbf{d}\eta + \int_{-\infty}^{0} U \mathbf{d}\Phi \, dY\,, \end{aligned} \tag{A.1}$$

since $\mathbf{d}\phi = \mathbf{d}\Phi - \Phi_Y \mathbf{d}\eta$ and so

$$\widetilde{\omega} = \mathbf{d}\widetilde{\alpha} = \mathbf{d}\widetilde{w} \wedge \mathbf{d}\eta + \int_{-\infty}^{0} \mathbf{d}U \wedge \mathbf{d}\Phi \, dY ,$$

recovering the system $(\widetilde{M}, \widetilde{\omega}, \widetilde{S})$ formulated by Baesens & MacKay [1992, §2].

Appendix B. Twist coefficient

Let $H(q,p) = H_2 + H_3$ be the Hamiltonian (2.15), where H_2 (resp. H_3) denotes the quadratic (resp. cubic) part of H. Let

$$\mathbf{J} = \begin{pmatrix} \mathbf{0} & \mathbf{I}_2 \\ -\mathbf{I}_2 & \mathbf{0} \end{pmatrix} .$$

The equations of motion are given by

$$\dot{X} = \mathbf{J}\nabla H(X) ,$$

where $X = (q,p)^T$.

Let us follow the normalization described by Kaufmann [1986]. First we normalize the linear part. In order to simplify the notations, $1 - F$ is denoted by f and $\frac{1}{3} - W$ is denoted by w. The quadratic part of the Hamiltonian H_2 can be written as

$$H_2 = \tfrac{1}{2}\langle X, \mathbf{A}X\rangle ,$$

where

$$\mathbf{A} = \begin{pmatrix} -4f & 0 & 0 & 0 \\ 0 & 2w & 1 & 0 \\ 0 & 1 & 0 & 0 \\ 0 & 0 & 0 & \frac{45}{2} \end{pmatrix} .$$

The eigenvalues μ of $\mathbf{JA}$, which are the roots of $|\mathbf{JA} - \mu\mathbf{I}_4|$, are purely imaginary: $\mu = \pm iK_1, \pm iK_2$, with K_1 and K_2 satisfying (4.2). The eigenvectors e_1 and e_2 corresponding to the eigenvalues iK_1 and iK_2 are

$$e_k = \begin{pmatrix} 1 \\ 0 \\ 0 \\ -\frac{2}{45}K_k \end{pmatrix} + i \begin{pmatrix} 0 \\ K_k \\ -4f/K_k \\ 0 \end{pmatrix} , \qquad k = 1, 2 .$$

Kaufmann defines the signature as the sign of

$$v_k = -\langle \mathrm{Re}\,(e_k), \mathbf{A}\,\mathrm{Re}\,(e_k)\rangle .$$

One finds that

$$v_1 = \tfrac{2}{45}K_1^2(K_2^2 - K_1^2) , \quad v_2 = \tfrac{2}{45}K_2^2(K_1^2 - K_2^2) .$$

It follows that

$$\mathrm{sgn}(v_1) = +1 , \ \ \mathrm{sgn}(v_2) = -1 , \quad |v_1| = \tfrac{2}{45}K_1^2(K_2^2 - K_1^2) , \ |v_2| = \tfrac{2}{45}K_2^2(K_2^2 - K_1^2) .$$

Define the transformation matrix $T = [t_1, t_2, t_3, t_4]$ where

$$\begin{aligned} t_1 &= \sqrt{K_1/|v_1|}\,\mathrm{Re}\,(e_1) , \\ t_2 &= \sqrt{K_2/|v_2|}\,\mathrm{Re}\,(e_2) , \\ t_3 &= -\mathrm{sgn}(v_1)\sqrt{K_1/|v_1|}\,\mathrm{Im}\,(e_1) , \\ t_4 &= -\mathrm{sgn}(v_2)\sqrt{K_2/|v_2|}\,\mathrm{Im}\,(e_2) . \end{aligned}$$

The new coordinates $x = (x_1, x_2, y_1, y_2)^T$ are defined as $X = Tx$. The quadratic part of the Hamiltonian is transformed to

$$\tilde{H}_2 = -\tfrac{1}{2}K_1(x_1^2 + y_1^2) + \tfrac{1}{2}K_2(x_2^2 + y_2^2)\,. \tag{B.1}$$

The cubic part of the Hamiltonian becomes

$$\tilde{H}_3 = \tfrac{1}{6}e_{111}x_1^3 + \tfrac{1}{2}e_{112}x_1^2x_2 + \tfrac{1}{2}e_{122}x_1x_2^2 + \tfrac{1}{6}e_{222}x_2^3\,.$$

The expressions for the coefficients e_{ijk} can be easily computed.

Then Kaufmann describes the transformation of a Hamiltonian $\tilde{H}$ (whose quadratic part has the form B.1) into its Birkhoff normal form $\hat{H}$, which is described by

$$\hat{H} = \hat{H}_2 + \hat{H}_4\,,$$

where $\hat{H}_2 = \tilde{H}_2$ and

$$\hat{H}_4 = \tfrac{1}{8}d_{11}(\tilde{x}_1^2 + \tilde{y}_1^2)^2 + \tfrac{1}{8}d_{22}(\tilde{x}_2^2 + \tilde{y}_2^2)^2 + \tfrac{1}{4}d_{12}(\tilde{x}_1^2 + \tilde{y}_1^2)(\tilde{x}_2^2 + \tilde{y}_2^2)\,.$$

Kaufmann provides the relations giving d_{ij} in terms of e_{ijk}. The twist coefficient is the determinant D defined as follows:

$$D = -K_2^2 d_{11} - K_1^2 d_{22} - 2K_1K_2d_{12}\,.$$

The coefficients d_{ij} found following the normal form approach are precisely the ones found in §4.

References

[1] C. Baesens & R. S. MacKay [1992], *Uniformly travelling water waves from a dynamical systems viewpoint: some insights into bifurcations from Stokes' family*, J. Fluid Mech. **241**, 333–347

[2] T. J. Bridges [1992], *Spatial Hamiltonian structure, energy flux and the water–wave problem*, Proc. Roy. Soc. Lond. **A439**, 297–315

[3] T. J. Bridges & A. Mielke [1995], *A proof of the Benjamin-Feir instability*, Arch. Rat. Mech. Anal. **133**, 145–198

[4] A. Deprit & A. Deprit–Bartholomé [1967], *Stability of the triangular Lagrangian points*, The Astronomical Journal **72**, 173–179

[5] G. Iooss & K. Kirchgässner [1992], *Water waves for small surface tension: an approach via normal form*, Proc. Roy. Soc. Edinburgh **122A**, 267–299

[6] G. Iooss & J. Los [1990], *Bifurcation of spatially quasi-periodic solutions in hydrodynamic stability problems*, Nonlinearity **3**, 851–871

[7] S. Kaufmann [1986], *An explicit formula for the discussion of stability in autonomous Hamiltonian systems*, ZAMP **37**, 735–753

[8] T. Kawahara [1972], *Oscillatory solitary waves in dispersive media*, J. Phys. Soc. Japan **33**, 260–264

[9] A. Mielke [1991], *Hamiltonian and Lagrangian flows on center manifolds with applications to elliptic variational problems*, Lect. Notes in Math. **1489**, Springer: Heidelberg

[10] R. Moeckel [1990], *Generic bifurcations of the twist coefficient*, Ergod. Th. & Dynam. Sys. **10**, 185–195

[11] J. K. Moser [1968], *Lectures on Hamiltonian systems*, Mem. Amer. Math. Soc. **81**, 1–60

[12] I. Percival [1974], *Variational principles for the invariant toroids of classical dynamics*, J. Phys. A: Math., Nucl. Gen. **7**, 794–802

[13] J. A. Zufiria [1987], *Symmetry breaking in periodic and solitary gravity–capillary waves on water of finite depth*, J. Fluid Mech. **184**, 183–206

Department of Mathematical and Computing Sciences, University of Surrey, Guildford, GU2 5XH, UK

E-mail address: t.bridges@mcs.surrey.ac.uk

Institut Non-Linéaire de Nice, UMR 129 CNRS–UNSA, 1361 route des Lucioles, F-06560 Valbonne

E-mail address: dias@ecu.unice.fr

Contemporary Mathematics
Volume **200**, 1996

On modulations of weakly nonlinear water waves

T. Colin, F. Dias, and J.-M. Ghidaglia

ABSTRACT. The authors recently studied rotational effects on the modulations of water waves (Colin, Dias & Ghidaglia [1995]). In the process of deriving the amplitude equations, they found a new singularity for which the derivation failed. In this note, the nature of this singularity is discussed. Moreover, it is shown that the well-known resonant interaction between a short capillary–gravity wave and a long gravity wave is present only for one-dimensional modulations.

1. Introduction

The authors recently studied the modulational stability of Stokes' waves in the context of rotational free-surface Euler equations (Colin, Dias & Ghidaglia [1995], referred to as CDG below). Their primary interest was to study rotational effects on the stability of water waves and we refer to their conclusion (Section 5 of CDG) for a discussion of these effects. However, in the process of deriving the amplitude equations, they found a new singularity for which the derivation failed. This singularity is not related to the introduction of rotational effects and is also present in the classical potential case, although it was never mentioned before. In this note, the nature of this singularity is discussed. The motivation for this discussion was raised by Pierce [1995], who believes that this singularity is artificial because we neglected the second order homogeneous solutions. The singularity appeared as we were trying to show (and not to assume, as is usually done) that the mean flow depends on the slow space and time variables x_1 and t_1 through $(x_1 - c_g t_1)$, where c_g is the group velocity. In this note, we show that indeed one can neglect the second order homogeneous solutions. Therefore, our proof that the mean flow only depends on $(x_1 - c_g t_1)$ still holds away from the singular case. However, the singularity found previously disappears if one takes the average of the amplitude equations with respect to $(x_1 + c_g t_1)$. In other words, the Davey–Stewartson equations are still valid in the singular case, but the mean flow must be replaced by its average value with respect to $(x_1 + c_g t_1)$.

In this note, we also show that the well-known resonant interaction between a short capillary–gravity wave and a long gravity wave is present only for one-dimensional modulations.

1991 *Mathematics Subject Classification.* Primary 76B15.

2. Formulation of the problem

In this section, we recall the problem and the ansatz used in order to derive the Davey–Stewartson equations. We work in the general framework of rotational motions, but, as mentioned in the introduction, the questions we want to address are independent of the fact that the flow is rotational or not.

The three-dimensional Euler equations describing the motion of an inviscid and incompressible liquid layer of mean depth h read as follows. Denoting by $\Omega(t)$ the volume occupied at time t by the liquid,

$$\Omega(t) = \left\{X = (x,y,z) \in \mathbb{R}^3, -h \le z \le \eta(x,y,t)\right\} , \tag{2.1}$$

and by $\underline{V} = (u,v,w)$ the velocity vector field, we have

$$\underline{V}_t + (\underline{V}\cdot\nabla)\underline{V} + \nabla p = \underline{g} \quad \text{in } \Omega(t) , \tag{2.2}$$

$$\nabla\cdot\underline{V} = 0 \quad \text{in } \Omega(t) , \tag{2.3}$$

together with the boundary conditions

$$w = 0 , \quad \text{for } z = -h , \tag{2.4}$$

$$w = \frac{\partial\eta}{\partial t} + u\frac{\partial\eta}{\partial x} + v\frac{\partial\eta}{\partial y} , \quad \text{for } z = \eta(x,y,t) , \tag{2.5}$$

$$p - p_0 = -T\kappa , \quad \text{for } z = \eta(x,y,t) , \tag{2.6}$$

where $\eta(x,y,t)$ is the elevation of the free surface, κ its mean curvature:

$$\kappa = \frac{\eta_{xx}(1+\eta_y^2) - 2\eta_x\eta_y\eta_{xy} + \eta_{yy}(1+\eta_x^2)}{(1+\eta_x^2+\eta_y^2)^{3/2}} , \tag{2.7}$$

p the pressure, p_0 the pressure above the liquid, g the acceleration due to gravity and T the surface tension per unit density of the liquid.

The state of rest ($\underline{V} = 0$, $p = p_0 - gz$, $\eta = 0$) is solution to (2.2)–(2.6). Its linear stability is governed by the following set of equations on the strip

$$S = \left\{X = (x,y,z) \in \mathbb{R}^3, -h \le z \le 0\right\} :$$

$$\tilde{\underline{V}}_t + \nabla\tilde{p} = 0 , \tag{2.8}$$

$$\nabla\cdot\tilde{\underline{V}} = 0 , \tag{2.9}$$

$$\tilde{w} = 0 , \quad \text{for } z = -h , \tag{2.10}$$

$$\tilde{w} = \frac{\partial\tilde{\eta}}{\partial t} , \quad \text{for } z = 0 , \tag{2.11}$$

$$-g\tilde{\eta} + \tilde{p} = -T\Delta\tilde{\eta} , \quad \left(\Delta = \frac{\partial^2}{\partial x^2} + \frac{\partial^2}{\partial y^2}\right) , \quad \text{for } z = 0 , \tag{2.12}$$

where

$$\underline{V} = \tilde{\underline{V}} + \cdots , p = p_0 - gz + \tilde{p} + \cdots , w = \tilde{w} + \cdots , \eta = \tilde{\eta} + \cdots .$$

The existence of plane wave solutions to (2.8)–(2.12) of the form

$$(\tilde{\underline{V}}, \tilde{p}, \tilde{w}, \tilde{\eta}) = e^{i(kx-\omega t)}(\hat{V}(z), \hat{p}(z), \hat{w}(z), \hat{\eta}) \tag{2.13}$$

is decided upon the dispersion relation

$$\omega^2 = k(g + Tk^2)\tanh(kh) . \tag{2.14}$$

Hence solutions to (2.8)–(2.12) are bounded and the state of rest is linearly stable. It is not asymptotically stable; therefore the study of the nonlinear stability of the state of rest cannot be decided at this stage.

In order to address the problem of nonlinear interactions of small amplitude waves, we assume that the initial data for (2.2)–(2.6) have the following form:

$$\underline{V}(t=0) = \sum_{n=1}^{3} \epsilon^n \underline{V}_n^0(x, \epsilon x, \epsilon^2 x, \epsilon y, \epsilon^2 y, z) + o(\epsilon^3)\,, \tag{2.15}$$

$$\eta(t=0) = \sum_{n=1}^{3} \epsilon^n \eta_n^0(x, \epsilon x, \epsilon^2 x, \epsilon y, \epsilon^2 y) + o(\epsilon^3)\,. \tag{2.16}$$

That is, the initial data are of size ϵ and depend also on the slow variables $(\epsilon x, \epsilon^2 x, \epsilon y, \epsilon^2 y)$. At this point we make the following ansatz on $\underline{V}$, p and η. For that, it is convenient to introduce several temporal scales: $t_0 = t, t_1 = \epsilon t, t_2 = \epsilon^2 t$ and we assume that $\underline{V}, \eta, p$ have uniformly valid asymptotic expansions in terms of ϵ:

$$\underline{V} = \sum_{n=1}^{3} \epsilon^n \underline{V}_n(x_0, x_1, x_2, y_1, y_2, z; t_0, t_1, t_2) + o(\epsilon^3)\,, \tag{2.17}$$

$$p = p_0 - gz + \sum_{n=1}^{3} \epsilon^n p_n(x_0, x_1, x_2, y_1, y_2, z; t_0, t_1, t_2) + o(\epsilon^3)\,, \tag{2.18}$$

$$\eta = \sum_{n=1}^{3} \epsilon^n \eta_n(x_0, x_1, x_2, y_1, y_2; t_0, t_1, t_2) + o(\epsilon^3)\,, \tag{2.19}$$

where $\underline{V} = (u, v, w)$, $\underline{V}_n = (u_n, v_n, w_n)$ and

$$x_0 = x,\ x_1 = \epsilon x,\ x_2 = \epsilon^2 x,\ y_1 = \epsilon y,\ y_2 = \epsilon^2 y, \tag{2.20}$$

denote the spatial scales introduced in (2.15). In the next section, we will apply the derivative expansion method to this problem (Davey & Stewartson [1974], Kawahara [1975], Djordjevic & Redekopp [1977]).

3. Three-dimensional flows: amplitude equations

In order to obtain the amplitude equations, we substitute the ansatz (2.17)–(2.19) into (2.2)–(2.6) and we expand (2.5)–(2.6) around $z = 0$. Equating coefficients of equal powers of ϵ yields three sets of equations up to the third order.

As solutions to the first order linear equations, we take the plane waves introduced in Section 2. More precisely:

$$u_1 = \mathrm{i}A(x_1, x_2, y_1, y_2; t_1, t_2) \cosh[k(z+h)]\mathrm{e}^{i\theta} + \text{c.c.}\,, \tag{3.1}$$

$$v_1 = 0\,, \tag{3.2}$$

$$w_1 = A(x_1, x_2, y_1, y_2; t_1, t_2) \sinh[k(z+h)]\mathrm{e}^{i\theta} + \text{c.c.}\,, \tag{3.3}$$

$$p_1 = \frac{\mathrm{i}\omega}{k} A(x_1, x_2, y_1, y_2; t_1, t_2) \cosh[k(z+h)]\mathrm{e}^{i\theta} + \text{c.c.}\,, \tag{3.4}$$

$$\eta_1 = \frac{\mathrm{i}}{\omega} A(x_1, x_2, y_1, y_2; t_1, t_2) \sinh(kh)\mathrm{e}^{i\theta} + \text{c.c.}\,, \tag{3.5}$$

where $\theta = kx_0 - \omega t_0$ and ω satisfies the dispersion relation (2.14)

$$\omega^2 = k(g + Tk^2)\tanh(kh)\,,$$

and c.c. denotes the complex conjugate of the preceding terms. The complex amplitude A depends only upon the slow scales x_1, x_2, y_1, y_2, t_1 and t_2 and the aim of the expansion is to determine the differential equation satisfied by A. We now substitute this first order solution (3.1)–(3.5) into the second order equations and we look for a solution in the form

$$\begin{aligned} \underline{V}_2 &= e^{2i\theta}\underline{V}_{22}(x_1,x_2,y_1,y_2,z;t_1,t_2) + e^{i\theta}\underline{V}_{21}(x_1,x_2,y_1,y_2,z;t_1,t_2) + \text{c.c.} \\ &\quad +\underline{V}_{20}(x_1,x_2,y_1,y_2,z;t_1,t_2)\,, \end{aligned} \tag{3.6}$$

$$\begin{aligned} p_2 &= e^{2i\theta}p_{22}(x_1,x_2,y_1,y_2,z;t_1,t_2) + e^{i\theta}p_{21}(x_1,x_2,y_1,y_2,z;t_1,t_2) + \text{c.c.} \\ &\quad +p_{20}(x_1,x_2,y_1,y_2,z;t_1,t_2)\,, \end{aligned} \tag{3.7}$$

$$\begin{aligned} \eta_2 &= e^{2i\theta}\eta_{22}(x_1,x_2,y_1,y_2;t_1,t_2) + e^{i\theta}\eta_{21}(x_1,x_2,y_1,y_2;t_1,t_2) + \text{c.c.} \\ &\quad +\eta_{20}(x_1,x_2,y_1,y_2;t_1,t_2)\,. \end{aligned} \tag{3.8}$$

One finds for the coefficient of $(e^{i\theta})^0$:

$$\left\{ \begin{array}{lcl} p_{20} &=& -|A|^2\cosh[2k(z+h)] + \zeta(x_1,x_2,y_1,y_2;t_1,t_2)\,, \\ w_{20} &=& 0\,, \\ \eta_{20} &=& \frac{1}{g}(\zeta - |A|^2)\,. \end{array} \right. \tag{3.9}$$

For the term in $(e^{i\theta})^1$, one obtains

$$\left\{ \begin{array}{lcl} u_{21} &=& (z+h)\sinh[k(z+h)]\frac{\partial A}{\partial x_1}\,, \\ v_{21} &=& \frac{1}{k}\cosh[k(z+h)]\frac{\partial A}{\partial y_1}\,, \\ w_{21} &=& -i(z+h)\cosh[k(z+h)]\frac{\partial A}{\partial x_1}\,, \\ p_{21} &=& \frac{\omega}{k}\Big((z+h)\sinh[k(z+h)]\frac{\partial A}{\partial x_1} - \frac{1}{k}\cosh[k(z+h)]\frac{\partial A}{\partial x_1} \\ && \qquad - \frac{1}{\omega}\cosh[k(z+h)]\frac{\partial A}{\partial t_1}\Big)\,, \\ \eta_{21} &=& \frac{1}{\omega}\Big(h\cosh(kh)\frac{\partial A}{\partial x_1} + \frac{\sinh(kh)}{\omega}\frac{\partial A}{\partial t_1}\Big)\,, \end{array} \right. \tag{3.10}$$

together with the non-secularity condition

$$\frac{\partial A}{\partial t_1} + c_g\frac{\partial A}{\partial x_1} = 0\,, \tag{3.11}$$

where c_g denotes the group velocity $d\omega/dk$. Moreover the higher harmonic terms are given by

$$\left\{ \begin{array}{lcl} u_{22} &=& -\mu A^2\cosh[2k(z+h)]\,, \\ v_{22} &=& 0\,, \\ w_{22} &=& i\mu A^2\sinh[2k(z+h)]\,, \\ p_{22} &=& -\mu\frac{\omega}{k}A^2\cosh[2k(z+h)] + \frac{1}{2}A^2\,, \\ \eta_{22} &=& -\frac{k}{2\omega^2}A^2\sinh(2kh) - \mu\frac{1}{2\omega}A^2\sinh(2kh)\,, \end{array} \right. \tag{3.12}$$

where

$$\mu = \frac{3k}{2\omega}\frac{g(1-\sigma^2)+Tk^2(3-\sigma^2)}{g\sigma^2+(\sigma^2-3)Tk^2}\,, \qquad \sigma = \tanh(kh)\,.$$

The denominator of μ vanishes when $g\sigma^2 + (\sigma^2-3)Tk^2 = 0$, which corresponds to the well-known second-harmonic resonance.

Note that, at this stage, one could add to the above given solution $(u_{21}, v_{21}, w_{21}, p_{21}, \eta_{21})$ an homogeneous solution (Pierce [1995]), i.e. terms like

$$
\begin{array}{c}
\mathrm{i}D(x_1, x_2, y_1, y_2; t_1, t_2)\cosh[k(z+h)]\,,\\
0\,,\\
D(x_1, x_2, y_1, y_2; t_1, t_2)\sinh[k(z+h)]\,,\\
\frac{\mathrm{i}\omega}{k}D(x_1, x_2, y_1, y_2; t_1, t_2)\cosh[k(z+h)]\,,\\
\frac{\mathrm{i}}{\omega}D(x_1, x_2, y_1, y_2; t_1, t_2)\sinh(kh)\,.
\end{array}
\tag{3.13}
$$

In CDG, we did not add these terms; however, the parameter ϵ that was introduced in order to apply the derivative expansion method is artificial and has no precise physical meaning. It is therefore possible to use $\tilde{A} = A + \epsilon D$ as a new amplitude. If we use this new amplitude, the non-secularity condition obtained for the term $e^{\mathrm{i}\theta}$ of the second order solution yields the equivalent of (3.11), that is

$$\frac{\partial \tilde{A}}{\partial t_1} + c_g \frac{\partial \tilde{A}}{\partial x_1} = 0\,.$$

This last equation yields with (3.11)

$$\frac{\partial D}{\partial t_1} + c_g \frac{\partial D}{\partial x_1} = 0\,. \tag{3.14}$$

Note that Djordjevic & Redekopp [1977] take into account the second order homogeneous solution, but they assume *a priori* that D satisfies (3.14). Again, we show that this is not an hypothesis.

Introducing (3.12) into the third order perturbation equations yields the following set of amplitude equations:

$$2\mathrm{i}\sinh(kh)\left(\frac{\partial A}{\partial t_2} + c_g\frac{\partial A}{\partial x_2}\right) + \sinh(kh)\frac{d^2\omega}{dk^2}\frac{\partial^2 A}{\partial \xi^2} + c_g\frac{\sinh(kh)}{k}\frac{\partial^2 A}{\partial y_1^2}$$

$$= \frac{k\omega}{g\cosh(kh)}A\zeta + \frac{2k^2 A}{\cosh(kh)}\int_{-h}^{0}\cosh[2k(z+h)]u_{20}(z)\,dz - \nu|A|^2 A\,, \tag{3.15}$$

$$\frac{\partial \zeta}{\partial x_1} + \frac{\partial u_{20}}{\partial t_1} = 0 \quad \text{in } S\,, \tag{3.16}$$

$$\frac{\partial \zeta}{\partial y_1} + \frac{\partial v_{20}}{\partial t_1} = 0 \quad \text{in } S\,, \tag{3.17}$$

$$\frac{\partial w_{30}}{\partial z} + \frac{\partial v_{20}}{\partial y_1} + \frac{\partial u_{20}}{\partial x_1} = 0 \quad \text{in } S\,, \tag{3.18}$$

$$w_{30} = 0 \quad \text{at } z = -h\,, \tag{3.19}$$

$$w_{30} - \frac{1}{g}\left(c_g\frac{\partial |A|^2}{\partial \xi} + \frac{\partial \zeta}{\partial t_1}\right) - \frac{\sinh(2kh)}{\omega}\frac{\partial |A|^2}{\partial \xi} = 0 \quad \text{at } z = 0\,, \tag{3.20}$$

and A depends on x_1 and t_1 only through $\xi = x_1 - c_g t_1$. Note that there is a contribution of D to equation (3.15), and this contribution is

$$2\mathrm{i}\sinh(kh)\left(\frac{\partial D}{\partial t_1} + c_g\frac{\partial D}{\partial x_1}\right),$$

which is zero according to (3.14).

The coefficient ν is given by

$$\begin{aligned}\nu &= \tfrac{1}{2}k^3\sigma^2\left[g^3(9-12\sigma^2+13\sigma^4-2\sigma^6)+k^2g^2(36-62\sigma^2+33\sigma^4-6\sigma^6)T\right.\\ &\quad \left.+gk^4(33-55\sigma^2+30\sigma^4-6\sigma^6)T^2+k^6(6-14\sigma^2+10\sigma^4-2\sigma^6)T^3\right]\\ &\quad \times\cosh(kh)\left[g\omega^3(\sigma^2-1)(g\sigma^2-3Tk^2+k^2\sigma^2T)\right]^{-1}.\end{aligned}$$

This value of ν corresponds to that given in Djordjevic & Redekopp [1977] after a correction by a multiplication factor due to the fact that we do not take the same linear solution as they do. The set of equations (3.15)–(3.20) governs the modulations of a small amplitude wave train.

Let us conclude this section by saying that the introduction of the term (3.13) does not have any influence on the amplitude equations (3.15)–(3.20).

The next step is to simplify this system and to clarify the situation concerning the singularity found in CDG.

4. On the singularity $gh = \frac{g^2\sinh^2(2kh)}{\omega^2}$.

For the convenience of the reader, we recall here how the Davey–Stewartson equations were derived from (3.15)–(3.20) in CDG, and how the new singularity appeared. We will show that the functions ζ and $\int_{-h}^{0}\cosh[2k(z+h)]u_{20}(z)\,dz$ depend on x_1 and t_1 only through $\xi = x_1 - c_g t_1$, except for a singular case. We first apply the differential operator

$$\left(\frac{\partial}{\partial t_1}+c_g\frac{\partial}{\partial x_1}\right)\frac{\partial}{\partial t_1}$$

on (3.15) and the fact that $\frac{\partial A}{\partial t_1}+c_g\frac{\partial A}{\partial x_1}=0$ yields

(4.1)

$$k\omega\left(\frac{\partial}{\partial t_1}+c_g\frac{\partial}{\partial x_1}\right)\frac{\partial\zeta}{\partial t_1}+2gk^2\int_{-h}^{0}\left(\frac{\partial}{\partial t_1}+c_g\frac{\partial}{\partial x_1}\right)\frac{\partial u_{20}}{\partial t_1}\cosh[2k(z+h)]\,dz=0\,.$$

Equation (3.16) gives

$$\frac{\partial\zeta}{\partial x_1}=-\frac{\partial u_{20}}{\partial t_1}\,.$$

Therefore, (4.1) leads to

$$\left(\frac{\partial}{\partial t_1}+c_g\frac{\partial}{\partial x_1}\right)\left(\frac{\partial}{\partial t_1}-\frac{g\sinh(2kh)}{\omega}\frac{\partial}{\partial x_1}\right)\zeta=0\,.$$

Hence ζ can be written in the form

$$\zeta = f_1(x_1-c_gt_1,y_1)+f_2(x_1-\beta t_1,y_1)\,, \tag{4.2}$$

where $\beta=-g\sinh(2kh)/\omega$ and f_1, f_2 are arbitrary functions. We have dropped the dependences in x_2, y_2, t_2 which are considered as parameters in this calculation. On the other hand, (3.16) and (3.17) imply

$$\frac{\partial^2\zeta}{\partial x_1^2}=-\frac{\partial^2u_{20}}{\partial t_1\partial x_1}\quad\text{and}\quad\frac{\partial^2\zeta}{\partial y_1^2}=-\frac{\partial^2v_{20}}{\partial t_1\partial y_1}\,, \tag{4.3}$$

while $\frac{\partial(3.18)}{\partial t_1}$ gives

$$\frac{\partial^2w_{30}}{\partial z\partial t_1}=-\frac{\partial^2v_{20}}{\partial t_1\partial y_1}-\frac{\partial^2u_{20}}{\partial t_1\partial x_1}\,. \tag{4.4}$$

Equations (4.3) and (4.4) lead to

$$\frac{\partial^2 w_{30}}{\partial t_1 \partial z} = \Delta_{x_1,y_1}\zeta\,.$$

The boundary condition (3.19) yields

$$\frac{\partial w_{30}}{\partial t_1} = (z+h)\Delta_{x_1,y_1}\zeta\,, \tag{4.5}$$

since ζ does not depend on z. We substitute (4.5) in $\frac{\partial(3.20)}{\partial t_1}$:

$$h\Delta_{x_1,y_1}\zeta - \frac{1}{g}\left(-c_g^2\frac{\partial^2|A|^2}{\partial\xi^2} + \frac{\partial^2\zeta}{\partial t_1^2}\right) + \frac{\sinh(2kh)}{\omega}c_g\frac{\partial^2|A|^2}{\partial\xi^2} = 0\,. \tag{4.6}$$

Combining (4.2) and (4.6) and denoting by Ξ the quantity $x_1 - \beta t_1$ yields

$$\begin{aligned} h\frac{\partial^2 f_1}{\partial\xi^2} + h\frac{\partial^2 f_1}{\partial y_1^2} + h\frac{\partial^2 f_2}{\partial\Xi^2} + h\frac{\partial^2 f_2}{\partial y_1^2} + \frac{c_g^2}{g}\frac{\partial^2|A|^2}{\partial\xi^2} & \\ -\frac{c_g^2}{g}\frac{\partial^2 f_1}{\partial\xi^2} - \frac{\beta^2}{g}\frac{\partial^2 f_2}{\partial\Xi^2} + \frac{c_g\sinh(2kh)}{\omega}\frac{\partial^2|A|^2}{\partial\xi^2} &= 0\,. \end{aligned} \tag{4.7}$$

It follows that

$$\frac{\partial^2 f_2}{\partial\Xi^2}\left(h - \frac{\beta^2}{g}\right) + h\frac{\partial^2 f_2}{\partial y_1^2} = \alpha(y_1)\,. \tag{4.8}$$

Since $\alpha(y_1) \to 0$ as $\Xi \to \infty$, it follows that $\alpha(y_1) \equiv 0$ and (4.8) reads

$$\frac{\partial^2 f_2}{\partial\Xi^2}\left(1 - \frac{\beta^2}{gh}\right) + \frac{\partial^2 f_2}{\partial y_1^2} = 0\,. \tag{4.9}$$

Three cases can occur:

(i) If $gh - \beta^2 > 0$: taking the Fourier transform in Ξ, y_1, we obtain that f_2 has to be linear in Ξ, y_1 and since f_2 tends to zero at infinity, $f_2 \equiv 0$.

(ii) If $gh - \beta^2 < 0$: f_2 satisfies a wave equation in the variables Ξ, y_1, hence it can be written as

$$f_2 = f_{2+}(\Xi + \mathcal{C}y_1) + f_{2-}(\Xi - \mathcal{C}y_1)\,,$$

where $\mathcal{C} = \sqrt{\beta^2/gh - 1}$. It follows that $f_{2+} = f_{2-} = 0$, since f_2 has to decay at infinity.

(iii) If $gh - \beta^2 = 0$: it is a singular case and one cannot conclude. This singularity occurs when

$$gh = \frac{g^2\sinh^2(2kh)}{\omega^2}\,. \tag{4.10}$$

In both cases (i) and (ii), we obtain that ζ depends on x_1 and t_1 only through ξ. Now, (3.15) implies that

$$k\omega\zeta + 2gk^2\int_{-h}^{0}\cosh[2k(z+h)]u_{20}\,dz$$

depends on x_1 and t_1 only through ξ and we denote its value by $2gk^2\mu(\xi, y_1)$. Hence

$$\int_{-h}^{0}\cosh[2k(z+h)]u_{20}\,dz = \mu(\xi,y_1) - \frac{\omega}{2gk}\zeta\,. \tag{4.11}$$

Differentiating (4.11) with respect to t_1 leads to

$$\int_{-h}^{0} \cosh[2k(z+h)] \frac{\partial u_{20}}{\partial t_1}(z)\, dz = -c_g \frac{\partial \mu}{\partial \xi} + \frac{c_g \omega}{2gk} \frac{\partial \zeta}{\partial \xi}. \tag{4.12}$$

Moreover, multiplying (3.16) by $\cosh[2k(z+h)]$ and integrating on $[-h;0]$ leads to

$$\int_{-h}^{0} \cosh[2k(z+h)] \frac{\partial u_{20}}{\partial t_1}(z)\, dz = -\frac{\partial \zeta}{\partial x_1} \frac{\sinh(2kh)}{2k}. \tag{4.13}$$

Combining (4.13) and (4.12) yields

$$\frac{1}{c_g}\left(\frac{c_g \omega}{2gk} + \frac{\sinh(2kh)}{2k}\right) \frac{\partial \zeta}{\partial \xi} = \frac{\partial \mu}{\partial \xi}.$$

It follows that

$$\mu = \frac{\zeta}{c_g}\left(\frac{c_g \omega}{2gk} + \frac{\sinh(2kh)}{2k}\right).$$

Hence (4.11) reads

$$\omega \zeta + 2gk \int_{-h}^{0} \cosh[2k(z+h)] u_{20}(z)\, dz = \frac{2gk}{c_g}\left(\frac{c_g \omega}{2gk} + \frac{\sinh(2kh)}{2k}\right) \zeta,$$

and (3.15) becomes

$$2i \sinh(kh)\left(\frac{\partial A}{\partial t_2} + c_g \frac{\partial A}{\partial x_2}\right) + \sinh(kh) \frac{d^2\omega}{dk^2} \frac{\partial^2 A}{\partial \xi^2} + \frac{c_g \sinh(kh)}{k} \frac{\partial^2 A}{\partial y_1^2}$$

$$= \frac{2k^2}{c_g \cosh(kh)}\left(\frac{c_g \omega}{2gk} + \frac{\sinh(2kh)}{2k}\right) \zeta A - \nu |A|^2 A. \tag{4.14}$$

and

$$(gh - c_g^2) \frac{\partial^2 \zeta}{\partial x_1^2} + gh \frac{\partial^2 \zeta}{\partial y_1^2} = -c_g^2 \left(1 + \frac{g \sinh(2kh)}{c_g \omega}\right) \frac{\partial^2 |A|^2}{\partial \xi^2}. \tag{4.15}$$

Equations (4.14) and (4.15) form the Davey–Stewartson system.

Concerning the new singularity that we have found, we notice that we can proceed following the idea of Pierce & Knobloch [1994]. Let us introduce the variable

$$\xi_1 = x_1 + c_g t_1.$$

Using ξ and ξ_1 instead of x_1 and t_1, and taking the mean value of equations (3.15)-(3.20) with respect to ξ_1 yields the following set of equations:

$$2i \sinh(kh)\left(\frac{\partial A}{\partial t_2} + c_g \frac{\partial A}{\partial x_2}\right) + \sinh(kh) \frac{d^2\omega}{dk^2} \frac{\partial^2 A}{\partial \xi^2} + c_g \frac{\sinh(kh)}{k} \frac{\partial^2 A}{\partial y_1^2}$$

$$= \frac{k\omega}{g \cosh(kh)} A \langle \zeta \rangle + \frac{2k^2 A}{\cosh(kh)} \int_{-h}^{0} \cosh[2k(z+h)] \langle u_{20} \rangle (z)\, dz - \nu |A|^2 A, \tag{4.16}$$

$$\frac{\partial\langle\zeta\rangle}{\partial\xi} - c_g\frac{\partial\langle u_{20}\rangle}{\partial\xi} = 0 \quad \text{in } S, \tag{4.17}$$

$$\frac{\partial\langle\zeta\rangle}{\partial y_1} - c_g\frac{\partial\langle v_{20}\rangle}{\partial\xi} = 0 \quad \text{in } S, \tag{4.18}$$

$$\frac{\partial\langle w_{30}\rangle}{\partial z} + \frac{\partial\langle v_{20}\rangle}{\partial y_1} + \frac{\partial\langle u_{20}\rangle}{\partial\xi} = 0 \quad \text{in } S, \tag{4.19}$$

$$\langle w_{30}\rangle = 0 \quad \text{at } z=-h, \tag{4.20}$$

$$\langle w_{30}\rangle - \frac{c_g}{g}\left(\frac{\partial|A|^2}{\partial\xi} - \frac{\partial\langle\zeta\rangle}{\partial\xi}\right) - \frac{\sinh(2kh)}{\omega}\frac{\partial|A|^2}{\partial\xi} = 0 \quad \text{at } z=0, \tag{4.21}$$

where $\langle.\rangle$ denotes the mean value in the ξ_1 component. It is easy to see that the same proof as above shows that A and $\langle\zeta\rangle$ satisfy the D–S equations.

In order to conclude these computations, let us note that in the case when $gh \neq \frac{g^2\sinh^2(2kh)}{\omega^2}$, we know that $\langle\zeta\rangle = \zeta$, and therefore A and ζ satisfy the D–S equations (4.14)-(4.15). But when $gh = \frac{g^2\sinh^2(2kh)}{\omega^2}$, we do not know whether $\langle\zeta\rangle = \zeta$ or not; however, A and $\langle\zeta\rangle$ still satisfy the D–S equations, i.e.

$$2i\sinh(kh)\left(\frac{\partial A}{\partial t_2} + c_g\frac{\partial A}{\partial x_2}\right) + \sinh(kh)\frac{d^2\omega}{dk^2}\frac{\partial^2 A}{\partial\xi^2} + \frac{c_g\sinh(kh)}{k}\frac{\partial^2 A}{\partial y_1^2}$$

$$= \frac{2k^2}{c_g\cosh(kh)}\left(\frac{c_g\omega}{2gk} + \frac{\sinh(2kh)}{2k}\right)\langle\zeta\rangle A - \nu|A|^2 A. \tag{4.22}$$

Since ζ depends only on ξ, (4.6) yields

$$(gh - c_g^2)\frac{\partial^2\langle\zeta\rangle}{\partial x_1^2} + gh\frac{\partial^2\langle\zeta\rangle}{\partial y_1^2} = -c_g^2\left(1 + \frac{g\sinh(2kh)}{c_g\omega}\right)\frac{\partial^2|A|^2}{\partial\xi^2}. \tag{4.23}$$

That is why, one can say that when $gh = \frac{g^2\sinh^2(2kh)}{\omega^2}$ the system is not resonant, but it is singular since the analysis is different for this value of the parameters.

5. On the long wave-short wave resonance

Djordjevic & Redekopp [1977] noticed that when $c_g^2 = gh$ the modulation equations that they found are singular. However, the form of the Davey–Stewartson equations that we have found here (equations (4.14)-(4.15)) is not singular when $c_g^2 = gh$. Indeed, they become

$$2i\sinh(kh)\left(\frac{\partial A}{\partial t_2} + c_g\frac{\partial A}{\partial x_2}\right) + \sinh(kh)\frac{d^2\omega}{dk^2}\frac{\partial^2 A}{\partial\xi^2} + \frac{c_g\sinh(kh)}{k}\frac{\partial^2 A}{\partial y_1^2}$$

$$= \frac{2k^2}{c_g\cosh(kh)}\left(\frac{c_g\omega}{2gk} + \frac{\sinh(2kh)}{2k}\right)\zeta A - \nu|A|^2 A, \tag{5.1}$$

and

$$gh\frac{\partial^2\zeta}{\partial y_1^2} = -c_g^2\left(1 + \frac{g\sinh(2kh)}{c_g\omega}\right)\frac{\partial^2|A|^2}{\partial\xi^2}. \tag{5.2}$$

These equations are similar to that obtained for pure capillary waves by Djordjevic & Redekopp [1977] (see their equations (A7) and (A8)), when exchanging the roles of the x and y variables. Therefore, there is no long wave-short wave resonance in 3D. That is why Grimshaw [1977], who studied the modulations of an internal gravity wave packet and considered the special case of the resonant interaction with a long wave, obtained in the 3D case the same amplitude equations for the resonance as that obtained by Djordjevic & Redekopp [1977] in the 2D resonant case! Of course, in the 2D case, when the functions do not depend on the variable y, equation (5.2) implies that $|A|^2$ does not depend on ξ, i.e. the train of wave is not modulated, and there is indeed a resonance. Therefore a different scaling must be introduced in order to study modulations near that resonance.

References

[1] Benney D. J. and Roskes G. J. (1969), *Wave instabilities*, Stud. Appl. Math. **48**, 377-85.

[2] Colin T., Dias F. and Ghidaglia J.-M. (1995), *On rotational effects in the modulations of weakly nonlinear water waves over finite depth*, Eur. J. Mech. B/Fluids **14**, 775-93.

[3] Davey A. and Stewartson K. (1974), *On three-dimensional packets of surface waves*, Proc. R. Soc. Lond. A **338**, 101-10.

[4] Djordjevic V. D. and Redekopp L. G. (1977), *On two-dimensional packets of capillary–gravity waves*, J. Fluid Mech. **79**, 703-14.

[5] Kawahara T. (1975), *Nonlinear self-modulation of capillary–gravity waves on a liquid layer*, J. Phys. Soc. Japan **38**, 265-70.

[6] Grimshaw R. H. J. (1977), *The modulation of an internal gravity-wave packet, and the resonance with the mean motion*, Stud. Appl. Math. **56**, 241-66.

[7] Pierce R. D. and Knobloch E. (1994), *On the modulational stability of traveling and standing waves*, Phys. Fluids **6**, 1177-90.

[8] Pierce R. D. (1995), Personal communication.

Mathématiques Appliquées de Bordeaux, CNRS ERS 123 and Université de Bordeaux I, 351 cours de la Libération, F-33405 Talence Cedex

E-mail address: colin@ceremab.u-bordeaux.fr

Institut Non-Linéaire de Nice, UMR 129 CNRS–UNSA, 1361 route des Lucioles, F-06560 Valbonne

E-mail address: dias@ecu.unice.fr

Centre de Mathématiques et de Leurs Applications, CNRS URA 1611 and Ecole Normale Supérieure de Cachan, 61 Av. du Président Wilson, F-94235 Cachan Cedex

E-mail address: jmg@cmla.ens-cachan.fr

Contemporary Mathematics
Volume **200**, 1996

Birkhoff normal forms for water waves

Walter Craig

To the memory of T. Brooke Benjamin

ABSTRACT. The water wave problem describes the free surface of a fluid which satisfies Euler's equations and is irrotational. It may be formulated as a hamiltonian system, where energy, mass and momentum are conserved integrals and the equilibrium solution is an elliptic stationary point of the flow. It is natural to seek a sequence of Birkhoff normal forms transformations for this system, despite the fact that it is a problem with infinitely many degrees of freedom. The third order normal form transformation is convergent in a neighborhood of the equilibrium solution; this is shown using an abstract form of the Cauchy-Kowalewsky theorem due to Nirenberg, and estimates of the Dirichlet-Neumann operator for the fluid domain. We also consider higher order normal forms transformations. A.I. Dyachenko and V.E. Zakharov have calculated for the deep water problem that all the coefficients of 'nonintegrable' quartic terms in the water wave hamiltonian vanish. We have verified this fact, and furthermore have shown that the phenomenon of integrability of the formal Birkhoff normal form in terms of the action - angle variables of the harmonic oscillator does not continue to quintic terms. We use the explicit form of the fifth order hamiltonian to discuss the periodic standing waves in the 1:2 resonant plane. There are four families of such orbits in a neighborhood of the equilibrium; two are stable, and two are unstable with at least one hyperbolic normal direction. In more generality, we show that finite mode approximations of a standing wave are unstable to modes which have four times the maximum represented wave number. This is proposed as the reason behind the difficulties with numerical methods for the time - periodic standing wave.

§1 INTRODUCTION

This article concerns the water wave problem, taking the point of view that some aspects of the transformation theory of finite dimensional hamiltonian mechanics can be successfully applied in problems with infinitely many degrees of freedom. Starting with the original fluid dynamics problem of water waves, posed in two dimensions, we introduce a version of Zakharov's hamiltonian [**17**] for the problem, which involves the Dirichlet - Neumann operator of the fluid domain [**4**].

1991 *Mathematics Subject Classification.* Primary 76B15; Secondary 34C15, 35L45.
The research of the author was supported in part by NSF Grant # DMS 94 01514.

Mass, momentum and total energy are conserved for the system, and if the added mass is normalized to zero, the equilibrium solution can be seen to be an elliptic stationary point for the flow. Thus it is natural to ask whether a Birkhoff normal forms transformation can be used to simplify the lower order terms of the hamiltonian. On a neighborhood of this equilibrium in a space of analytic functions, the third order normal forms transformation exists, eliminating from the transformed hamiltonian all cubic terms. When the fluid is chosen to be infinitely deep, one can continue the procedure, at least on a formal level, and there are certain interesting aspects of the higher order normal forms transformations. It is remarkable that in the fourth order transformed hamiltonian, every coefficient of a 'nonintegrable' resonant monomial vanishes exactly (V. E. Zakharov [**18**], A. I. Dyachenko & V. E. Zakharov [**6**] and W. Craig & P. Worfolk [**5**]). Although it is tempting to speculate that this presages the vanishing of the coefficients of all higher order terms of a normal form which are not functions of the action variables of the quadratic hamiltonian, this turns out to not be true, and indeed in [**5**] and in [**7**] nonvanishing cases are given. The explicit form of the hamiltonian lends itself readily to calculation; in this paper we study the periodic standing waves in the 1:2 resonant plane, and show that in a neighborhood of the equilibrium there are four families of periodic solutions. Two are families of stable elliptic orbits, and the other two are unstable hyperbolic periodic orbits, with an asymptotic bound on their Lyapunov exponents. In more generality, we show that no finite mode truncation of the water wave problem at fifth order is invariant for the full evolution. As a particular case, we show that because of the 1:2 resonance, any finite mode approximation of the time - periodic standing wave is modified by a nontrivial correction, under refinements which include modes of four times the previously represented wave number. In many cases we show that this approximate standing wave is dynamically unstable, with a positive Lyapunov exponent. This result is proposed as an explanation of the recognized difficulty of numerical calculations of standing waves.

There are well known discussoins of the classical standing wave problem, for example see I. Tadjbakhsh and J. B. Keller [**16**]. Our point of view on this problem is a natural continuation of the analysis of [**4**]. It is also related to the interesting work of P. J. Bryant and M. Stiassnie [**1**] on the subject of deep water standing waves, in which they consider the 1:2:3 resonant subspace (corresponding to wave numbers 1:4:9) without performing normal forms transformations. However in their paper they set all of the five wave interactions to zero, and study a system with sixth order contributions that is intended to approximate standing waves. Our work contrasts to this, and in fact we have found that there are nontrivial five wave interactions, and therefore our description of the structure of the small amplitude periodic standing waves and their stability differs from the account in [**1**].

The main disadvantage of this approach to the water wave problem is that (as far as I know) the rigorous analysis is limited to the category of analytic data. On the other hand it has a certain elegance of point of view, and it should encourage one to take on other problems with these techniques. Indeed, in J. Shatah [**15**] and in H. P. McKean and J. Shatah [**11**], the analysis of normal forms transformations plays a role in the existance theory for the nonlinear Schrödinger and Klein - Gordon equations. Furthermore, Birkhoff normal forms have been used by S. Kuksin and J. Pöschel in their paper [**10**] on invariant finite dimensional tori for the nonlinear Schrödinger equation, in order to show that the problem satisfies a condition of genuine nonlinearity. In the case of the water wave problem, Y. Kodama [**8**][**9**] has proposed a normal form in terms of nonlinear differential operators (local operators) in expansions in the long wave scaling regime. This is quite different from the present approach, and his method is much more difficult to justify by rigorous analysis.

§2 The water wave problem

This is a classical problem for an ideal fluid with a free surface, being acted upon by gravity. We will consider the fluid motion to be two dimensional, with the fluid domain to be defined as $S(\eta) = \{(x,y) \ : \ -h < y < \eta(x,t), \ 0 < x < 2\pi\}$, which has a horizontal bottom boundary. We include the possibility that $h = +\infty$, so that the fluid region may be infinitely deep. The fluid itself is taken to be incompressible and irrotational, so that the Euler equations give rise to a potential flow;

$$\begin{aligned} u &= \nabla\varphi \\ \Delta\varphi &= 0 \ , \qquad \text{for } (x,y) \text{ in } S(\eta) \ , \end{aligned} \tag{1}$$

therefore

$$\nabla \cdot u = 0 \ , \qquad \nabla \wedge u = 0 \ .$$

Boundary conditions on solid, impermeable boundaries such as $\{(x,y) \ : \ y = -h\}$ are that

$$\partial_n \varphi = 0 \ . \tag{2}$$

Top boundary conditions introduce all of the time dependence and nonlinearity into the problem; on the free surface $\{y = \eta(x)\}$ they are

$$\begin{aligned} \partial_t \eta &= \partial_y \varphi - \partial_x \eta \partial_x \varphi \\ \partial_t \varphi &= -g\eta - \tfrac{1}{2}(\nabla\varphi)^2 \ . \end{aligned} \tag{3}$$

Boundary conditions on the lateral sides of the fluid domain also have to be specified. We will be interested in two possible choices, periodic boundary conditions

$$\eta(x+2\pi,t) = \eta(x,t) \ , \qquad \varphi(x+2\pi,y,t) = \varphi(x,y,t) \ , \tag{4}$$

or possibly Neumann boundary conditions, in case that the fluid is taken to occupy a region with solid vertical walls,

$$\begin{aligned} \partial_x \eta(0,t) &= 0 = -\partial_x \eta(\pi,t) \\ \partial_x \varphi(0,y,t) &= 0 = -\partial_x \varphi(\pi,y,t) \ . \end{aligned} \tag{5}$$

Sometimes one adds the restoring force of surface tension to these equations, but this article will not discuss this more complicated situation. It is also the case that these equations could be given in three (or higher) dimensional settings, with for example $x \in \mathbf{T}^{n-1}$ an $(n-1)$-dimensional torus, but again this article only considers the two-dimensional case, with the dimension of the surface $n = 1$.

These equations can be written as a hamiltonian system, which is the starting point of our work. It is straightforward that the choice of hamiltonian function should be the total energy of the system. The kinetic energy is the Dirichlet integral for the fluid domain $S(\eta)$;

$$K = \int_0^{2\pi} \int_{-h}^{\eta(x)} \tfrac{1}{2}(\nabla\varphi)^2 \, dydx$$

and the potential energy is

$$V = \int_0^{2\pi} \tfrac{1}{2} \, g\eta^2(x) \, dx \, .$$

Thus the total energy is

$$H = K + V \, . \tag{6}$$

The subtle question is as to the correct choice of canonical variables. This was given in a paper of Zakharov [**17**] to be

$$(\eta(x), \; \xi(x) = \varphi(x, \eta(x))) \, . \tag{7}$$

This merits a remark on the Dirichlet problem for the fluid domain $S(\eta)$. Given $\eta(x)$ defining the domain, the boundary value problem (1)(2), for Dirichlet data given on the top free surface $\{y = \eta(x)\}$ is well posed. That is, for given $\eta(x)$ the boundary values $\xi(x)$ parametrize the solutions $\varphi(x,y) = \varphi[\eta,\xi](x,y)$ of (1)(2). It is then ξ that plays the role of the second set of canonical variables for the water wave problem. One central question is how this relationship between $\xi(x)$ and $\varphi(x,y)$ depends upon η.

Expressing the hamiltonian H in terms of (η, ξ),

$$\begin{aligned} H(\eta,\xi) &= \int_0^{2\pi} \int_{-h}^{\eta(x)} \tfrac{1}{2}(\nabla\varphi)^2 \, dydx + \int_0^{2\pi} \tfrac{1}{2} \, g\eta^2(x) \, dx \\ &= - \int_0^{2\pi} \int_{-h}^{\eta(x)} \tfrac{1}{2}\varphi\Delta\varphi \, dydx + \int_{y=-h} \varphi\nabla\varphi \cdot ndS \\ &\quad + \int_{y=\eta(x)} \varphi\nabla\varphi \cdot ndS + \int_0^{2\pi} \tfrac{1}{2} \, g\eta^2(x) \, dx \, , \end{aligned} \tag{8}$$

where the second expression follows from integration by parts. The first two terms of the resulting expression vanish, as φ is harmonic and satisfies the bottom boundary conditions, and we are left with the statement that the hamiltonian is

$$H(\eta,\xi) = \int_0^{2\pi} \tfrac{1}{2}\xi G(\eta)\xi \, dx + \int_0^{2\pi} \tfrac{1}{2} \, g\eta^2 \, dx \, . \tag{9}$$

The operator $G(\eta)$ appearing in (9) is the Dirichlet - Neumann operator for the fluid domain. It is a linear operator, giving from the data ξ the normal derivative of the solution φ of the boundary value problem (1)(2), with

$$\varphi(x, \eta(x)) = \xi(x) \, . \tag{10}$$

In other words, when φ satisfies (1)(2), with Dirichlet data (10), then

$$G(\eta)\xi(x) \, dx = (\nabla\varphi \cdot n)(x) \, dS(x) \, . \tag{11}$$

THEOREM 1. *The Dirichlet - Neumann operator satisfies $G(\eta) \geq 0$ and it is symmetric. Furthermore it is a bounded mapping $G(\eta) : H^1 \to L^2$ which for some R is analytic in the class of operators from H^1 to L^2 for $\eta \in B_R(0) \subseteq Lip([0, 2\pi])$.*

The analyticity result comes essentially from a result in harmonic analysis by R. Coifman and Y. Meyer [**3**] on the boundedness of Cauchy integrals with Lipschitz coefficients.

The system (1)(2) and (3) is rewritten as

$$\begin{aligned} \partial_t \eta &= \delta_\xi H = G(\eta)\xi \ , \\ \partial_t \xi &= -\ \delta_\eta H = -g\eta - \delta_\eta K \ . \end{aligned} \tag{12}$$

We will return to the calculation of the more complicated expression $\delta_\eta K$.

The analyticity of the operator $G(\eta)$ means that we may express it in terms of its Taylor series expansion

$$G(\eta) = \sum_{j \geq 0} G_j(\eta) \ , \tag{13}$$

with each term $G_j(\eta)$ homogeneous of degree j in $\eta(x)$. Placing (13) into the hamiltonian, we write that

$$\begin{aligned} H(\eta, \xi) &= H_2(\eta, \xi) + \sum_{j \geq 3} H_j(\eta, \xi) \\ &= \int_0^{2\pi} \tfrac{1}{2}(\xi G_0 \xi + g\eta^2) \ dx + \sum_{j \geq 3} \int_0^{2\pi} \tfrac{1}{2} \xi G_{j-2}(\eta) \xi \ dx \ . \end{aligned} \tag{14}$$

The basic conservation laws for the water wave problem are that mass and momentum are conserved, as well as the hamiltonian itself. The expression for the mass is

$$M \ = \ \int_0^{2\pi} \eta(x) \ dx \tag{15}$$

and the momentum (or impulse, as Brooke Benjamin would insist) is

$$I \ = \ \int_0^{2\pi} \eta(x) \partial_x \xi(x) \ dx \ . \tag{16}$$

Additionally, the average of the velocity potential is conserved,

$$\Xi \ = \ \int_0^{2\pi} \xi(x) \ dx \ . \tag{17}$$

The fact that they are conserved is seen by the computation of their Poisson bracket with H;

$$\{M, H\} = 0 \ , \qquad \{I, H\} = 0 \ , \qquad \{\Xi, H\} = 0 \ ,$$

where we define the Poisson bracket between two functionals $A(\eta, \xi), B(\eta, \xi)$ in the standard way

$$\{A, B\}(\eta, \xi) = \int_0^{2\pi} \delta_\xi A(x) \delta_\eta B(x) - \delta_\eta A(x) \delta_\xi B(x) \ dx \ . \tag{18}$$

Since level surfaces of M and Ξ are invariant under the flow of (12), we make it a point to restrict ourselves to the subspace in which both $M = 0$ and $\Xi = 0$.

For any initial surface $\eta(x)$ the first can be implemented by a possible change of definition of the depth h, and the latter can be assigned arbitrarily, as potentials are physically relevant only up to choice of sign.

These expressions for the equations of motion for the water wave problem in terms of the Dirichlet - Neumann operator and its Taylor expansion have been used in a number of publications, for example see [**4**].

§3 Linear theory

Zero is an equilibrium solution of equations (12), and the equations linearized about zero can be expressed in terms of the quadratic part of the hamiltonian

$$H_2(\eta,\xi) = \int_0^{2\pi} \tfrac{1}{2}(\xi G_0 \xi + g\eta^2)\, dx \ . \tag{19}$$

It is easy to see that the operator G_0, the Dirichlet - Neumann operator for the rectangle, is given by the Fourier multiplier

$$G_0 = D\tanh(hD) \ . \tag{20}$$

Indeed, when $\xi(x) = \exp(ikx)$, then the harmonic extension of ξ to the rectangle $\{(x,y) \ : \ 0 < x < 2\pi, \ -h < y < 0\}$, which also satisfies the bottom boundary conditions, is $\varphi(x,y) = (\cosh(k(y+h))/\cosh(kh))\exp(ikx)$. Its exterior normal derivative on the surface $\{y = 0\}$ is $k\tanh(hk)\exp(ikx)$, and using the notation that $D = \frac{1}{i}\partial_x$, the expression (20) follows.

The quadratic part of the hamiltonian can be expressed in terms of the Fourier series for the functions η and ξ. If we write that $\eta(k)$ is the $k-th$ Fourier coefficient of $\eta(x)$, and similarly for $\xi(x)$, then the Plancherel identity gives that

$$H_2 = \sum_{k\in\mathbf{Z}\backslash\{0\}} \tfrac{1}{2}(k\tanh(hk)|\xi(k)|^2 + g|\eta(k)|^2) \ . \tag{21}$$

Proposition 2. *The water wave problem linearized about equilibrium is a harmonic oscillator of infinitely many degrees of freedom, with frequencies*

$$\omega(k) = \sqrt{gk\tanh(hk)} \ , \qquad k \in \mathbf{Z}\backslash\{0\} \ . \tag{22}$$

In the case of deep water, $h = +\infty$ and the frequencies are

$$\omega(k) = \sqrt{g|k|} \ , \qquad k \in \mathbf{Z}\backslash\{0\} \ . \tag{23}$$

One recognizes the well known dispersion for linear water waves in the frequencies. Our restriction to the subspace $\{M(\eta,\xi) = \Xi(\eta,\xi) = 0\}$ has eliminated the only non-elliptic eigenspaces of the quadratic form H_2, and allows us to avoid the singularity of the deep water dispersion relation (23) at $k = 0$.

We still need to express all of the other Taylor coefficients of $G(\eta)$, in order to fully describe the nonlinear problem.

Proposition 3. *(i) An expression for the variation of the Dirichlet integral with respect to the upper boundary is given by*

$$\delta_\eta \int_0^{2\pi} \xi G(\eta)\xi \ dx = \frac{1}{(1+(\partial_x\eta)^2)}\left((\partial_x\xi)^2 - (G(\eta)\xi)^2 - 2\partial_x\eta\,\partial_x\xi\, G(\eta)\xi\right) \ . \tag{24}$$

(ii) There is a recursion formula for the Taylor coefficients of $G(\eta)$. We give the first several Taylor coefficients as examples.

$$
\begin{aligned}
G_0 &= D\tanh(hD) \ , \\
G_1(\eta) &= D\eta(x)D - G_0(D)\eta(x)G_0(D) \ , \\
G_2(\eta) &= -\tfrac{1}{2}(D^2\eta^2(x)G_0(D) + G_0(D)\eta^2(x)D^2 \\
&\quad - 2G_0(D)\eta(x)G_0(D)\eta(x)G_0(D)) \ , \\
G_3(\eta) &= \tfrac{1}{6}D\eta^3(x)D^3 - \tfrac{1}{6}G_0(D)\eta^3(x)D^2G_0(D) \\
&\quad - \tfrac{1}{2}G_1(\eta)\eta^2(x)D^2 - G_2(\eta)\eta(x)G_0(D) \ .
\end{aligned} \tag{25}
$$

The formula (24) is related to the Hadamard variational formula for the variation of the Green's function in terms of the domain, and in fact one derivation of (24) comes from this. This recursion formula of (ii) is quite simple, and higher Taylor coefficients can be readily calculated, by hand or by computer algebra. A complete description of this recursion formula appears in [**4**].

§4 Normal forms

A normal forms transformation can be naively considered as an attempt to compare the nonlinear hamiltonian system

$$\partial_t u = J\delta_u H(u) = J\delta_u(H_2(u) + H_3(u) + \cdots) \tag{26}$$

with its linear part

$$\partial_t U = J\delta_u H_2(U) \ . \tag{27}$$

For this the standard method in the subject (see [**2**] is to perform canonical transformations $u = F_N(v)$ such that in the new variables the system (26) is given in the form

$$\partial_t v = J\delta_v(H_2(v) + \sum_{3\le n\le N} H_n^{res}(v) + R_{N+1}(v)) \tag{28}$$

where the terms $H_n^{res}(v)$ are the so - called resonant terms which cannot be removed by such considerations, and $R_{N+1}(v)$ is the error, of order at least $N+1$. In neighborhoods of an elliptic stationary point there is a straightforward criterion for a term to be identified as resonant. To describe this it is convenient to work in complex coordinates, in the Fourier transform variables. Define

$$z(k) = \frac{\text{re } \eta(k)}{\sqrt{(\omega(k)/2g)}} + i\frac{\text{re } \xi(k)}{\sqrt{g/2\omega(k)}} \tag{29}$$

for the cosine series in x, and

$$w(k) = \frac{\text{im } \eta(k)}{\sqrt{(\omega(k)/2g)}} + i\frac{\text{im } \xi(k)}{\sqrt{g/2\omega(k)}} \tag{30}$$

for the sine series. Denote the pair by $Z(k) = (z(k), w(k))$.

DEFINITION 4. A monomial

$$Z^P\overline{Z}^Q = \prod_k Z(k)^{p(k)}\overline{Z}(k)^{q(k)} \tag{31}$$

of degree $N = |P| + |Q| = \sum_k p(k) + q(k)$ is *resonant* if

$$0 = \langle P - Q, \omega\rangle = \sum_k (p(k) - q(k))\omega(k)\ . \tag{32}$$

With regard to resonance in the water wave problem, there are the following elementary facts.

PROPOSITION 5. *(i) The only monomials $Z^P\overline{Z}^Q$ which appear in the hamiltonian for the water wave problem have the property that*

$$0 = \langle P - Q, k\rangle = \sum_k (p(k) - q(k))k\ . \tag{33}$$

(ii) There are no solutions of the third order resonance relations (32)(34), that is, of the relations

$$\omega(k) \pm \omega(\ell) \pm \omega(m) = 0, \qquad k \pm \ell \pm m = 0\ . \tag{34}$$

PROOF. Statement (i) follows from the conservation of momentum (16) for the water wave problem. One can easily see (ii) from the facts that the dispersion relation $\omega(k)$ is increasing and concave in $k \geq 0$. □

To state a theorem of content regarding the normal forms transformations for the water wave problem, take $r > 1$ and define the analytic classes of sequences $h_{(r,\rho)}$ to be those sequences $\{Z(k)\}_{k\in\mathbf{Z}}$ such that the following exponentially weighted norm is finite,

$$\|Z\|^2_{(r,\rho)} = \sum_{k\in\mathbf{Z}} \langle k\rangle^{2r} \exp(2\rho|k|)|Z(k)|^2\ . \tag{35}$$

THEOREM 6. *The third order normal forms transformation exists for the water wave problem. That is, there is a constant $R > 0$ such that in the ball $B_R(0) \subseteq h_{(r,\rho)}$, there is a $0 < \rho_1 \leq \rho$ and a canonical transformation*

$$F_3\ : B_R(0) \subseteq h_{(r,\rho)} \to h_{(r,\rho_1)} \tag{36}$$

which removes the cubic terms from the water wave hamiltonian.

PROOF. From the cubic contribution to (14)

$$H_3(Z,\overline{Z}) = \sum_{|P|+|Q|=3} h_3(P,Q)Z^P\overline{Z}^Q\ , \tag{37}$$

we form an auxiliary hamiltonian

$$K_3(Z,\overline{Z}) = \sum_{|P|+|Q|=3} \frac{h_3(P,Q)}{i\langle\omega, Q-P\rangle} Z^P\overline{Z}^Q\ . \tag{38}$$

One then proves that the vector field $J\delta_{\overline{Z}}K_3$ satisfies Cauchy estimates with respect to the scale of Hilbert spaces $h_{(r,\sigma)}$, $0 < \sigma < \sigma_0$. This allows one to use the abstract version of the Cauchy-Kowalewski theorem, due to L. Nirenberg (see **[13]**, **[14]** or

[12]) to define the transformation as the time - 1 hamiltonian flow for K_3. The result then follows from comparison of Taylor coefficients. □

§5 Higher order normal forms

In the case of deep water, for which $h = +\infty$ and the linear frequencies are $\{\omega(k) = \sqrt{g|k|}\}$, we are able to do the algebra for higher order normal forms, at least on a formal level. For ease of notation, take $g = 1$. For a fourth order normal form there are two types of resonant monomials. The first type is generic at even orders in all hamiltonians which satisfy Proposition 5 (i);

$$\omega(k) + \omega(\ell) - \omega(k) - \omega(\ell) = 0 \ . \tag{39}$$

These are usually considered benign, as the associated resonant monomial can be written in terms of the integrals of motion of the quadratic hamiltonian (at least in the absence of other 1:1 resonances). In fact these terms play a role in the nondegeneracy conditions of KAM theory. There is however a second type of fourth order resonance for the deep water problem, exhibited by the example of the resonant quartet

$$1 : -2 : -2 : 3 \quad (\omega's) \ , \qquad 1 : 4 : 4 : -9 \quad (k's) \ ,$$

which also satisfies (32)(33) with $N = 4$. The class of these fourth order non-generic resonant quartets appears in the study of the modulational instability of deep water waves, which is known as the Benjamin - Feir instability. For this reason we call these the *Benjamin - Feir* resonances. It is straightforward to classify them all, in a two integer parameter family.

Proposition 7. *For the deep water problem, the only resonant quartets are the generic ones of (39), or else for $p, m \in \mathbf{Z}$ they are of the form*

$$p \times (m^2, (m+1)^2, m^2(m+1)^2, (m^2+m+1)^2) \ . \tag{40}$$

Resonant terms correspond to essential nonlinearities that appear in a hamiltonian system. Resonant terms which correspond to benign resonances (39) generally give rise to contributions to the full hamiltonian which are functions of integrals of the quadratic hamiltonian H_2. It is reasonable to approximate the dynamical system by truncating the hamiltonian in a Birkhoff normal form, to some given finite order. If the only nontrivial resonant terms contributing to this approximation are benign, then the resulting approximate system is completely integrable, in terms of the integrals of the quadratic hamiltonian. Because of the presence of resonance relations (40) which are not benign at fourth order, and this question of complete integrability of the Birkhoff normal form approximation, I think that the following calculation is quite an interesting result.

Theorem 8 *(Dyachenko and Zakharov* [6], *Craig and Worfolk* [5]*). In the case $h = +\infty$, the normal forms transformation which removes the cubic terms of the water wave hamiltonian H also removes all of the Benjamin - Feir quartic terms.*

Proof. In outline, we have a hamiltonian

$$H = H_2 + H_3 + H_4 + R_5 \ ,$$

which under the third order normal forms transformation becomes

$$H^{(1)} = H_2 + (H_3 + \{K_3, H_2\}) + (H_4 + \tfrac{1}{2}\{K_3, H_3\}) + R_5^{(1)} . \tag{41}$$

Since these represent long calculations, both papers [**5**] and [**6**] used numerical algebra in the calculations of $H_4^{(1)} = (H_4 + \frac{1}{2}\{K_3, H_3\})$. Focussing upon the coefficients of the Benjamin - Feir resonant monomials, by these explicit calculations one shows that their coefficients vanish in (41). Under a further (formal) transformation which removes all nonresonant monomials, the remaining hamiltonian $H^{(2)}$ has quartic term which only contains contributions from the benign resonant terms associated with the resonant quartets of (39). □

The vanishing of these non - benign resonant terms implies that the resulting fourth order normal forms approximation to the water wave problem is completely integrable. The structure of this system is analysed in [**5**], and will be discussed briefly in the following section of this paper. It is enough of a coincidence that Dyachenko and Zakharov [**6**] make the conjecture that, in carrying out the calculations to higher order, the coefficients of all nongeneric resonant monomials will vanish as well.

At fifth order there are many resonances, as we can easily compute. The simplest is a polynomial in only two indices, involving $(Z(p), \overline{Z}(p), Z(4p), \overline{Z}(4p))$, and the resonant quintet is

$$1 : 1 : 1 : -1 : -2 \, , \qquad 1 : 1 : 1 : 1 : -4 \, . \tag{42}$$

THEOREM 9 *(Craig and Worfolk* [**5**]*). The coefficients of the* $(1 : 1 : 1 : 1 : 2)$ *fifth order resonant terms in the water wave hamiltonian* $H^{(3)}$ *are non-zero.*

PROOF. The proof is again by numerical algebra, where we have to compute the resonant component of the fifth order term

$$H_5^{(3)} = (H_5 + \{K_3, H_4\} + \tfrac{1}{3}\{K_3, \{K_3, H_3\}\}) \, , \tag{43}$$

at least for those monomials consisting only of the variables $(z(p), \overline{z}(p), w(p), \overline{w}(p), z(4p), \overline{z}(4p), w(4p), \overline{w}(4p))$. We find explicitly that the resonant polynomial is

$$\begin{aligned} \frac{p^4}{12\pi}\sqrt[4]{\frac{4p}{\pi^2}} \; \mathrm{re} \; \Big(& z(4p)\big(z(p)\overline{z}(p)^3 + w(p)\overline{w}(p)^3 - 3\overline{z}(p)\overline{w}(p)(z(p)\overline{w}(p) + \overline{z}(p)w(p))\big) \\ & - w(4p)\big(z(p)\overline{w}(p)^3 - \overline{z}(p)^3 w(p) - 3\overline{z}(p)\overline{w}(p)(z(p)\overline{z}(p) - w(p)\overline{w}(p))\big)\Big) \, . \end{aligned} \tag{44}$$

□

The conclusion is that the water wave problem is not an integrable system, at least not with respect to the action variables of the quadratic part of the hamiltonian. Subsequent to [**5**], Dyachenko, Lvov and Zakharov [**7**] have also calculated resonant quintets which contribute nontrivial fifth degree terms to the hamiltonian in Birkhoff normal form. These can be quite complicated, as the following proposition shows.

PROPOSITION 10. *The resonant quintets which involve only two indices are all integer multiples of the quintet given in (42). The only resonant quintets which involve three distinct indices are integer multiples of the quintets in the following list:*

$$\begin{aligned}&(1:1:2:2:4)\\&(1:1:3:3:4)\\&(1:1:9:9:16)\\&(9:9:9:16:25)\ .\end{aligned} \tag{45}$$

There are exactly 19 families of resonant quintets which involve four distinct indices (which I will not list here), and a sampling of resonant quintets which include five distinct indices is

$$\begin{aligned}&(1:4:16:25:36)\\&(4:9:16:25:36)\\&(4:25:36:49:64)\\&(1:4:16:100:121)\ .\end{aligned}$$

THEOREM 11 *(Dyachenko, Lvov and Zakharov* [**7**]*). There is a family of resonant quintets in the form*

(46)

$$(k_1,k_2,k_3,k_4,k_5) =$$

$$a\times(((p-1)^2-q^2)^2,((p+1)^2-q^2)^2,(p^2+4-(q+1)^2)^2,p^2+4-(q-1)^2)^2,16),$$

where p,q are rational, $0<|p|,|q|,|p\pm q|<1$, and a is chosen so that the k_ℓ are integers. The coefficient of the associated resonant quintic monomial is

$$\frac{2}{\pi^{3/2}}\sqrt{\frac{\omega(k_1)\omega(k_2)\omega(k_3)}{\omega(k_4)\omega(k_5)}}\frac{k_1k_2k_3k_4k_5}{\max(k_1,k_2,k_3)} \tag{47}$$

For the result of a complicated calculation, this is a remarkably simple result.

§6 THE DYNAMICAL SYSTEM AT FOURTH AND FIFTH ORDERS

Given the above normal forms calculations, it is natural that we use the truncated hamiltonian to define a dynamical system which will approximate the evolution of the water wave problem. First of all we remark that the water wave equations, as do many other hamiltonian systems which come from partial differential equations, have infinitely many 1:1 resonances. We have mostly considered the case of periodic boundary conditions, for which both of the Fourier modes $\cos(kx)$, $\sin(kx)$ evolve in time under the linear equations (27) with the same temporal frequencies $\omega(k)$. This is to say that the four - dimensional plane described by $P(k)=\{(\text{ re }\eta(k),\text{ re }\xi(k),\text{ im }\eta(k),\text{ im }\xi(k))\not\equiv 0\ :\ \textit{with all others zero}\ \}$, is a resonant plane, in which all orbits of the linearized equations have the same frequency. The Birkhoff normal forms transformations are near identity transformations, so the same is true for the four - dimensional plane spanned by setting all but $(z(k),\overline{z}(k),w(k),\overline{w}(k))$ to zero, in the transformed variables. Due to the 1:1 resonance, within this plane the correct action variables are not $z(k)\overline{z}(k)$ and $w(k)\overline{w}(k)$, but rather

$$\begin{aligned}I_1(k)&=\tfrac{1}{2}(z(k)\overline{z}(k)+w(k)\overline{w}(k))\\I_2(k)&=\tfrac{1}{2i}(z(k)\overline{w}(k)-\overline{z}(k)w(k))\ ,\end{aligned} \tag{48(i)}$$

with canonical conjugate angle variables

$$\tag{48(ii)}\begin{aligned}\varphi_1(k) &= -\tfrac{1}{2}\arctan\left(\frac{\text{im }(z^2(k)+w^2(k))}{\text{re }(z^2(k)+w^2(k))}\right)\\ \varphi_2(k) &= -\tfrac{1}{2}\arctan\left(\frac{z(k)\overline{w}(k)+\overline{z}(k)w(k)}{z(k)\overline{z}(k)-w(k)\overline{w}(k)}\right).\end{aligned}$$

§**6.1 Fourth order Birkhoff normal form.** Imagine having performed the normal forms transformations such that the water wave hamiltonian is free of non-resonant quartic terms (I will avoid a discussion of the analysis of this transformation, however it can be considered on a formal level, and is calculated on individual monomials). We can then write the water wave hamiltonian in action - angle variables as

$$\tag{49}\begin{aligned}H^{(2)} =&\sum_{k>0}\sqrt{k}I_1(k)-\frac{k^3}{8\pi}(I_1^2(k)-3I_2^2(k))+\sum_{0<k<\ell}\frac{k^2\ell}{\pi}I_2(k)I_2(\ell)+R_5^{(2)}\\ =&h^{(4)}(I_1,I_2)+R_5^{(2)}.\end{aligned}$$

Upon truncation of the remainder term $R_5^{(2)}$ of (49), the resulting hamiltonian $h^{(4)}=h^{(4)}(I_1,I_2)$ gives a fourth order approximation to the water waves problem with cyclic angle variables (φ_1,φ_2). Thus $h^{(4)}$ defines an integrable system, such that all of its orbits are almost periodic,

$$\tag{50}\begin{aligned}\partial_t I_1(k) =&0\,, \qquad \partial_t\varphi_1(k) = -\delta_{I_1(k)}h^{(4)}\\ \partial_t I_2(k) =&0\,, \qquad \partial_t\varphi_2(k) = -\delta_{I_2(k)}h^{(4)}\,.\end{aligned}$$

The four-dimensional symplectic planes discussed above are $P(k)=\{(I_1(k),I_2(k),\varphi_1(k),\varphi_2(k))\not\equiv 0$, *all other coordinates vanishing*$\}$, and are invariant under the hamiltonian flow of $h^{(4)}$. Initial data in $P(k)$ gives rise to three classes of wave motions; (i) travelling waves, which are time - periodic as well as spatially periodic, (one in each direction), (ii) standing waves, which are also time - periodic, and (iii) time quasi-periodic waves, which have two temporal quasi - frequencies. Each solution lies on a 2 - torus of solutions identical up to translation, parametrized by the angles $(\varphi_1(k),\varphi_2(k))$, as the hamiltonian $h^{(4)}$ is cyclic with respect to these variables. Their frequencies depend explicitly upon the action variables $(I_1(k),I_2(k))$. In the case of time - periodic standing waves within $P(k)$, which are asked to satisfy the Neumann boundary conditions (5), one takes only the cosine series in x; therefore $w(k)=0$ and $I_2(k)=0$, $I_1(k)=z(k)\overline{z}(k)$ remains constant in time, and $\varphi_1(k)$ increases linearly in t with frequency $(-\sqrt{k}+(k^3/4\pi)I_1(k))$. All solutions of the fourth order system are nonlinear superpositions of these three basic solution classes available for each k.

One can also characterize the bifurcations of standing waves and travelling waves of the integrable hamiltonian system (50). A graphic description of the infinite - dimensional phase space, the three basic classes of solutions and the standing and travelling waves is reported in [**5**].

§**6.2 Fifth order Birkhoff normal form.** It is also tempting to study the approximation to the water wave problem given through the hamiltonian which includes only resonant terms up to fifth order, after a third hypothetical normal forms transformation. According to the normal forms procedure, the fifth order contribution will be

$$H_5^{(3)} = (H_5 + \{K_3, H_4\} + \tfrac{1}{3}\{K_3, \{K_3, H_3\}\})^{res} \ . \tag{51}$$

We will denote the resulting hamiltonian by

$$h^{(5)} = h^{(4)} + H_5^{(3)} \ . \tag{52}$$

This is quite a complicated quantity, but it can be studied in detail on certain subspaces of our infinite dimensional phase space. Choose a finite set of indices $\vec{k} = (k_1, k_2, \dots k_n)$, and define the coordinate subspaces of periodic functions $P(\vec{k}) = \{(z, \overline{z}, w, \overline{w}) \ : \ z(k) = 0 = w(k) \textit{ unless } k = k_\ell, \textit{ for some } \ell = 1, \dots n\}$. In an analogous way define the finite dimensional coordinate subspaces of functions satisfying Neumann boundary conditions (5); $N(\vec{k}) = P(\vec{k}) \cap \{w(k) = 0 \ \forall k \in \mathbf{Z}\}$.

PROPOSITION 12. *(i) The infinite dimensional subspace given by $N = \{w(k) = 0 \ , \forall k \in \mathbf{Z}\}$ is invariant under the normal forms transformations, and is invariant under the hamiltonian flows of H, $H^{(n)}$ and $h^{(n)}$. All solutions in this subspace are even in x about the points $x = 0, \pi$, and therefore the solutions satisfy the Neumann boundary conditions (5).*
(ii) For any choice of indices $\vec{k}$, the subspaces $P(\vec{k})$ and $N(\vec{k})$ are invariant under the hamiltonian flow of $h^{(4)}$.
(iii) For no finite choice of indices $\vec{k}$ are either of the coordinate subspaces $P(\vec{k})$ or $N(\vec{k})$ invariant under the hamiltonian flow of $h^{(5)}$.

PROOF. The case (i) of the proposition follows by symmetry under reflection in $x = 0$ and $x = \pi$, and (ii) is evident from the form of the integrals of the hamiltonian $h^{(4)}$, so we will focus on statement (iii). Suppose that $\vec{k} = (k_1, k_2 \dots k_n)$ is our choice of indices (in increasing order), with $k_n = m$. The coordinate subspace $P(\vec{k})$ would be invariant if there were no non-zero normal component to the hamiltonian vector field on $P(\vec{k})$, in some coordinate $z(q)$ and/or $w(q)$ direction, for $q \neq k_1, \dots m$. These can arise from some non-zero quintic polynomial of the hamiltonian $h^{(5)}$ which is linear in the coordinates $(z(q), \overline{z}(q), w(q), \overline{w}(q))$, such that it otherwise only depends upon $(z(k_1), \overline{z}(k_1), \dots w(m), \overline{w}(m))$. Looking through the list in Proposition 10, there are a number of candidates, including the polynomial (44), whose gradient is nonvanshing on $P(\vec{k})$, giving rise to a non-zero component in the $4m$ coordinate direction. The same considerations hold for the subspace $N(\vec{k})$. Notice that in the list supplied by Proposition 10 and reference [**5**] there are no resonant quintic monomials other than this $(1:1:1:1:2)$ case which, when considered on $P(\vec{k})$, give a contribution to the hamiltonian vector field with a non-zero $k = 4m$ component. □

The remainder of this section will concentrate on the dynamics of the fifth order approximation to the water wave hamiltonian $h^{(5)}$, restricted to finite dimensional coordinate subspaces $N(\vec{k})$ of the standing wave problem, where $\vec{k} = (k_1, \dots k_{n-1}, m)$. The point of this discussion is that it is very natural to approximate the water wave problem by restricting the evolution to finite dimensional coordinate subspaces, so that only finitely many Fourier modes are permitted to be non-zero. The significance of Proposition 12 is that this is never really enough, as the evolution dictated by the full system, among other things, has a dominant contribution which drives the component with wave number $k = 4m$ to be non-zero. We will show below that the $m \times (1:2)$ resonance particularly affects the time - periodic solutions of approximations of the water wave problem. In many cases we will show that upon admitting Fourier modes with wave numbers $4m$, all periodic solutions within the approximating subspace $N(\vec{k})$ bifurcate to unstable periodic solutions within $N(\vec{k}, 4m)$, and these new unstable solutions have Lyapunov exponent which grows strongly with m (they are in fact of order $m^{17/4}$).

We start with $\vec{k} = (p, 4p)$, where the dynamics describes a two - mode approximation to the standing wave problem in exact 1:2 resonance, with motion of the other modes simply supressed. On this resonant subspace the term (44) simplifies substantially, and the hamiltonian is given by

$$
\begin{aligned}
h^{(5)}(p,4p)(I_1(p), I_1(4p), \varphi_1(p), \varphi_1(4p)) &= \sqrt{p}(I_1(p) + 2I_1(4p)) \\
- \frac{p^3}{8\pi}(I_1^2(p) + 64I_1^2(4p)) &+ \frac{2p^4}{3\pi}\sqrt[4]{\frac{p}{\pi^2}} I_1^2(p) I_1^{\frac{1}{2}}(4p)\cos(\varphi_1(4p) - 2\varphi_1(p)) \; .
\end{aligned} \tag{53}
$$

THEOREM 13. *Consider the hamiltonian vector field of (53), which is that of hamiltonian (52) restricted to the 4 - dimensional subspace $N(p, 4p)$. There are two Poisson commuting integrals of motion, $h^{(5)}(p, 4p)$ itself and $J_1 = (I_1(p) + 2I_1(4p))/\sqrt{5}$, and the evolution is completely integrable. On each surface $\{J_1 = \text{const.}\}$ there are at least two time - periodic orbits, given by $(J_2, \psi_2) = (J_2^{(1)}, \psi_2)$ and $(J_2^{(2)}, n\pi/\sqrt{5})$, n even, and these orbits are elliptic (that is to say, stable). In addition there is a constant $c_1(p)$ such that for $0 < J_1 < c_1(p)$ and n odd there are two more time - periodic orbits, $(J_2, \psi_2) = (J_2^{(3)}, n\pi/\sqrt{5})$ and $(J_2^{(4)}, n\pi/\sqrt{5})$, with $J_2^{(3)} < J_2^{(4)}$. These orbits are hyperbolic (that is, unstable) with Lyapunov exponents respectively*

$$
\begin{aligned}
\lambda^{(3)}(p) &\sim \Lambda^{(3)}(p) J_1^{3/2} \; , \qquad \Lambda^{(3)}(p) = O(p^{17/4}) \; , \\
\lambda^{(4)}(p) &\sim \Lambda^{(4)}(p) J_1^{5/4} \; ,
\end{aligned} \tag{54}
$$

for small J_1. Finally, there is a constant $c_2(p) > c_1(p)$ such that for larger amplitudes $J_1 > c_2(p)$ there again appear two more time periodic orbits of the hamiltonian flow.

PROOF. This result follows from explicit calculations of the hamiltonian flow within the 1:2 resonant plane $N(p, 4p)$. We change the problem to rotating coordinates with the canonical transformation

$$
\begin{pmatrix} I_1(p) \\ I_1(4p) \end{pmatrix} = \frac{1}{\sqrt{5}} \begin{pmatrix} 1 & -2 \\ 2 & 1 \end{pmatrix} \begin{pmatrix} J_1 \\ J_2 \end{pmatrix} \tag{55}
$$

$$
\begin{pmatrix} \varphi_1(p) \\ \varphi_1(4p) \end{pmatrix} = \frac{1}{\sqrt{5}} \begin{pmatrix} 1 & -2 \\ 2 & 1 \end{pmatrix} \begin{pmatrix} \psi_1 \\ \psi_2 \end{pmatrix} \; . \tag{56}
$$

Since $I_1(p), I_1(4p) \geq 0$, the range of the transformation is $J_1 \geq 0$ and $-2I_1(p)/\sqrt{5} \leq J_2 \leq I_1(4p)/\sqrt{5}$. The hamiltonian (53) becomes

$$
\begin{aligned}
h^{(5)}(p,4p)(J,\psi) =& \sqrt{5p}J_1 - \frac{p^3}{40\pi}(257J_1^2 + 252J_1J_2 + 68J_2^2) \\
&+ \frac{2p^4}{15\pi}\sqrt[4]{\frac{p}{5\pi^2}}(J_1 - 2J_2)^2(2J_1 + J_2)^{1/2}\cos(\sqrt{5}\psi_2) \; ,
\end{aligned} \tag{57}
$$

giving rise to the hamiltonian system

$$
\begin{aligned}
\partial_t J_1 &= 0 \ , \qquad & \partial_t \psi_1 &= -\delta_{J_1} h^{(5)}(p,4p) \\
\partial_t J_2 &= \delta_{\psi_2} h^{(5)}(p,4p) \ , \qquad & \partial_t \psi_2 &= -\delta_{J_2} h^{(5)}(p,4p) \ .
\end{aligned} \tag{58}
$$

The hamiltonian (57) is cyclic with respect to ψ_1, therefore J_1 is a conserved quantity of (58) in addition to the hamiltonian $h^{(5)}$, and it can be used as a parameter in the description of the vector field of the system in the variables (J_2, ψ_2);
(59)

$$
\begin{pmatrix} \delta_{\psi_2} h^{(5)}(p,4p) \\ -\delta_{J_2} h^{(5)}(p,4p) \end{pmatrix} = \begin{pmatrix} -\frac{p^4}{15\pi}\sqrt[4]{\frac{p}{5\pi^2}}(J_1 - 2J_2)^2(2J_1+J_2)^{1/2}\sin(\sqrt{5}\psi_2) \\ \frac{P^3}{10\pi}(63J_1 + 34J_2) + \frac{p^4}{3\pi}\sqrt[4]{\frac{p}{5\pi^2}}\frac{(J_1-2J_2)(3J_1+2J_2)}{(2J_1+J_2)^{1/2}}\cos(\sqrt{5}\psi_2) \end{pmatrix} .
$$

Stationary points of (59) are critical points of the hamiltonian $h^{(5)}(p,4p)(J_2,\psi_2)$, and correspond to periodic orbits of the system $\partial_t z(k) = i\delta_{\overline{z}(k)} h^{(5)}(p,4p)$, $k = p, 4p$. Examining (59) on the singular subvarieties of our polar coordinates, one observes that (59) is not stationary on $N(p) = \{0 = I_1(4p) = (2J_1 + J_2)/\sqrt{5}\}$, reflecting the fact that $N(p)$ is not an invariant set for the restricted problem. On the subspace $N(4p) = \{0 = I_1(p) = (J_1 - 2J_2)/\sqrt{5}\}$, the hamiltonian $h^{(5)}(p,4p)$ coincides with $h^{(4)}$ up to quadratic order, giving a family of periodic solutions with frequency close to $\sqrt{4p}$, given by $J_2^{(1)} = \frac{1}{2}J_1$. Other than on $N(4p)$, the first component of the vector field (59) can vanish only when $\sqrt{5}\psi_2 = n\pi$. For $\cos(\sqrt{5}\psi_2) = (-1)^n$, the zeros of $-\delta_{J_2} h^{(5)}(p,4p)(J_2,\psi_2)$ correspond to the roots of

$$
(63J_1 + 34J_2)(2J_1 + J_2)^{1/2} + (-1)^n \frac{10p}{3}\sqrt[4]{\frac{p}{5\pi^2}}(J_1 - 2J_2)(3J_1 + 2J_2) = 0 \tag{60}
$$

over the range of the transformation. These are obtained by elementary inspection. There is always one root $J_2^{(2)}$ when n is even, within the interval $-(63/34)J_1 < J_2^{(2)} < -(3/2)J_1$. In addition, for n odd and small, *i.e.* $0 < J_1 < c_1(p)$, there are two other roots $J_2^{(3)} \le J_2^{(4)}$ in the range $-2J_1 < J_2^{(3)}, J_2^{(4)} < -(63/34)J_1$, which converge to $J_2^{(3)} = -2J_1$ and $J_2^{(4)} = -(63/34)J_1$ respectively, as J_1 tends to zero, and converge to each other as J_1 tends toward $c_1(p)$. In fact, one finds from (60) that for small J_1

$$
(2J_1 + J_2^{(3)}) = O(J_1^2) \ , \quad (63/34)J_1 - J_2^{(4)} = O(J_1) \ . \tag{61}
$$

From the asymptotic (61) we see that the family of solutions $(J_2^{(3)}, \psi_2 = n\pi/\sqrt{5})$, n odd, is the most relevant to the standing wave problem, for it is the bifurcation branch of periodic orbits of frequency close to $\sqrt{p}$ which is tangent to the standing wave solutions (ii) of hamiltonian (49).

Finally, there is a second constant $c_2(p) > c_1(p)$ such that for all $J_1 > c_2(p)$, for n odd, there are two other stationary points of (59), in the region $-(3/2)J_1 < J_2 < (1/2)J_1$.

To understand the stability of these periodic solutions one studies the linearized vector field of (59) about the above stationary points. For fixed J_1 the restricted hamiltonian $H^{(5)}(p,4p)$ takes its minimum within $N(4p)$, so that the periodic orbits in $N(4p)$ are elliptic. Off of this coordinate plane, one studies the linearized vector field

$$
J\delta^2 h^{(5)}(p,4p) = \begin{pmatrix} \partial_{J_2}\partial_{\psi_2} h^{(5)}(p,4p) & \partial^2_{\psi_2} h^{(5)}(p,4p) \\ -\partial^2_{J_2} h^{(5)}(p,4p) & -\partial_{J_2}\partial_{\psi_2} h^{(5)}(p,4p) \end{pmatrix} , \tag{62}
$$

evaluated at the stationary points. For small enough $0 < J_1$ this can be seen to have two pure imaginary eigenvalues for n even, and two real eigenvalues for n odd, with opposing signs. At the stationary point $(J_2, \psi_2) = (J_2^{(3)}, n\pi/\sqrt{5})$, n odd, the real eigenvalues are $\pm\lambda^{(3)}(J_1,p) \sim \Lambda^{(3)}(p)J_1^{3/2}$, and at the stationary point $(J_2, \psi_2) = (J_2^{(4)}, n\pi/\sqrt{5})$, n odd, they are $\pm\lambda^{(4)}(J_1,p) \sim \Lambda^{(4)}(p)J_1^{5/4}$. This gives the statement of instability in the theorem. We have not looked at the stability of these time - periodic solutions for values of J_1 which are large, as this is somewhat more complicated and it is less relevant to the water wave problem. □

Now let us consider the water wave hamiltonian (52), restricted to an arbitrary finite dimensional coordinate subspace $N(\vec{k})$, $\vec{k} = (k_1, \dots k_{n-1}, m)$; we will call the restricted hamiltonian $h^{(5)}(\vec{k})$. Evaluated on $N(\vec{k})$, the only component of the hamiltonian vector field of the full hamiltonian (52) tangent to the subspace $N(4m)$ comes from (44). We will observe the changes under the extension of the restricted system to $N(\vec{k}, 4m)$. There are two cases to discuss, determined by the resonance relations among the indices $(k_1, \dots m)$.

Case (i): Suppose that for any multiindices $P = (p(k_1), \dots p(k_{n-1}), p(m))$, $Q = (q(k_1), \dots q(k_{n-1}), q(m))$ such that $|P| + |Q| = 5$, $0 = \langle k, P - Q\rangle = \sum_{\ell=1}^{n} k_\ell(p(k_\ell) - q(k_\ell))$ and $0 = \langle \omega, P - Q\rangle = \sum_{\ell=1}^{n} \sqrt{k_\ell}(p(k_\ell) - q(k_\ell))$, then $p(m) = q(m) = 0$. That is to say, there is no fifth order resonance relation among $\{k_1, \dots k_{n-1}, m\}$ which involves m. Then $J_1 = (I_1(m) + 2I_1(4m))/\sqrt{5}$ and $J_2 = (I_1(4m) - 2I_1(m))/\sqrt{5}$ are integrals of motion of $h^{(5)}(\vec{k}, 4m)$, and the coordinates $(I_1(m), I_1(4m), \varphi_1(m), \varphi_1(4m))$ obey the system of equations given by the hamiltonian vector field (59). In this case every periodic solution within $N(\vec{k})$ bifurcates to a periodic solution in $N(\vec{k}, m)$ which has $I_1(4m) = O(J_1^2)$. Furthermore, all of these solutions have at least one positive Lyapunov exponent $\lambda(m) \sim \Lambda(m)J_1^{3/2}$, with a non-zero component of its unstable direction in the plane $N(4m)$. We point out that the constant $\Lambda(m)$ diverges strongly with m, with the order as given in (54).

Case (ii): It may also be the case that there are fifth order resonance relations which involve the wave number m. Still the resulting hamiltonian system has two integrals; $h^{(5)}(\vec{k}, 4m)$ itself, and $J_1 = (\sum_{\ell=1}^{n} \sqrt{k_\ell} I_1(k_\ell) + \sqrt{4m} I_1(4m))$. Level surfaces $\{J_1 = R^2\} = S^{2n+1}(R)$ are $2n+1$-dimensional spheres, on which critical points of $h^{(5)}(\vec{k}, 4m)$ correspond to perodic orbits. All critical points are pushed off of $N(\vec{k}) \cap S^{2n+1}(R)$ by the presence of the $m \times (1:2)$ fifth order resonant term, by an amount which is at least $I_1(4m) \geq O(I_1(m)^2)$. Furthermore the hamiltonian $h^{(5)}(\vec{k}, 4m)$ is invariant under the hamiltonian flow of J_1, so by a Lusternik - Schnirlman category argument, there are at least $(n+1)$ distinct critical points. A minimum (at least local) of $h^{(5)}(\vec{k}, 4m)$ on $S^{2n+1}(R)$ occurs in $N(4m)$, for $I_1(4m) = J_1/\sqrt{4m}$. By examining the vector field we conclude that the maximum and minimum of $h^{(5)}(\vec{k}, 4m)$ on $S^{2n+1}(R)$ must occur for $I_1(4m) > O(J_1)$. This suggests strongly that the rest of the critical points possess at least one hyperbolic direction, including especially those periodic orbits satisfying $dist.(I_1(4m), N(\vec{k})) = O(J_1^2)$, analogous to the features of the model problem that have been described in Theorem 13.

§**6.3 Conclusions.** The technique of Birkhoff normal forms is an infinite dimensional transformation theory with which we study the initial value problem for the water wave problem. The method is uniform in wave number since the transformations include all Fourier coefficients of the solutions, and while it is an expansion in amplitudes, it does not rely upon an Ansatz as to the form of the solution. After a (formal) fourth order normal forms transformation, [**5**] describes the resulting completely integrable system. The set of periodic standing wave solutions for the water wave equations truncated to this order is invariant under change of phase, and these time - periodic solutions are stable within the phase space of standing waves N. When five wave interactions are included, the system is more complicated, however the Hamiltonian is still explicit on the resonant subspace $N(p, 4p)$, where its restriction is shown to have four families of time - periodic standing waves, whose stability is determined explicitly. In particular, the most relevant periodic solutions are the ones which bifurcate from the one mode standing waves of the system at fourth order Birkhoff normal form; this family of solutions is shown to be unstable, with positive Lyapunov exponent. Clearly under further extension of the perturbation analysis this instability is expected to persist, at least for small amplitudes.

The paper of Bryant and Stiassnie [**1**] also discusses time - periodic standing waves using a perturbation regime for small amplitudes (which are not normal forms transformations), and it presents certain statements about their instability. The authors however have restricted themselves to subspaces $N(\vec{k})$ with $\vec{k} = (p_1^2, p_2^2, \dots p_N^2)$, an Ansatz which already omits certain relevant four wave interactions. In addition the five wave interactions within $N(\vec{k})$ are set to zero, thus the resonant standing waves within $N(k, 4k, 9k)$ which they discuss (see [**1**], formulae (2.14) through (2.17)) do not have correctly chosen phases ([**1**] (2.20)) and the fourth order perturbation calculation of their amplitude is not correct. Furthermore the five wave resonant interactions which are missing are the source of the instability of the solution branch of standing waves that is described in section 6.2 above.

The instability of the time - periodic standing waves within $N(k_1, k_2 \dots k_N)$ to perturbations within $N(k_1, k_2 \dots k_N, 4k_N)$ is a serious problem for certain methods of approximating standing waves. In particular, calculations by shooting methods, which would essentially be based on a Poincaré section, would be highly sensitive to error, under increases of resolution which include four times a previously represented wave number. One should rather choose methods which avoid treating the problem dynamically.

There is no rigorous mathematical proof of the existance of time - periodic standing waves, and due to the presence of small divisors in any formal perturbation expansion this is a challenging analytical problem. The rigorous results presented in this paper on the third order Birkhoff normal form can conceivably play a role in the answer to this question, as it shows that the 'Arnold condition', or condition of genuine nonlinearity, is satisfied by the water wave problem in deep water.

References

[1] P. J. Bryant and M. Stiassnie, *Different forms for nonlinear standing waves in deep water*, J. Fluid Mechanics **272** (1994), 135 – 156.

[2] R. C. Churchill, M. Kummer and D. L. Rod, *On averaging, reduction and symmetry in Hamiltonian systems*, J. Differential Equations **49** (1983), 359 – 414.

[3] R. Coifman and Y. Meyer, *Nonlinear harmonic analysis and analytic dependence*, Proc. Sympos. Pure. Math., vol. 43, Amer. Math. Soc., Providence, RI, 1985, pp. 71 – 78.

[4] W. Craig and C. Sulem, *Numerical simulation of gravity waves*, J. Comp. Physics **108** (1993), 73 – 83.

[5] W. Craig and P. Worfolk, *An integrable normal form for water waves in infinite depth*, Physica D **84** (1995), 513 – 531.

[6] A. I. Dyachenko and V. E. Zakharov, *Is free surface hydrodynamics an integrable system?*, Physics Letters A **190** (1994), 144 – 148.

[7] A. I. Dyachenko, Y. V. Lvov and V. E. Zakharov, *Five-wave interaction on the surface of deep fluid*, Physica D **87** (1995), 233 – 261.

[8] Y. Kodama, *Normal forms and solitons*, Topics in soliton theory and exactly solvable nonlinear equations, World Science Publ. Co., Singapore, 1987, pp. 319 – 339.

[9] Y. Kodama, *Normal forms and solitons in the shallow water waves*, Nonlinear water waves: IUTAM Symposium / Tokyo 1987, Springer Verlag, Berlin Heidelberg, 1988, pp. 85 –91.

[10] S. B. Kuksin and J. Pöschel, *Invariant Cantor manifolds of quasiperiodic oscillations for a nonlinear Schrödinger equation*, Annals of Math. (1995), (to appear).

[11] H. P. McKean and J. Shatah, *Nonlinear Schrödinger and heat equations*, Commun. Pure Applied Math. **44** (1991), 1067 – 1080.

[12] J. Moser (1979), lectures in the course 'nonlinear functional analysis', Courant Institute.

[13] L. Nirenberg, *An astract form of the Cauchy - Kowalewski theorem*, J. Differential Geom. **6** (1972), 561 – 576.

[14] T. Nishida, *A note on a theorem of Nirenberg*, J. Differential Geom. **12** (1977), 629 – 633.

[15] J. Shatah, *Normal forms and quadratic nonlinear Klein - Gordon equations*, Commun. Pure Applied Math. **38** (1985), 685 – 696.

[16] I. Tadjbakhsh and J. B. Keller, *Standing surface waves of finite amplitude*, J. Fluid Mechanics **8** (1960), 442 – 451.

[17] V. E. Zakharov, *Stability of periodic waves of finite amplitude on the surface of deep fluid*, J. Appl. Mech. Tech. Physics **2** (1968), 190 – 194.

[18] V. E. Zakharov, *Space-time spectra of sea waves in the weak turbulent approximation*, seminar talk, MSRI Berkeley, 7 February 1994.

Mathematics Department and Lefschetz Center for Dynamical Systems, Brown University, Providence, RI 02912, USA

CMLA, Ecole Normale Supérieure, 94235 Cachan cedex, France
E-mail address: craigw@math.brown.edu

Contemporary Mathematics
Volume **200**, 1996

Remarks on the Stability of Generalized KP Solitary Waves

A. de Bouard and J.C. Saut

ABSTRACT. We study the stability of the set of ground state solitary waves solutions of a generalized Kadomtsev-Petviashvili equation, when the critical exponent in the nonlinear term is less than 4/3. In the case when it is larger than 4/3, we make some remarks about a proof of instability which was given in [**19**].

1. Introduction

The existence and stability of solitary waves is one of the main mathematical questions arising when one considers a nonlinear conservative dispersive equation. A method for studying orbital stability, based on the fact that solitary waves for these equations are usually critical points of some functional constructed with the help of invariant quantities (energy and charge) of the evolution equation, was first introduced by Benjamin [**1**] in the context of the Korteweg-de vries equation, and then extended by many authors to other nonlinear dispersive models (see [**3**], [**11**], [**20**]). This method uses hypothesis on the spectrum of the second derivative of the above mentioned functional, that are however not always easy to establish. On the other hand, this technical part may be avoided when it is possible to show directly that the solitary waves considered (the ground states in general) are minima of energy constrained by constant charge, provided that this minimum is unique (see [**5**] in the context of nonlinear Schrödinger equations). When the minimum is not unique, one only gets the stability of the set of minima.

Here, we are concerned with the stability properties of solitary waves of the generalized Kadomtsev-Petviashvili (KP) equations

$$\left\{ \begin{array}{l} u_t + u^p u_x + u_{xxx} - v_y = 0 \\ v_x = u_y \end{array} \right. \tag{1.1}$$

where $u = u(t, x, y)$, $(x, y) \in \mathbb{R}^2$ and $t \geq 0$. Existence of solitary waves propagating along the $x-$direction for (1.1) was proved in [**7**] (see also [**19**]), while decay and symmetry properties of those peculiar solutions was proved in [**8**]. The initial value problem for (1.1) with $u(0) = u_0$ prescribed, has been studied by many authors ([**9**], [**15**], [**17**], [**18**], see also [**10**], [**21**] for results using inverse scattering techniques

1991 *Mathematics Subject Classification.* Primary 35Q53, 35B35; Secondary 76B15.

when $p = 1$). In order to give a theorem stating the main results related to the initial value problem for (1.1), we define for $s \geq 0$:

$$X_s = \left\{ f \in H^s(\mathbb{R}^2), \mathcal{F}^{-1}\left(\frac{\hat{f}}{\xi_1}\right) \in H^s(\mathbb{R}^2) \right\}$$

where $H^s(\mathbb{R}^2)$ is the usual Sobolev space of order s and $\mathcal{F}^{-1}(f)$, $\hat{f}$ denote respectively the inverse and direct Fourier transform of f. We also define as in [**7**] and [**8**], the energy space Y as the closure of $\partial_x \mathcal{C}_0^\infty(\mathbb{R}^2)$ for the norm

$$\|\partial_x \varphi\|_Y = \left(|\nabla\varphi|_{L^2}^2 + |\partial_x^2 \varphi|_{L^2}^2\right)^{1/2}.$$

Note that if $u \in Y$, then $v = \partial_x^{-1} u_y$ is well defined (see [**7**]). For a proof of the following theorem, see [**15**], [**17**], [**18**].

THEOREM 1.1. *Let $u_0 \in X_s$, $s \geq 3$, such that $\partial_y^2 u_0 = \partial_x^2 \psi$ for some $\psi \in L^2(\mathbb{R}^2)$. There exists $T > 0$ such that (1.1) has a unique solution*

$$u \in \mathcal{C}([0,T], H^s(\mathbb{R}^2)) \cap \mathcal{C}^1([0,T], H^{s-3}(\mathbb{R}^2))$$

$$v \in \mathcal{C}([0,T], H^{s-1}(\mathbb{R}^2)).$$

In addition, $\int_{\mathbb{R}^2} |u(t,x,y)|^2 dxdy$ and

$$E(u(t)) = \frac{1}{2}\int_{\mathbb{R}^2} (u_x^2 + v^2)dxdy - \frac{1}{(p+1)(p+2)}\int_{\mathbb{R}^2} u^{p+2} dxdy$$

are independent of $t \in [0,T]$, and if $1 \leq p < 4/3$, then $\sup_{t\in[0,T]} \|u(t)\|_Y \leq C(|u_0|_{L^2}, E(u_0))$.

In [**7**], we proved the existence of finite energy solitary waves propagating along the x-direction with a positive velocity, i.e. solutions of (1.1) of the form $u(x-ct, y)$, with $u \in Y$ and $c > 0$, when $p < 4$. If we set $v = \partial_x^{-1} u_y$, then (u, v) satisfies the system of equations

$$\left\{ \begin{array}{l} -cu_x + u^p u_x + u_{xxx} - v_y = 0 \\ v_x = u_y \end{array} \right. \tag{1.2}$$

or equivalently if $p < 4$, u satisfies the following equation in Y' (the dual space of Y)

$$-cu + \frac{u^{p+1}}{p+1} + u_{xx} - \partial_x^{-1} v_y = 0 \tag{1.3}$$

where $v = \partial_x^{-1} u_y \in L^2(\mathbb{R}^2)$ and $\partial_x^{-1} v_y \in Y'$. More precisely, concerning equation (1.3), we have

THEOREM 1.2 ([**7**] [**8**]). *Equation (1.3) has a nontrivial solution $u \in Y$ if and only if $1 \leq p < 4$. Any solution $u \in Y$ is continuous and satisfies $(x^2+y^2)u \in L^\infty(\mathbb{R}^2)$, and $(x^2+y^2)^\delta \nabla u$, $(x^2+y^2)^\delta u_{xx} \in L^2(\mathbb{R}^2)$ for any δ, $0 < \delta < 1$. In the case where p is an integer, $u \in H^\infty(\mathbb{R}^2) = \cap_{s\geq 0} H^s(\mathbb{R}^2)$.*

The solutions of Theorem 1.2, when $1 \leq p < 4$, were obtained as minima of the problem

$$I_{\lambda^*} = \inf \left\{ \int_{\mathbb{R}^2} [u_x^2 + cu^2 + (\partial_x^{-1} u_y)^2] dxdy, \ , u \in Y, \ , \int_{\mathbb{R}^2} u^{p+2} dxdy = \lambda^* \right\}$$

for some $\lambda^* > 0$. It was also proved in [8] that those minima are exactly the ground state solutions of equation (1.3), that is the solutions of (1.3) that minimize the action

$$S(u) = E(u) + \frac{c}{2} \int_{\mathbb{R}^2} |u(x,y)|^2 dxdy$$

among all other solutions of (1.3) (see [8], Lemma 2.1). We will from now on denote by $\mathcal{G}_c$ the set of the ground state solutions of (1.3). The natural question arizing concerning those solutions is then the question of their orbital stability, in the following sense.

DEFINITION 1.3. Let φ_c be a solitary wave of (1.3). Then φ_c is orbitally stable if for all $\varepsilon > 0$, there is a $\delta > 0$ such that for any $u_0 \in X_s$, $s \geq 3$, with $\|u_0 - \varphi_c\|_Y \leq \delta$, the corresponding solution $u(t)$ of (1.1) with $u(0) = u_0$ satisfies

$$\sup_{t \geq 0} \inf_{(r,s) \in \mathbb{R}^2} \|u(t) - \varphi_c(. - r, . - s)\|_Y \leq \varepsilon.$$

In Section 2, use is made of a new characterization of the ground states in the case $1 \leq p < 4/3$, to show that the set of ground states is stable for these values of p (see Theorem 2.1). However, we have not been able to prove that the ground states themselves are orbitally stable, because we do not know whether or not they are isolated points in Y, up to spatial translations (i.e. whether they are unique up to translations in a sufficiently small neighborhood of their orbit). In the complementary case $4/3 < p < 4$, a proof of instability of the solitary waves was given in [19], by applying a general theory of [16]. The aim of Section 3 is to justify several technical points which were neglected in [19].

2. Stability of the set of ground states

It was proved in [8] that the ground states of equation (1.3) were exactly the solutions of the minimization problem I_{λ^*} defined in the introduction. We will now use another characterization of the ground states solutions for $p < 4/3$, in order to show that the set of those solutions is stable when p is less then the critical value $p_c = 4/3$. More precisely, let $\mathcal{G}_c$ be the set of the ground state solutions of equation (1.3). Then, the stability result is the following :

THEOREM 2.1. *Let $p < 4/3$, $c > 0$ and $\varphi_c \in \mathcal{G}_c$; then for all positive ε, there is a positive δ such that if $u_0 \in X_s$, $s \geq 3$, and $\|u_0 - \varphi_c\|_Y \leq \delta$, then the solution $u(t)$ of (1.1) with $u(0) = u_0$ satisfies*

$$\sup_{t \geq 0} \inf_{\psi \in \mathcal{G}_c} \|u(t) - \psi\|_Y \leq \varepsilon.$$

Note that if $\varphi \in \mathcal{G}_c$, then all the translates $\varphi_c(. - x, . - y)$, with $(x, y) \in \mathbb{R}^2$ are also elements of $\mathcal{G}_c$. On the other hand, if for some cylindrically symmetric function $\varphi_c \in Y$, $\mathcal{G}_c$ is reduced to the set of all the spatial translates of φ_c, then Theorem 2.1

says that the solution $\varphi_c(x-ct,y)$ of equation (1.1) is orbitally stable. However, we do not know whether the ground state solution of equation (1.3) (or equivalently the solution of the minimization problem I_{λ^*}) is unique up to translations or not.

The key point in the proof of Theorem 2.1 is the fact that for $p < 4/3$, the ground states solutions are characterized as minimizers of energy constrained by constant L^2-norm (note that any solitary wave $\varphi(x-ct,y)$, with φ solution of equation (1.3) is a critical point of the functional $S(\varphi)$). This characterization was first used for the ground states of the subcritical nonlinear Schrödinger equation by Cazenave and Lions ([**5**], see also [**4**]) to prove the stability of those solutions. It is stated in the following lemma.

LEMMA 2.2. *There is a positive number μ^* such that*

(i) *The minimization problem*

$$J = \inf\left\{E(u),\ u \in Y,\ \int_{\mathbb{R}^2} |u|^2 dxdy = \mu^*\right\}$$

has at least one solution.

(ii) *Any minimizing sequence u_n for J is relatively compact in Y up to translations, i.e. there are subsequences u_{n_k}, and $(x_{n_k}, y_{n_k}) \in \mathbb{R}^2$ such that $u_{n_k}(\cdot - x_{n_k}, \cdot - y_{n_k})$ converges strongly in Y to some u which is a minimum of J.*

(iii) *The set of minima of J is exactly equal to $\mathcal{G}_c$.*

PROOF. (i) and (ii) are proved by making use of the concentration-compactness principle of P.L. Lions ([**13**]). Note that we may assume that $c = 1$, since the scale change $\tilde{u}(x,y) = c^{-1/p}u(x/\sqrt{c}, y/c)$ transforms any solution u of equation (1.3) into a solution $\tilde{u}$ of the same equation with $c = 1$, and furthermore,

$$E(\tilde{u}) = c^{-\frac{2}{p}+\frac{1}{2}}E(u)$$

and

$$|\tilde{u}|_{L^2}^2 = c^{-\frac{2}{p}+\frac{3}{2}}|u|_{L^2}^2.$$

Let us denote by J_μ, for $\mu > 0$, the minimization problem

$$J_\mu = \inf\left\{E(u),\ u \in Y,\ \int_{\mathbb{R}^2} |u|^2 dxdy = \mu\right\}.$$

Then, for any $\mu > 0$, one has $-\infty < J_\mu < 0$ and $J_\mu = \mu^{\frac{4-p}{4-3p}}J_1$: indeed, setting $u_\mu(x,y) = \mu^{2/(4-3p)}u(\mu^{p/(4-3p)}x, \mu^{2p/(4-3p)}y)$, one can easily show that for $\mu > 0$, $J_\mu = \mu^{\frac{4-p}{4-3p}}J_1$. Assume now that $J_\mu \geq 0$ for some $\mu > 0$, then $J_\mu \geq 0$ for any $\mu > 0$, and thus, $E(u) \geq 0$ for any $u \in Y$. But it is clear that if u^* is a solution of the minimization problem I_{λ^*} (defined in the introduction), then $E(\lambda u^*) < 0$ for sufficiently large λ, leading thus to a contradiction.

On the other hand, from the inequality (see [**2**])

$$\begin{aligned}\frac{1}{(p+1)(p+2)}\int_{\mathbb{R}^2}|u|^{p+2} &\leq C_1|u|_{L^2}^{(4-p)/2}|u_x|_{L^2}^{p}|\partial_x^{-1}u_y|_{L^2}^{p/2} \\ &\leq \tfrac{1}{4}\textstyle\int_{\mathbb{R}^2}\left(u_x^2 + (\partial_x^{-1}u_y)^2\right) + C_2|u|_{L^2}^{2(4-p)/(4-3p)},\end{aligned}$$

one can deduce that

$$E(u) \geq \frac{1}{4}\int_{\mathbb{R}^2} (u_x^2 + (\partial_x^{-1}u_y)^2) - C_3|u|_{L^2}^{2(4-p)/(4-3p)}$$

so that $J_\mu \geq -C_3\mu^{2(4-p)/(4-3p)} > -\infty$ for any $\mu > 0$.

Now, if u_n is a minimizing sequence for J_μ, and if $\varphi_n \in L^2_{loc}(\mathbb{R}^2)$ with $\partial_x\varphi_n = u_n$ and $\partial_x^{-1}u_{ny} = \partial_y\varphi_n$, then u_n is a bounded sequence in Y by the use of the preceding inequality, and there is a subsequence, still denoted by u_n such that $\|u_n\|_Y^2 \to \beta > 0$. We then apply the concentration compactness lemma [**13**] with $\rho_n = \|u_n\|_Y^2$.

The "vanishing" case is avoided because in this case, u_n tends to 0 in $L^q(\mathbb{R}^2)$ for any q with $2 < q < 6$ (see [**7**]) and thus $\int u_n^{p+2}dxdy$ tends to 0 as n goes to infinity. But this contradicts the fact that $J_\mu < 0$.

In the "dichotomy" case, one would show as in [**7**] that for some α with $0 < \alpha < \beta$, and for all $\varepsilon > 0$, there exist $\delta(\varepsilon) > 0$ (with $\delta(\varepsilon) \to 0$ as $\varepsilon \to 0$), two sequences u_n^1 and u_n^2 in Y, and an integer $n_0 \geq 0$ such that for $n \geq n_0$,

$$\begin{aligned} \|u_n^1 + u_n^2 - u_n\|_Y &\leq \delta(\varepsilon) \\ |\|u_n^1\|_Y^2 - \alpha| &\leq \delta(\varepsilon) \\ |\|u_n^2\|_Y^2 - (\beta-\alpha)| &\leq \delta(\varepsilon) \\ |\int_{\mathbb{R}^2}(u_n)^2 - \int_{\mathbb{R}^2}(u_n^1)^2 - \int_{\mathbb{R}^2}(u_n^2)^2| &\leq \delta(\varepsilon). \end{aligned}$$

But this would then lead to a contradiction, due to the subadditivity property

$$J_\mu < J_\nu + J_{\mu-\nu} \text{ for any } \nu, \quad 0 < \nu < \mu$$

which follows from the equality $J_\mu = \mu^{(4-p)/(4-3p)}J_1$ and the fact that $J_1 < 0$.

The only remaining possibility is then the "concentration" of the sequence u_n, up to translations. Using the relative compactness of the injection $Y \subset L^q_{loc}$, one then shows that some subsequence of $u_n(.-x_n, .-y_n)$, with $(x_n, y_n) \in \mathbb{R}^2$, converges strongly in $L^2(\mathbb{R}^2)$ and $L^{p+2}(\mathbb{R}^2)$ to some $u \in Y$, which is a minimum of J_μ (see [**7**]). It follows that it converges in fact to u strongly in Y. This achieves to prove (i) and (ii) which are actually true for any positive value of μ^*.

In order to prove (iii), we fix λ^* such that the solutions of I_{λ^*} are exactly the ground state solutions of equation (1.3) (see [**8**], Lemma 2.1) and set $\mu^* = |u^*|_{L^2}^2$ where u^* is any minimum of I_{λ^*}. First, μ^* does not depend on the choice of u^* : indeed, by the computations of [**7**], Section 2, one can prove that $\int(u^*)^2 = \frac{4-p}{2(p+2)}I_{\lambda^*}$ for any minimum u^* of I_{λ^*}, or equivalently for any ground state solution of equation (1.3). In the same way, the energy E takes the same value on any ground state, and in order to show that the minima of J_{μ^*} are exactly the ground states of (1.3), it suffices to prove that if v is a minimum of J_{μ^*}, then v solves equation (1.3). But then, for some $\theta \in \mathbb{R}$, v satisfies

$$-v_{xx} + \partial_x^{-2}v_{yy} + \theta v - \frac{v^{p+1}}{p+1} = 0,$$

and by the same computations as in [**7**], Section 2, one has $0 > E(v) = \frac{\theta(3p-4)}{2(4-p)}|u^*|_{L^2}^2 = \theta E(u^*)$, hence $\theta \geq 1$. On the other hand, $\tilde{v}(x,y) = \theta^{-1/p}v(x/\sqrt{\theta}, y/\theta)$ solves (1.3) (with $c = 1$), and using the inequality $S(\tilde{u}) \geq S(u^*)$, one shows that $\theta \leq 1$. This achieves to prove (iii). □

PROOF OF THEOREM 2.1. Theorem 2.1 follows from Lemma 2.2 by using classical arguments that we repeat here for the sake of clarity. Assume that $\mathcal{G}_c$ is not stable, then one can find $\varphi \in \mathcal{G}_c$, $u_n^0 \in X_s$, $t_n \geq 0$ and $\varepsilon > 0$ with $\|u_n^0 - \varphi\|_Y \to 0$ as $n \to \infty$ and such that if $u_n(t)$ is the maximal solution of (1.1) with $u_n(0) = u_n^0$, then

$$\inf_{(x,y)\in\mathbb{R}^2} \inf_{u\in\mathcal{G}_c} \|u_n(t_n) - u(.-x,.-y)\|_Y \geq \varepsilon,$$

or equivalently, since $\mathcal{G}_c$ is stable under spatial translations,

$$\inf_{u\in\mathcal{G}_c} \|u_n(t_n) - u\|_Y \geq \varepsilon.$$

But $|u_n(t_n)|_{L^2}^2 = |u_n^0|_{L^2}^2 \to |\varphi|_{L^2}^2 = \mu^*$ and $E(u_n(t_n)) = E(u_n^0) \to E(\varphi) = J$ as n goes to infinity; hence $\frac{\mu^*}{|u_n(t_n)|_{L^2}^2} u_n(t_n)$ is a minimizing sequence for J, and by Lemma 2.2, for some $(x_{n_k}, y_{n_k}) \in \mathbb{R}^2$, $u_{n_k}(t_{n_k}, .-x_{n_k}, .-y_{n_k})$ converges strongly in Y to a ground state of equation (1.3), contradicting the fact that $\inf_{u\in\mathcal{G}_c} \|u_n(t_n) - u\|_Y \geq \varepsilon$.

3. Remarks on the instability of cylindrically symmetric solitary waves

In the case $p > 4/3$, considering a solitary wave φ_c of (1.3) with $c > 0$, Wang et. al. have constructed in [**19**] a curve χ_ω, with $\chi_c = \varphi_c$, on which the L^2-norm is constant while the energy E has a local maximum at $\omega = c$. This shows that φ_c is a saddle point of energy constrained by constant L^2-norm. The preceding authors then claimed that the instability of the solitary wave φ_c followed as in the theorem of Shatah and Strauss [**16**] by constructing a Lyapunov functional $A(u(t))$ with the help of $Y_c = \int_{-\infty}^x y_c dx$, where $y_c = \left.\frac{\partial \chi_\omega}{\partial \omega}\right|_{\omega=c}$. However, it is not clear whether $A(u(t))$ is bounded independently of time as is required in the proof of Shatah and Strauss, because Y_c has no reason to be in $L^2(\mathbb{R}^2)$ (note that taking for φ_c the lump soliton of the KPI equation (see [**14**]) leads to $Y_c \notin L^2(\mathbb{R}^2)$). The aim of this section is to give indications of the fact that it is however possible to justify the proof of instability, at least when φ_c is cylindrically symmetric - which is the case for the ground states, as was proved in [**8**]. We will thus prove the following theorem.

THEOREM 3.1. *Let $4/3 < p < 4$ and φ_c be a cylindrically symmetric solitary wave of (1.3). Then φ_c is orbitally unstable.*

We assume now that $4/3 < p < 4$ and we fix $c > 0$ and φ_c a cylindrically symmetric solitary wave solution of (1.3). For $\varepsilon > 0$, let

$$U_\varepsilon = \{u \in Y, \inf_{(x,y)\in\mathbb{R}^2} \|u - \varphi_c(.-x,.-y)\|_Y < \varepsilon\}.$$

The following lemma shows that the minimum in this definition is attained and that if u is cylindrically symmetric, i.e. $u(x,-y) = u(x,y)$, then one only has to translate u in the x variable. The lemma is proved by making use of the implicit function theorem and the fact that φ_c is cylindrically symmetric (see [**3**], [**6**], [**11**]). The integrals in the lemma are well defined, as can be seen in the proof of [**8**], Theorem 1.1.

LEMMA 3.2. *For $\varepsilon > 0$ sufficiently small, there is a C^1 curve $\alpha : U_\varepsilon \to \mathbb{R}^2$ such that for all $u \in U_\varepsilon$,*

$$\int_{\mathbb{R}^2} u(\mathbf{x}+\alpha(u))\partial_x\varphi_c(\mathbf{x})d\mathbf{x} = \int_{\mathbb{R}^2} u(\mathbf{x}+\alpha(u))\partial_y\varphi_c(\mathbf{x})d\mathbf{x} = 0$$

Furthermore, if $u(x,y) = u(x,-y)$ for all $(x,y) \in \mathbb{R}^2$, then $\alpha(u) = (\alpha_1(u), 0)$ and

$$\alpha_1'(u) = \frac{\partial_x\varphi_c(. - \alpha(u))}{\int_{\mathbb{R}^2} u(\mathbf{x})\partial_x^2\varphi_c(x-\alpha_1(u),y)dxdy}.$$

Wang et. al. defined a curve χ_ω, $\omega > 0$ by setting

$$\chi_\omega(x,y) = \varphi_\omega\left(\frac{x}{\sigma(\omega)}, \frac{y}{\sigma^2(\omega)}\right)$$

where $\sigma(\omega) = [\int |\varphi_c|^2 dxdy / \int |\varphi_\omega|^2 dxdy]^{1/3}$ is choosen in order that $\int |\chi_\omega|^2 dxdy = \int |\varphi_c|^2 dxdy$, and where φ_ω is the solitary wave of speed ω obtained by an appropriate rescaling of φ_c. Then, they proved the following :

LEMMA 3.3 ([**19**]). *In the case $4/3 < p < 4$, $E(\chi_\omega) < E(\varphi_c)$ for $\omega \neq c$ and ω near c.*

In order to define rigorously the Lyapunov functional $A(u(t))$ and to compute $\frac{dA}{dt}$, we need the following lemma.

LEMMA 3.4. *$y_c \in Y$*

PROOF. Using the decay properties of φ_c obtained in [**8**], we easily get $y_c \in L^2(\mathbb{R}^2)$. Since for some constants α_c, β_c and γ_c, y_c expresses as

$$y_c = \alpha_c\varphi_c + \beta_c x\partial_x\varphi_c + \gamma_c y\partial_y\varphi_c,$$

one shows easily that $y_c \in Y$, provided that $y\partial_x\partial_y\varphi_c$ and $y\partial_y(\partial_x^{-1}\partial_y\varphi_c)$ are in L^2. Let $\psi = \partial_x^{-1}\partial_y\varphi_c$, then ψ satisfies

$$\psi_y = -c\varphi_{cx} + \varphi_{cxxx} + \varphi_c^p\varphi_{cx}.$$

Multiplying this equation by $y^2\psi_y$ and integrating by parts leads to

(3.1)

$$\int_{\mathbb{R}^2} y^2\psi_y^2 dxdy = \int_{\mathbb{R}^2}\left[-cy^2\varphi_{cy}^2 + c\varphi_c^2 - y^2\varphi_{cyy}\varphi_{cxx} - \frac{1}{p+1}y^2{\varphi_c}^{p+1}\varphi_{cyy}\right]dxdy.$$

Again, by the decay properties obtained in [**8**], all the terms in the right hand side, except maybe the third one, are finite. In order to treat the third term, one shows that $y^\delta\varphi_{cyy} \in L^2$ for $\delta \in (0,1)$, by multiplying the equation

$$\varphi_{cyy} = -c\varphi_{cxx} + \varphi_{cxxxx} + \left(\frac{\varphi_c^{p+1}}{p+1}\right)_{xx}$$

by $y^{2\delta}\varphi_{cyy}$, integrating by parts and using again the decay properties of [**8**]. Since by Lemma 3.5 of [**8**], $y^{2-\delta}u_{xx} \in L^2$, it follows that the third term in (3.1) is also finite and that $y\psi_y \in L^2(\mathbb{R}^2)$. The fact that $y\partial_x\partial_y\varphi_c \in L^2(\mathbb{R}^2)$ is proved in the same way. □

We now choose ω sufficiently close to c, so that Lemma 3.3 holds; Since $\chi_\omega \in Y$, χ_ω is cylindrically symmetric, and

$$\partial_x^{-2}\partial_y^2\chi_\omega = -\frac{c}{\sigma^2}\chi_\omega + \frac{1}{\sigma^4}\chi_{\omega xx} + \frac{1}{(p+1)\sigma^{2(p+1)}}\chi_\omega^{p+1} \quad \in L^2(\mathbb{R}^2),$$

we may choose $u_0 \in \partial_x C_0^\infty(\mathbb{R}^2)$, with $\partial_x^{-2}\partial_y^2 u_0 \in L^2(\mathbb{R}^2)$, u_0 cylindrically symmetric, $|u_0|_{L^2}^2 = |\chi_\omega|_{L^2}^2 = |\varphi_c|_{L^2}^2$, and u_0 sufficiently close to χ_ω in Y-norm, so that $E(\varphi_c) - E(u_0) > 0$ (note that $\partial_x C_0^\infty(\mathbb{R}^2)$ is dense in Y). Since $u_0 \in X_s$ with $s \geq 3$, it follows that the maximal solution $u(t)$ given by Theorem 1.1 is in $\mathcal{C}([0,T^*], X_s)$, i.e. $\partial_x^{-1}u(t) \in \mathcal{C}([0,T^*], H^s(\mathbb{R}^2))$. We then fix $\varepsilon > 0$, and define, as long as $u(t)$ stays in U_ε

$$A(u(t)) = -\int_{\mathbb{R}^2} y_c(x - \alpha_1(u(t)), y)\partial_x^{-1}u(t,x,y)dxdy. \tag{3.2}$$

LEMMA 3.5. *Let $u(t) \in \mathcal{C}([0,T^*], X_s)$ be a cylindrically symmetric solution of (1.1) and $A(u(t))$ be defined by (3.2). Then there is a constant δ with $0 < \delta < 1$ and for any $T \in [0,T^*]$, there is a positive constant C depending only on $\sup_{0\leq t\leq T}\|u(t)\|_Y$ such that*

$$|A(u(t))| \leq C(1 + t^\delta), \quad 0 \leq t \leq T.$$

PROOF. If $S(t)$ is the group associated with the linear part of (1.1), then

$$\partial_x^{-1}u(t) = S(t)\partial_x^{-1}u_0 + \int_0^t S(t-s)[u^{p+1}(s)]ds. \tag{3.3}$$

Since $S(t)$ is unitary in X_s and satisfies (see [**15**])

$$|S(t)v|_{L^\infty} \leq \frac{C}{|t|}|v|_{L^1}, \text{ for any } v \in L^1(\mathbb{R}^2) \cap X_0,$$

it follows from (3.3) that if $r' \in (1,2) \cap [\frac{2}{p+1}, \frac{6}{p+1}] \neq \emptyset$, and if $\frac{1}{q'} = \frac{1}{r'} - \frac{1}{2}$, (i.e. $q' > 1$), then

$$\begin{aligned} |\partial_x^{-1}u(t)|_{L^2} &\leq C(|\partial_x^{-1}u_0|_{L^2} + |u^{p+1}|_{L^{q'}(0,t,L^{r'})}) \\ &\leq C(|\partial_x^{-1}u_0|_{L^2} + t^{1/q'}\sup_{0\leq s\leq T}|u(s)|_{L^{(p+1)r'}}^{p+1}) \end{aligned} \tag{3.4}$$

and by the injection $Y \subset L^{(p+1)r'}$, one has

$$|A(u(t))| \leq C|y_c|_{L^2}|\partial_x^{-1}u(t)|_{L^2} \leq C(1 + t^{1/q'})$$

where C only depends on u_0 and on $\sup_{0\leq s\leq T}\|u(s)\|_Y$. □

The last lemma we need is proved as in [**3**].

LEMMA 3.6. *For $u \in U_\varepsilon$, let*

$$B(u) = y_c(. - \alpha(u)) + \frac{\langle u, y_c(. - \alpha(u))\rangle}{\langle u, \partial_x^2\varphi_c(. - \alpha(u))\rangle}\partial_x^2\varphi_c(. - \alpha(u)).$$

Then B is $\mathcal{C}^1$ from $U_\varepsilon^c = \{v \in U_\varepsilon,\ v$ is cylindrically symmetric $\}$ into Y and there is a $\mathcal{C}^1$ functional $\Lambda : \{v \in U_\varepsilon^c,\ \int |v|^2 = \int |\varphi_c|^2\} \to \mathbb{R}$ such that if $v \in U_\varepsilon^c$ with $\int |v|^2 = \int |\varphi_c|^2$ and v is not a translate of φ_c, then

$$E(\varphi_c) < E(v) + \Lambda(v)\langle E'(v), B(v)\rangle.$$

We now end the proof of Theorem 3.1 : having choosen u_0 as before, and defined $A(u(t))$ as in (3.2), we assume that $u(t) \in U_\varepsilon^c$ for $t \in [0,T]$, and we formally compute

$$\frac{d}{dt}A(u(t)) = \langle \alpha_1'(u(t)), u_t\rangle\langle \partial_x y_c(.-\alpha(u(t)), \partial_x^{-1}u(t)\rangle - \langle y_c(.-\alpha(u(t))), \partial_t\partial_x^{-1}u(t)\rangle.$$

But we have $\alpha_1'(u(t)) \in \mathcal{C}([0,T], L^2)$ by Lemma 3.2, $\partial_t u \in \mathcal{C}([0,T], L^2)$, $\partial_x^{-1}u \in \mathcal{C}([0,T], L^2)$, and finally $\partial_t\partial_x^{-1}u(t) = \partial_x^{-1}u_t(t) = E'(u(t)) \in \mathcal{C}([0,T], Y')$, so that the preceding computations are justified, and give

$$\frac{d}{dt}A(u(t)) = \langle B(u(t)), E'(u(t))\rangle. \tag{3.5}$$

The proof is now ended as in the classical case (see [**3**], [**16**]) : choosing ε sufficiently small, we may assume that $|\Lambda(v)| \le 1$ for all $v \in U_\varepsilon^c$, where Λ is the functional defined in Lemma 3.6. We then have for $0 \le t \le T$,

$$0 < E(\varphi_c) - E(u_0) = E(\varphi_c) - E(u(t)) < |\Lambda(u(t))\langle B(u(t)), E'(u(t))\rangle| \le \left|\frac{d}{dt}A(t)\right|$$

by (3.5) and Lemma 3.6. From this and Lemma 3.5, we deduce that T is necessarily finite i.e. that $u(t)$ must leave U_ε^c in a finite time, showing thus that φ_c is orbitally unstable.

References

[1] T.B. Benjamin, *The stability of solitary waves.* Proc. Roy. Soc. Lond. A 328 (1972), 153-183.

[2] O.V. Besov, V.P. Il'in, S.M. Nikolskii, *Integral Representations of Functions and Imbeddings Theorems*, vol. I, J. Wiley, 1978.

[3] J.L. Bona, P.E. Souganidis, W.A. Strauss, *Stability and instability of solitary waves of Korteweg-de Vries type*, Proc. Roy. Soc. Lond. A 411 (1987), 395-412.

[4] T. Cazenave, *An Introduction to Nonlinear Schrödinger Equations.* Textos de Métodos Matematicos 26, Instituto de Metematica-UFRJ Rio de Janeiro, 1993.

[5] T. Cazenave, P.-L. Lions, *Orbital stability of standing waves for some nonlinear Schrödinger equations*, Comm. Math. Phys. 85 (1982), 549-561.

[6] A. de Bouard, *Stability and instability of some nonlinear dispersive solitary waves in higher dimension*, Proc. Roy. Soc. Edinburgh 126 A (1996), 89-112.

[7] A. de Bouard, J.C. Saut, *Solitary waves of generalized Kadomtsev-Petviashvili equations*, Annales I.H.P. Analyse Non Linéaire, to appear.

[8] A. de Bouard, J.C. Saut, *Symmetries and decay of the generalized KP solitary waves*, Preprint, 1995.

[9] A. V. Faminskii, *The Cauchy problem for the generalized Kadomtsev-Petviashvili equation*, Siberian Math. J. 33 N1 (1992), 133-143.

[10] A.S. Fokas, L.Y. Sung, *On the solvability of the N-wave, Davey-Stewartson and Kadomtsev-Petviashvili equations*, Inverse Problems 8 (1992), 673-708.

[11] M. Grillakis, J. Shatah, W.A. Strauss, *Stability theory of solitary waves in the presence of symmetry I*, J. Funct. Anal. 74 (1987), 160-197.

[12] B.B. Kadomtsev, V.I. Petviashvili, *On the stability of solitary waves in weakly dispersing media*, Sov. Phys. Doklady 15 (1970), 539-541.

[13] P.-L. Lions, *The concentration-compactness principle in the calculus of variations, The locally compact case, I*, Annales I.H.P. Analyse Non Linéaire I, (1984), 104-145.

[14] V.I. Petviashvili, V.V. Yankov, *Solitons and turbulence*, in Review of Plasma Physics XIV, Kadomtsev ed. (1989) 1-62.

[15] J.-C. Saut, *Remarks on the generalized Kadomtsev-Petviashvili equations*, Indiana Math. J. 42 (1993), 1011-1026.

[16] J. Shatah, W.A. Strauss, *Instability of nonlinear bound states*, Comm. Math. Phys. 100 (1985), 173-190.

[17] M.M. Tom, *On a generalized Kadomtsev-Petviashvili equation*, Preprint, 1996.

[18] S. Ukaï, *Local solutions of the Kadomtsev-Petviashvili equation*, J. Fac. Sci. Univ. Tokyo, Sect. IA Math. 36 (1989), 193-209.

[19] X.P. Wang, M.J. Ablowitz, H. Segur, *Wave collapse and instability of solitary waves of a generalized Kadomtsev-Petviashvili equation*, Physica D 78 (1994), 97-113.

[20] M.I. Weinstein, *Lyapunov stability of ground states of nonlinear dispersive evolution equations*, Comm. Pure Appl. Math. 39 (1986), 51-68.

[21] X. Zhou, *Inverse scattering transform for the time dependent Schrödinger equation with applications to the KPI equation*, Comm. Math. Phys. 128 (1990), 551-564.

CNRS et Université Paris-Sud, URA 760, Bât. 425, Université de Paris-Sud, 91405 ORSAY CEDEX, FRANCE

E-mail address: Anne.Bouard@math.u-psud.fr

CNRS et Université Paris-Sud, URA 760, Bât. 425, Université de Paris-Sud, 91405 ORSAY CEDEX, FRANCE

E-mail address: Jean-Claude.Saut@math.u-psud.fr

Contemporary Mathematics
Volume **200**, 1996

On the Asymptotic Integrability of a Generalized Burgers Equation

Thanasis Fokas and Laihan Luo

1. Introduction

Certain nonlinear partial differential equations are generic, namely they can be derived, using some physically meaningful asymptotic expansion, from very large classes of nonlinear PDEs. It turns out that for many important physical systems these equations are **integrable** (see [1] for a discussion of this remarkable fact). In this case we say that the underlying physical system is **asymptotically integrable** up to $O(\varepsilon)$, where ε is the small expansion parameter. In certain cases when one formally extends the asymptotic integrability of the system up to $O(\varepsilon^2)$, one can find an explicit transformation which maps the equation describing the physical system to some integrable equation in this higher asymptotic order. The first example of such a transformation was found by Kodama [2], and was used to map the equation describing unidirectional small amplitude, long, water waves correct up to $O(\varepsilon^2)$, to the integrable equation obtained by combining the Korteweg-de Vries (KdV) equation with its first commuting flow.

One of the authors and Q.M. Liu [3] have generalized the transformations found by Kodama, and furthermore have shown that the concept of mastersymmetries (introduced in [4]) provides an algorithmic approach for finding such transformations. Using appropriate mastersymmetries, it is for example, possible to map the equation describing unidirectional water waves correct up to $O(\varepsilon^2)$, to the KdV itself. And it is also possible to establish formally the asymptotic integrability of two dimensional water waves up to $O(\varepsilon^2)$ **without** the unidirectionalization assumption.

In this paper we present a methodology for studying these transformations rigorously. Also we apply this methodology to the rigorous analysis of the following formal result [5]:

Let $v(x,t)$ satisfy the Burgers equation

$$v_t = v_{xx} + 2vv_x. \tag{1.1}$$

1991 *Mathematics Subject Classification.* Primary 36; Secondary 46, 72.

The authors were supported in part by EPSRC GR/J71885. This is part of a larger project undertaken jointly with Professor Jerry Bona.

Let $u(x,t)$ be defined by

$$u = v + \varepsilon P(v),$$

where $P(v)$ is given by

$$P(v) \doteq \frac{1}{2}(2A-\Gamma)v^2 + \frac{3A-B}{2}v_x\partial_x^{-1}v - \frac{A}{2}x(v_{xx}+2vv_x). \tag{1.2}$$

In (1.2) ∂_x^{-1} denotes integration with respect to x and A, B, Γ are constants. Then u satisfies the generalized Burgers equation

$$u_t = u_{xx} + 2uu_x + \varepsilon\left[Au_{xxx} + Buu_{xx} + \Gamma u_x^2 + \left(\Gamma + \frac{B}{2} - \frac{3A}{2}\right)u^2u_x\right] + O(\varepsilon^2). \tag{1.3}$$

Suppose that

$$u(x,0) = u_0(x) + \varepsilon u_1(x). \tag{1.4}$$

Our rigorous methodology for studying the initial value problem (1.3)-(1.4) involves the following steps:

(i) Given u_0 and u_1, define $\tilde{v}(x,0,\varepsilon)$ and $v(x,0,\varepsilon)$ by

$$\tilde{v}(x,0,\varepsilon) \doteq u_0(x) + \varepsilon[u_1(x) - P(u_0(x))], \tag{1.5}$$

and by

$$u_0(x) + \varepsilon u_1(x) = v(x,0,\varepsilon) + \varepsilon P(v(x,0,\varepsilon)), \tag{1.6}$$

respectively.

(ii) Define $\tilde{v}(x,t,\varepsilon)$ and $v(x,t,\varepsilon)$ as the solutions of the integrable equation (in our case the Burgers equation) satisfying $\tilde{v}(x,0,\varepsilon)$ and $v(x,0,\varepsilon)$ respectively. Using the fact that $\tilde{v}(x,0,\varepsilon)$ and $v(x,0,\varepsilon)$ are close in the L_∞-norm, establish that

$$\begin{gathered} |(v-\tilde{v})(\cdot,t,\varepsilon)|_\infty \le C\varepsilon^2, \quad |(v_x-\tilde{v}_x)(\cdot,t,\varepsilon)|_\infty \le C\varepsilon^2 \\ \text{and} \quad |(v-\tilde{v})(\cdot,t,\varepsilon)|_1 \le C\varepsilon^2, \quad \text{for all} \quad t>0, \end{gathered} \tag{1.7}$$

where C is some time independent constant (see Lemma 2).

(iii) Define $u(x,t,\varepsilon)$ and $\tilde{u}(x,t,\varepsilon)$ by

$$u(x,t,\varepsilon) = v(x,t,\varepsilon) + \varepsilon P(v(x,t,\varepsilon)) \tag{1.8}$$

and by

$$\tilde{u}(x,t,\varepsilon) = \tilde{v}(x,t,\varepsilon) + \varepsilon P(\tilde{v}(x,t,\varepsilon)), \tag{1.9}$$

respectively. Show that u and $\tilde{u}$ are well defined, and furthermore obtain the large time behavior of $|u|_\infty$ and $|\tilde{u}|_\infty$ (see Lemmas 3, 4 and Corollary 1). Using these results show that

$$|(u-\tilde{u})(\cdot,t,\varepsilon)|_\infty \le C\varepsilon^2, \quad \text{for all} \quad t>0, \tag{1.10}$$

where C is a time-independent constant (see Theorem 1).

We note that unfortunately there exists a linear relation among the coefficients of the $O(\varepsilon)$ terms of equation (1.3). This is to be contrasted with the cases of generalized KdV, and generalized nonlinear Schrödinger, where the coefficients of the $O(\varepsilon)$ terms are arbitrary. Further discussion of this linear constraint among the coefficients of (1.3) can be found in [6].

The generalized Burgers equation (without a linear constraint among its coefficients) can be obtained from a certain asymptotic expansion of the acoustic waves,

$$\begin{aligned}\rho_t = \ & \rho_{xx} + 2\rho\rho_x \\ & + \varepsilon[\tfrac{\rho_{xxx}}{4} - \tfrac{1}{\gamma+1}\rho\rho_{xx} + \tfrac{\gamma-1}{2(\gamma+1)}\rho_x^2 + (3\tfrac{\delta+\gamma}{(\gamma+1)^2} - 1)\rho^2\rho_x] + O(\varepsilon^2),\end{aligned} \tag{1.11}$$

where ρ is the perturbation of the density,

$$\gamma = \frac{P_0''\rho_0}{2a_0^2}, \qquad \delta = \frac{P_0'''\rho_0^2}{6a_0^2}, \qquad \varepsilon = -\frac{\mu}{2\rho_0 a_0 l},$$

$P_0 = P(\rho_0)$ is the equilibrium pressure, a_0 is the speed of sound, l is a typical length, prime denotes differentiation and μ is viscosity. The constraint satisfied by the coefficients of equation (1.3) corresponds to

$$8\delta + 6\gamma + 1 - 3\gamma^2 = 0. \tag{1.12}$$

Notation The L_q-norm of a function f which is qth-power absolutely integrable on $\mathbb{R}$ will be denoted by $|f|_q$, for $1 \leq q < \infty$. The L_∞-norm of a function f will be denoted by $|f|_\infty$. If $m \geq 0$ is an integer, $H^m(\mathbb{R})$ will denote the Sobolev space consisting of those $L_2(\mathbb{R})$-functions whose first m generalized derivatives lie in $L_2(\mathbb{R})$. The space $H^m(\mathbb{R})$ will be equipped with the usual norm,

$$||f||_{H^m(\mathbb{R})} = ||f||_m = \sum_{k=0}^{m} |f^{(k)}|_2.$$

In addition, $\mathcal{H}_n^m(\mathbb{R})$ will denote the weighted subspace of $H^m(\mathbb{R})$ with the weight $(1+x^2)^{n/2}$, for $n \geq 0$ and with the norm

$$||f||^2_{\mathcal{H}_n^m(\mathbb{R})} = ||f||^2_{m,n} = \sum_{k=0}^{m} \int_{-\infty}^{\infty} |f^{(k)}(x)|^2 (1+x^2)^{\frac{n}{2}}\, dx.$$

2. Main Results

Let $u_0(x)$ and $u_1(x)$ be in $\mathcal{H}_4^5(\mathbb{R})$ and $\mathcal{H}_2^3(\mathbb{R})$, respectively. Define $\tilde{v}(x,0,\varepsilon)$ by

$$\begin{aligned}\tilde{v}(x,0,\varepsilon) \ & \equiv u_0 + \epsilon\{u_1 - \tfrac{1}{2}(2A-\Gamma)u_0^2 - \tfrac{3A-B}{2}(u_0)_x\partial_x^{-1}u_0 \\ & \quad + \tfrac{A}{2}x[(u_0)_{xx} + 2u_0(u_0)_x]\} \\ & = u_0(x) + \epsilon\,(u_1(x) - P(u_0(x))),\end{aligned} \tag{2.1}$$

where ∂_x^{-1} denotes integration with respect to x. Then it follows from (2.1) that

$$\tilde{v}(x,0,\varepsilon) \in \mathcal{H}_2^3(\mathbb{R}). \tag{2.2}$$

Define $v(x,0,\varepsilon)$ by

$$u_0(x) + \varepsilon u_1(x) = v(x,0,\varepsilon) + \varepsilon P\left(v(x,0,\varepsilon)\right). \tag{2.3}$$

Then using (2.1) it follows that for some suitable small $\epsilon > 0$,

$$\begin{aligned} \sup_{x\in\mathbb{R}} |v(x,0,\varepsilon) - \tilde{v}(x,0,\varepsilon)| &\leq |R(u_0,u_1)|_\infty \varepsilon^2 + 0(\varepsilon^3) \\ &\leq \varepsilon^2 C(||u_0||_{5,4}, ||u_1||_{3,2}) + 0(\varepsilon^3), \end{aligned} \tag{2.4}$$

where $R(u_0,u_1)$ is given by

$$\begin{aligned} R(u_0,u_1) = \quad & (2A-\Gamma)u_0[u_1 - P(u_0)] \\ & + \tfrac{3A-B}{2}\left[(u_0)_x \partial_x^{-1}(u_1 - P(u_0)) + (u_1 - P(u_0))_x \partial_x^{-1} u_0\right] \\ & - \tfrac{A}{2} x\left[(u_1 - P(u_0))_{xx} + 2\left((u_1 - P(u_0))u_0\right)_x\right]. \end{aligned} \tag{2.5}$$

Lemma 1. *Let $v_0(x)$ be in $L_1(\mathbb{R}) \cap H^3(\mathbb{R})$. Then the solution of the initial-value problem*

$$v_t = v_{xx} + 2vv_x \tag{2.6}$$

$$v(x,0) = v_0(x) \tag{2.7}$$

has the following properties:

$$\begin{aligned} &|v(\cdot,t)|_1 \leq C, \quad |v(\cdot,t)|_\infty \leq C(1+t)^{-\frac{1}{2}}, \quad |v(\cdot,t)|_2 \leq C(1+t)^{-\frac{1}{4}}, \\ &|v_x(\cdot,t)|_2 \leq C(1+t)^{-\frac{3}{4}}, \quad |v_{xx}(\cdot,t)|_2 \leq C(1+t)^{-\frac{5}{4}}, \end{aligned} \tag{2.8}$$

for all $t > 0$.

Proof: It is straightforward to obtain that $|v(\cdot,t)|_1 \leq C$ and $|v(\cdot,t)|_2 \leq C(1+t)^{-\frac{1}{4}}$ (see [7]). Let

$$\phi(x,t) = e^{\int_{-\infty}^x v(x,t)dx}.$$

Then ϕ satisfies $\phi_t = \phi_{xx}$ with initial condition

$$\phi_0 = \phi(x,0) = e^{\int_{-\infty}^x v_0 dx}.$$

Following the line in [8], one can easily show the decay of ϕ_x, ϕ_{xx} and ϕ_{xxx} in L_2-nrom. Because $v = \frac{\phi_x}{\phi}$ and $\phi > 0$ for all $t > 0$, one can obtain the decay estimates stated in (2.8) for solution v. $\diamondsuit$

Lemma 2. *Let $v(x,t,\varepsilon)$ and $\tilde{v}(x,t,\varepsilon)$ be the solutions of equation (2.6), where the initial conditions $v(x,0,\epsilon)$ and $\tilde{v}(x,0,\varepsilon)$ are defined by equations (2.3) and (2.1) respectively. Assume that $v(x,0,\varepsilon)$ and $\tilde{v}(x,0,\varepsilon) \in \mathcal{H}_2^3(\mathbb{R})$. Then*

$$\begin{aligned} &|(v-\tilde{v})(\cdot,t,\varepsilon)|_\infty \leq C\epsilon^2, \ |(v_x - \tilde{v}_x)(\cdot,t,\varepsilon)|_\infty \leq C\epsilon^2, \\ &\text{and } |(v-\tilde{v})(\cdot,t,\varepsilon)|_1 \leq C\epsilon^2, \quad \text{for all } \ t > 0, \end{aligned} \tag{2.9}$$

where the constant C in (2.9) depends on $||u_0||_{6,4}$ and $||u_1||_{4,2}$.

Proof: Suppose v and $\tilde{v}$ both are solutions of (2.6) with initial conditions

$$v(x,0,\varepsilon) = v_0 \quad \text{and} \quad \tilde{v}(x,0,\varepsilon) = \tilde{v}_0,$$

respectively. Let

$$\begin{aligned} \psi(x,t) &= \exp(\textstyle\int_{-\infty}^{x} v(x,t,\varepsilon)dx) \qquad \text{and} \\ \phi(x,t) &= \exp(\textstyle\int_{-\infty}^{x} \tilde{v}(x,t,\varepsilon)dx). \end{aligned} \tag{2.10}$$

Then both ψ and ϕ satisfy the heat equation

$$\Psi_t = \Psi_{xx} \tag{2.11}$$

with the initial conditions

$$\begin{aligned} \psi(x,0) &= \psi_0 = \exp\left(\textstyle\int_{-\infty}^{x} v_0 dx\right) \qquad \text{and} \\ \phi(x,0) &= \phi_0 = \exp\left(\textstyle\int_{-\infty}^{x} \tilde{v}_0 dx\right), \end{aligned} \tag{2.12}$$

respectively. First note that since v_0 and $\tilde{v}_0$ lie in $\mathcal{H}_2^3(\mathbb{R})$, they certainly lie in $L_1(\mathbb{R}) \cap H^3(\mathbb{R})$. Then by Lemma 1, $|v|_1$ and $|\tilde{v}|_1$ are bounded for all $t > 0$. Hence $\psi(x,t)$ and $\phi(x,t)$ are positive and bounded for any $t > 0$. Furthermore, both ψ_x and ϕ_x satisfy (2.11) with initial conditions

$$\begin{aligned} \psi_x(x,0) &= (\psi_0)_x = v_0 \exp\left(\textstyle\int_{-\infty}^{x} v_0 dx\right) \quad \text{and} \\ \phi_x(x,0) &= (\phi_0)_x = \tilde{v}_0 \exp\left(\textstyle\int_{-\infty}^{x} \tilde{v}_0 dx\right), \end{aligned} \tag{2.13}$$

respectively. The functions ψ, ψ_x, ϕ, and ϕ_x can be found explicitly,

$$\begin{aligned} \psi &= \textstyle\int_{-\infty}^{\infty} G(x,t) * \psi_0(x)dx, \qquad & \phi &= \textstyle\int_{-\infty}^{\infty} G(x,t) * \phi_0(x)dx, \\ \psi_x &= \textstyle\int_{-\infty}^{\infty} G(x,t) * (\psi_0)_x(x)dx, & \phi_x &= \textstyle\int_{-\infty}^{\infty} G(x,t) * (\phi_0)_x(x)dx, \end{aligned} \tag{2.14}$$

where

$$G(x,t) = \frac{1}{\sqrt{4\pi t}} \exp\left(-\frac{|x|^2}{4t}\right). \tag{2.15}$$

Note that

$$v = \frac{\psi_x}{\psi} \qquad \text{and} \qquad \tilde{v} = \frac{\phi_x}{\phi}. \tag{2.16}$$

Then there exists a constant C which only depends on the initial data ψ_0 and ϕ_0 such that for all $t > 0$,

$$\begin{aligned} |v(x,t,\epsilon) - \tilde{v}(x,t,\epsilon)| &= \left|\left(\tfrac{\psi_x}{\psi} - \tfrac{\phi_x}{\phi}\right)(x,t)\right| \\ &= \left|\left(\tfrac{(\psi_x - \phi_x)\phi + (\phi - \psi)\phi_x}{\psi\phi}\right)(x,t)\right| \\ &\le C(|(\psi_x - \phi_x)(x,t)| + |(\psi - \phi)\phi_x(x,t)|, \end{aligned} \tag{2.17}$$

where we have used the fact that ψ and ϕ are positive and bounded for all $t > 0$. Using equations (2.14) it follows that

$$\begin{aligned}
&|(\psi_x - \phi_x)(\cdot, t)|_\infty \\
&= \sup_{x\in\mathbb{R}} \left| \int_{-\infty}^{\infty} G(x-\xi, t)(\psi_0 - \phi_0)_\xi(\xi) d\xi \right| \\
&\leq C|(\psi_0 - \phi_0)_x(\cdot)|_\infty \\
&= C \sup_{x\in\mathbb{R}} \left| v_0(x) e^{\int_{-\infty}^{x} v_0 dx} - \tilde{v}_0(x) e^{\int_{-\infty}^{x} \tilde{v}_0 dx} \right| \\
&= C \sup_{x\in\mathbb{R}} \left| v_0(x) \left(e^{\int_{-\infty}^{x} v_0 dx} - e^{\int_{-\infty}^{x} \tilde{v}_0 dx} \right) + (v_0(x) - \tilde{v}_0(x)) e^{\int_{-\infty}^{x} \tilde{v}_0 dx} \right| \\
&\leq C \sup_{x\in\mathbb{R}} \left| \exp(\varepsilon^2 \textstyle\int_{-\infty}^{x} R(u_0(x), u_1(x)) dx) - 1 \right| + \varepsilon^2 |R(u_0(\cdot), u_1(\cdot))|_\infty \\
&\leq C\varepsilon^2 (|R(u_0(\cdot), u_1(\cdot))|_1 + |R(u_0(\cdot), u_1(\cdot))|_\infty) \leq \varepsilon^2 C(||u_0||_{5,4}, ||u_1||_{3,2}).
\end{aligned} \tag{2.18}$$

Similarly, one can obtain

$$\begin{aligned}
|(\psi - \phi)(\cdot, t)|_\infty &= \sup_{x\in\mathbb{R}} \left| \int_{-\infty}^{\infty} G(x-\xi, t)(\psi_0 - \phi_0)(\xi) d\xi \right| \\
&\leq C|(\psi_0 - \phi_0)(\cdot)|_\infty \\
&= C \sup_{x\in\mathbb{R}} \left| e^{\int_{-\infty}^{x} v_0 dx} - e^{\int_{-\infty}^{x} \tilde{v}_0 dx} \right| \\
&\leq C \sup_{x\in\mathbb{R}} \left| \exp(\varepsilon^2 \textstyle\int_{-\infty}^{x} R(u_0(x), u_1(x)) dx) - 1 \right| \\
&\leq C\varepsilon^2 |R(u_0(\cdot), u_1(\cdot))|_1 \leq \varepsilon^2 C(||u_0||_{5,4}, ||u_1||_{3,2}).
\end{aligned} \tag{2.19}$$

Using (2.18) and (2.19) in (2.17) and noticing

$$|\phi_x(\cdot, t)|_\infty \leq |\phi_x(\cdot, t)|_2^{\frac{1}{2}} |\phi_{xx}(\cdot, t)|_2^{\frac{1}{2}} \leq C(||\phi_0(\cdot)||_3) \leq C(||u_0||_{5,4}, ||u_1||_{3,2}),$$

one shows that

$$\begin{aligned}
\sup_{x\in\mathbb{R}} |(v - \tilde{v})(x, t, \varepsilon)| &\leq C[|(\psi_x - \phi_x)(\cdot, t)|_\infty + |(\psi - \phi)\phi_x(\cdot, t)|_\infty] \\
&\leq \varepsilon^2 C(||u_0||_{5,4}, ||u_1||_{3,2}), \quad \text{for all } t > 0.
\end{aligned}$$

Following the steps from (2.17) to (2.19), one can also show that

$$\sup_{x\in\mathbb{R}} |(v_x - \tilde{v}_x)(x, t, \varepsilon)| \leq \varepsilon^2 C(||u_0||_{6,4}, ||u_1||_{4,2}), \quad \text{for all } t > 0.$$

To show the third estimate in (2.9), it follows from (2.17) that

$$\int_{-\infty}^{\infty} |(v - \tilde{v})(x, t, \varepsilon)| \, dx \leq C \left[|(\psi_x - \phi_x)(\cdot, t)|_1 + |(\psi - \phi)\phi_x(\cdot, t)|_1 \right]. \tag{2.20}$$

Using equation (2.14) again, it follows that

$$\begin{aligned}
&|(\psi_x - \phi_x)(\cdot,t)|_1 \\
&= \int_{-\infty}^{\infty} \left| \int_{-\infty}^{\infty} G(x-\xi,t)(\psi_0-\phi_0)_\xi(\xi)d\xi \right| dx \\
&\le C|(\psi_0-\phi_0)_x(\cdot)|_1 \\
&= C\int_{-\infty}^{\infty} \left| v_0(x)\left(e^{\int_{-\infty}^x v_0 dx} - e^{\int_{-\infty}^x \tilde v_0 dx}\right) + (v_0(x)-\tilde v_0(x))\, e^{\int_{-\infty}^x \tilde v_0 dx} \right| dx \\
&\le C|v_0(\cdot)|_1 \sup_{x\in\mathbb{R}} \left| e^{\int_{-\infty}^x v_0 dx} - e^{\int_{-\infty}^x \tilde v_0 dx} \right| + C\,|(v_0-\tilde v_0)(\cdot)|_1 \\
&\le C\varepsilon^2 |R(u_0(\cdot),u_1(\cdot))|_1 \le \varepsilon^2 C(||u_0||_{5,4}, ||u_1||_{3,2}).
\end{aligned} \tag{2.21}$$

Note that $\phi_x = \tilde v \exp\left(\int_{-\infty}^x \tilde v dx\right)$ and $|\tilde v(\cdot,t,\varepsilon)|_1 \le C$ for all $t>0$. It follows that

$$|\phi_x(\cdot,t)|_1 \le C, \qquad \text{for all } \ t>0. \tag{2.22}$$

Using (2.19) and (2.22), one shows that

$$|(\psi-\phi)\phi_x(\cdot,t)|_1 \le |\phi_x(\cdot,t)|_1 |(\psi-\phi)(\cdot,t)|_\infty \le \varepsilon^2 C(||u_0||_{5,4}, ||u_1||_{3,2}). \tag{2.23}$$

Using (2.21) and (2.23) in (2.20), one obtains

$$|(v-\tilde v)(\cdot,t,\varepsilon)|_1 \le \varepsilon^2 C(||u_0||_{5,4}, ||u_1||_{3,2}). \qquad \diamondsuit$$

Lemma 3. *Let $v(x,t)$ be the solution of equation (2.6) with the initial data $v_0(x)$ in $\mathcal{H}_2^3(\mathbb{R})$. Then*

$$u(x,t) = v + \varepsilon \left\{ \frac{1}{2}(2A-\Gamma)v^2 + \frac{3A-B}{2} v_x \partial_x^{-1} v - \frac{A}{2} x(v_{xx} + 2vv_x) \right\} \tag{2.24}$$

is well-defined for all $t>0$. Furthermore, the L_∞-norm of u is bounded above as

$$|u(\cdot,t)|_\infty \le C_1 + C_2\varepsilon,$$

for any $t>0$, where C_1 and C_2 are constants depending on $v_0 \in L_1(\mathbb{R}) \cap L_\infty(\mathbb{R})$ and $v_0 \in \mathcal{H}_2^3(\mathbb{R})$, respectively.

Proof: Since v_0 lies in $\mathcal{H}_2^3(\mathbb{R})$, it certainly lies in $L_1(\mathbb{R}) \cap H^3(\mathbb{R})$. Lemma 1 implies that the solution v of (2.6) with the initial data v_0 exists for all t in $L_1(\mathbb{R}) \cap H^3(\mathbb{R})$. Moreover, the solution v has the decay properties stated in (2.8).

Let $w(x,t) = \exp(\int_{-\infty}^x v(x,t)dx)$. Because $|v(\cdot,t)|_1 \le C$ for all $t>0$, it follows that

$$w(x,t) > 0, \qquad \text{for all } \ t>0.$$

Also w satisfies

$$w_t = w_{xx}$$

with initial data

$$w(x,0) = w_0 = \exp\left(\int_{-\infty}^x v_0 dx\right).$$

Let $\psi = w_x$. Then ψ satisfies

$$\psi_t = \psi_{xx} \tag{2.25}$$

$$\psi(x,0)=\psi_0=v_0\exp\left(\int_{-\infty}^{x}v_0dx\right). \tag{2.26}$$

Multiplying (2.25) by $2x^2\psi$, and integrating over $\mathbb{R}\times[0,t]$, we find

$$\begin{aligned}&\textstyle\int_{-\infty}^{\infty}x^2\psi^2(x,t)dx+2\int_0^t\int_{-\infty}^{\infty}x^2\psi_x^2(x,\tau)dxd\tau\\ &=\textstyle\int_{-\infty}^{\infty}x^2\psi_0^2dx+2\int_0^t|\psi(\cdot,\tau)|_2^2d\tau.\end{aligned} \tag{2.27}$$

Note that

$$\psi(x,t)=w_x(x,t)=v(x,t)\exp\left(\int_{-\infty}^{x}v(x,t)dx\right). \tag{2.28}$$

Hence

$$|\psi(\cdot,t)|_2^2=\int_{-\infty}^{\infty}v^2(x,t)e^{2\int_{-\infty}^{x}v(x,t)dx}dx\le C|v(\cdot,t)|_2^2. \tag{2.29}$$

Using this estimate and (2.8), one can show that

$$\int_0^t|\psi(\cdot,\tau)|_2^2\,d\tau\le C\int_0^t|v(\cdot,\tau)|_2^2\,d\tau\le C\int_0^t\frac{d\tau}{\sqrt{1+\tau}}\le C(1+t)^{\frac12}. \tag{2.30}$$

Using (2.30) in (2.27) it follows that

$$\int_{-\infty}^{\infty}x^2v^2(x,t)dx\le C\int_{-\infty}^{\infty}x^2w_x^2(x,t)dx\le C(1+t)^{\frac12}. \tag{2.31}$$

Let $\phi(x,t)=w_{xx}(x,t)$. Then ϕ satisfies

$$\phi_t=\phi_{xx} \tag{2.32}$$

with the initial data

$$\phi(x,0)=\phi_0=\left((v_0)_x+v_0^2\right)e^{\int_{-\infty}^{x}v_0dx}. \tag{2.33}$$

Multiplying (2.32) by $2x^2\phi$, integrating over $\mathbb{R}\times[0,t]$, and integrating by parts, one finds

$$\begin{aligned}&\textstyle\int_{-\infty}^{\infty}x^2\phi^2(x,t)dx+2\int_0^t\int_{-\infty}^{\infty}x^2\phi_x^2(x,\tau)dxd\tau\\ &=\textstyle\int_{-\infty}^{\infty}x^2\phi_0^2dx+2\int_0^t|\phi(\cdot,\tau)|_2^2d\tau.\end{aligned} \tag{2.34}$$

Note that

$$\phi=w_{xx}=(v_x+v^2)e^{\int_{-\infty}^{x}vdx}. \tag{2.35}$$

Hence using (2.8) and the inequality

$$|v(\cdot,t)|_4^4\le|v(\cdot,t)|_2^3|v_x(\cdot,t)|_2, \tag{2.36}$$

it follows that

$$|\phi(\cdot,t)|_2^2\le C\int_{-\infty}^{\infty}(v_x^2+v^4)(x,t)dx\le C(1+t)^{-\frac32}. \tag{2.37}$$

Using (2.37) in (2.34), one finds

$$\int_{-\infty}^{\infty}x^2\phi^2(x,t)dx=\int_{-\infty}^{\infty}x^2w_{xx}^2(x,t)dx\le C, \tag{2.38}$$

where C is a constant which depends on $||v_0||_{\mathcal{H}_2^1(\mathbb{R})}$.

Similarly, following the steps from (2.32) to (2.38), one obtains that

$$\int_{-\infty}^{\infty}x^2w_{xxx}^2(x,t)dx\le C_1\qquad\text{and}\qquad\int_{-\infty}^{\infty}x^2w_{xxxx}^2(x,t)dx\le C_2, \tag{2.39}$$

where the C's are constants which depend on $||v_0||_{\mathcal{H}_2^3(\mathbb{R})}$.

Since v_0 lies in $\mathcal{H}_2^3(\mathbb{R})$, it follows from Lemma 1 that

$$|v(\cdot,t)|_\infty \leq C(||v_0||_{L_1(\mathbb{R})\cap L_\infty(\mathbb{R})}). \tag{2.40}$$

Thus,

$$|v^2(\cdot,t)|_\infty \leq C, \quad \text{and} \quad |v_x\partial_x^{-1}v(\cdot,t)|_\infty \leq C, \quad \text{for all} \quad t>0, \tag{2.41}$$

where we have used $|v(\cdot,t)|_1 \leq C$ (see (2.8)).

By definition, one has

$$v = \frac{w_x}{w}, \tag{2.42}$$

and hence

$$v_x = \frac{w_{xx}w - w_x^2}{w^2}. \tag{2.43}$$

Note that w is positive and bounded in $L_\infty(\mathbb{R})$ for all $t > 0$. Then by using the relation (2.43), it follows that

$$\int_{-\infty}^{\infty} x^2v_x^2 \leq C\int_{-\infty}^{\infty} x^2(w_{xx}^2 + w_x^4)dx. \tag{2.44}$$

Using (2.8) and (2.31) we find

$$\begin{aligned} \textstyle\int_{-\infty}^{\infty} x^2w_x^4(x,t)dx &\leq |w_x(\cdot,t)|_\infty^2 \textstyle\int_{-\infty}^{\infty} x^2w_x^2(x,t)dx \\ &\leq C|v(\cdot,t)|_\infty^2\sqrt{1+t} \leq C, \quad \text{for all} \quad t>0. \end{aligned} \tag{2.45}$$

Equations (2.38) and (2.45), imply

$$\int_{-\infty}^{\infty} x^2v_x^2(x,t)dx \leq C_1\int_{-\infty}^{\infty} x^2(w_{xx}^2 + w_x^4)dx \leq C_2, \quad \text{for all} \quad t>0. \tag{2.46}$$

Note that

$$v_{xx} = \frac{w_{xxx}w^2 - 3w_{xx}w_xw + 2w_x^3}{w^3}. \tag{2.47}$$

Thus

$$\int_{-\infty}^{\infty} x^2v_{xx}^2(x,t)dx \leq C_1\int_{-\infty}^{\infty} x^2(w_{xxx}^2 + w_{xx}^2 + w_x^6)dx \leq C_2, \tag{2.48}$$

since again w is positive and bounded for all $t > 0$. Using (2.39) and (2.46) in (2.48) it follows that

$$\int_{-\infty}^{\infty} x^2v_{xx}^2(x,t)dx \leq C, \tag{2.49}$$

for some constant C depending on $||v_0||_{\mathcal{H}_2^2(\mathbb{R})}$. Hence using (2.46), (2.49), and the inequality

$$\begin{aligned} &\sup_{x\in\mathbb{R}} |xv_x(x,t)| \\ &\leq \left(\textstyle\int_{-\infty}^{\infty} x^2v_x^2(x,t)dx\right)^{\frac{1}{4}} \left(\textstyle\int_{-\infty}^{\infty}[(xv_x)_x]^2(x,t)dx\right)^{\frac{1}{4}} \\ &\leq 2\left(\textstyle\int_{-\infty}^{\infty} x^2v_x^2(x,t)dx\right)^{\frac{1}{4}} \left(|v_x(\cdot,t)|_2^2 + \textstyle\int_{-\infty}^{\infty} x^2v_{xx}^2(x,t)dx\right)^{\frac{1}{4}}, \end{aligned} \tag{2.50}$$

we conclude that

$$\sup_{x\in\mathbb{R}} |xv_x(x,t)| \le C(||v_0||_{\mathcal{H}_2^2(\mathbb{R})}), \quad \text{for all} \quad t>0. \tag{2.51}$$

Finally, note that

$$v_{xxx} = \frac{w_{xxxx}w^3 - 4w_{xxx}w_xw^2 + 12w_{xx}w_x^2w - 3w_{xx}^2w^2 - 6w_x^4}{w^4}. \tag{2.52}$$

Hence,

$$\int_{-\infty}^{\infty} x^2v_{xxx}^2(x,t)dx \le C\int_{-\infty}^{\infty} x^2(w_{xxxx}^2 + w_{xxx}^2 + w_{xx}^2 + w_x^8)dx. \tag{2.53}$$

Using (2.38), (2.39) and (2.45) it follows that

$$\int_{-\infty}^{\infty} x^2v_{xxx}^2(x,t)dx \le C(||v_0||_{\mathcal{H}_2^3(\mathbb{R})}). \tag{2.54}$$

Using these estimates together with (2.49) and (2.50), we find

$$\begin{aligned} &\sup\nolimits_{x\in\mathbb{R}} |xv_{xx}(x,t)| \\ &\le \left(\int_{-\infty}^{\infty} x^2v_{xx}^2(x,t)dx\right)^{\frac14}\left(\int_{-\infty}^{\infty}[(xv_{xx})_x]^2(x,t)dx\right)^{\frac14} \\ &\le 2\left(\int_{-\infty}^{\infty} x^2v_{xx}^2(x,t)dx\right)^{\frac14}\left(|v_{xx}(\cdot,t)|_2^2 + \int_{-\infty}^{\infty} x^2v_{xxx}^2(x,t)dx\right)^{\frac14} \\ &\le C(||v_0||_{\mathcal{H}_2^3(\mathbb{R})}). \end{aligned} \tag{2.55}$$

Combining equations (2.40), (2.41), (2.51) and (2.55) concludes the proof of the lemma. ◇

Next we shall derive some temporal decay properties of the solutions v of (2.6)-(2.7). But first we prove a useful lemma.

Lemma 4. *Let $v_0 \in \mathcal{H}_2^3(\mathbb{R})$. Then for all $t>0$, the initial-value problem (2.23)-(2.24) with such initial data has the following properties:*

$$\sup_{x\in\mathbb{R}} |xw_x(x,t)| \le C, \tag{2.56}$$

$$\sup_{x\in\mathbb{R}} |xw_{xx}(x,t)| \le C(1+t)^{-\frac12}, \tag{2.57}$$

$$\sup_{x\in\mathbb{R}} |xw_{xxx}(x,t)| \le C(1+t)^{-1}. \tag{2.58}$$

Proof: Note that from (2.23)-(2.24) it follows that

$$\begin{aligned} w_x(x,t) &= \int_{-\infty}^{\infty} G(x,t) * \left(v_0e^{\int_{-\infty}^{x} v_0dx}\right)dx \\ &= \frac{1}{\sqrt{4\pi t}}\int_{-\infty}^{\infty} e^{-\frac{(x-\xi)^2}{4t}}v_0(\xi)e^{\int_{-\infty}^{\xi} v_0(\eta)d\eta}d\xi. \end{aligned} \tag{2.59}$$

Hence from (2.59) one finds

$$\begin{aligned}
|xw_x(x,t)| &= \frac{1}{\sqrt{4\pi t}}\left|\int_{-\infty}^{\infty} xe^{-\frac{(x-\xi)^2}{4t}} v_0(\xi)e^{\int_{-\infty}^{\xi} v_0(\eta)d\eta}d\xi\right| \\
&\leq \frac{1}{\sqrt{4\pi t}}\left|\int_{-\infty}^{\infty}(x-\xi)e^{-\frac{(x-\xi)^2}{4t}} v_0(\xi)e^{\int_{-\infty}^{\xi} v_0(\eta)d\eta}d\xi\right| \\
&\quad +\frac{1}{\sqrt{4\pi t}}\left|\int_{-\infty}^{\infty} \xi e^{-\frac{(x-\xi)^2}{4t}} v_0(\xi)e^{\int_{-\infty}^{\xi} v_0(\eta)d\eta}d\xi\right| \\
&\leq C\int_{-\infty}^{\infty}|v_0(\xi)|d\xi + C\sup_{\xi\in\mathbb{R}}|\xi v_0(\xi)|\int_{-\infty}^{\infty} e^{-\xi^2}d\xi \\
&\leq C(|v_0(\cdot)|_1 + \sup_{\xi\in\mathbb{R}}|\xi v_0(\xi)|) \\
&\leq C(||v_0||_{\mathcal{H}_2^3(\mathbb{R})}).
\end{aligned} \tag{2.60}$$

Using this inequality one obtains that

$$\sup_{x\in\mathbb{R}}|xw_x(x,t)| \leq C(||v_0||_{\mathcal{H}_2^3(\mathbb{R})}), \quad \text{for all} \quad t>0.$$

Similarly, one can show that

$$\begin{aligned}
|xw_{xx}(x,t)| &= \frac{1}{\sqrt{4\pi t}}\left|\int_{-\infty}^{\infty} \frac{-x(x-\xi)}{2t}e^{-\frac{(x-\xi)^2}{4t}} v_0(\xi)e^{\int_{-\infty}^{\xi} v_0(\eta)d\eta}d\xi\right| \\
&\leq \frac{C}{\sqrt{4\pi t}}\int_{-\infty}^{\infty} \frac{(x-\xi)^2}{4t}e^{-\frac{(x-\xi)^2}{4t}}|v_0(\xi)|d\xi \\
&\quad +\frac{C}{\sqrt{4\pi t}}\int_{-\infty}^{\infty} \frac{|x-\xi|}{2t}e^{-\frac{(x-\xi)^2}{4t}}|\xi v_0(\xi)|d\xi \\
&\leq \frac{C}{\sqrt{t}}\left[\int_{-\infty}^{\infty}|v_0(\xi)|d\xi + \sup_{\xi\in\mathbb{R}}|\xi v_0(\xi)|\int_{-\infty}^{\infty}|\xi|e^{-\xi^2}d\xi\right] \\
&\leq \frac{C}{\sqrt{t}}\left[|v_0(\cdot)|_1 + \sup_{\xi\in\mathbb{R}}|\xi v_0(\xi)|\right] \\
&\leq \frac{C(||v_0||_{\mathcal{H}_2^3(\mathbb{R})})}{\sqrt{t}}.
\end{aligned} \tag{2.61}$$

Also

$$\begin{aligned}
&|xw_{xxx}(x,t)| \\
&= \frac{1}{\sqrt{4\pi t}}\left|\int_{-\infty}^{\infty} \frac{-x}{2t}e^{-\frac{(x-\xi)^2}{4t}} v_0(\xi)e^{\int_{-\infty}^{\xi} v_0(\eta)d\eta}d\xi\right| \\
&\quad +\frac{1}{\sqrt{4\pi t}}\left|\int_{-\infty}^{\infty} \frac{x(x-\xi)^2}{4t^2}e^{-\frac{(x-\xi)^2}{4t}} v_0(\xi)e^{\int_{-\infty}^{\xi} v_0(\eta)d\eta}d\xi\right| \\
&\leq \frac{C}{t}\int_{-\infty}^{\infty}\left[\frac{|x-\xi|}{\sqrt{4t}}e^{-\frac{(x-\xi)^2}{4t}}|v_0(\xi)| + \frac{1}{\sqrt{4t}}e^{-\frac{(x-\xi)^2}{4t}}|\xi v_0(\xi)|\right]d\xi \\
&\quad +\frac{C}{t}\int_{-\infty}^{\infty}\left[\frac{|x-\xi|^3}{(4t)^{\frac{3}{2}}}e^{-\frac{(x-\xi)^2}{4t}}|v_0(\xi)| + \frac{|x-\xi|^2}{(4t)^{\frac{3}{2}}}e^{-\frac{(x-\xi)^2}{4t}}|\xi v_0(\xi)|\right]d\xi \\
&\leq \frac{C}{t}\left[|v_0(\cdot)|_1 + \sup_{\xi\in\mathbb{R}}|\xi v_0(\xi)|\right] \leq \frac{C(||v_0||_{\mathcal{H}_2^3(\mathbb{R})})}{t}.
\end{aligned} \tag{2.62}$$

Hence one concludes that there are constants depending on $||v_0||_{\mathcal{H}_2^3(\mathbb{R})}$ such that

$$\sup_{x\in\mathbb{R}} |xw_{xx}(x,t)| \leq Ct^{-\frac{1}{2}} \text{ and } \sup_{x\in\mathbb{R}} |xw_{xxx}(x,t)| \leq Ct^{-1}, \quad \text{for all } t > 0. \qquad \diamondsuit$$

Corollary 1. *Let $v_0 \in \mathcal{H}_2^3(\mathbb{R})$. Let $v(x,t)$ be the solution (2.6) with initial data $v_0(x)$. Then u defined by (2.24) has the following property:*

$$|u(\cdot,t)|_\infty \leq C_1(1+t)^{-\frac{1}{2}} + \varepsilon C_2(1+t)^{-1}, \quad \text{for all } t \geq 0, \tag{2.63}$$

where C_1 and C_2 are constant depending $||v_0||_{L_1(\mathbb{R})\cap L_\infty(\mathbb{R})}$ and $||v_0||_{\mathcal{H}_2^3(\mathbb{R})}$ respectively.

Proof: Recall the relation (2.43) between a solution w of (2.11) and a solution v of (2.6), i.e.,

$$v_x = \frac{w_{xx}w - w_x^2}{w^2}.$$

From this relation since w is positive and bounded for all $t > 0$, it follows that

$$|xv_x(x,t)| \leq C(|xw_{xx}(x,t)| + |xw_x^2(x,t)|). \tag{2.64}$$

Lemma 1 implies

$$|w_x(\cdot,t)|_\infty \leq C|v(\cdot,t)|_\infty \leq C(1+t)^{-\frac{1}{2}}. \tag{2.65}$$

Using Lemma 4 and (2.65), one can show that

$$\begin{aligned} \sup\nolimits_{x\in\mathbb{R}} |xw_x^2(x,t)| &\leq |w_x(\cdot,t)|_\infty \sup\nolimits_{x\in\mathbb{R}} |xw_x(x,t)| \\ &\leq C(1+t)^{-\frac{1}{2}}. \end{aligned} \tag{2.66}$$

The use of Lemma 4 and (2.66) in (2.64) implies

$$\sup_{x\in\mathbb{R}} |xv_x(x,t)| \leq C(1+t)^{-\frac{1}{2}}. \tag{2.67}$$

Using (2.67) and (2.65) we find

$$\begin{aligned} \sup\nolimits_{x\in\mathbb{R}} |xv_x(x,t)v(x,t)| &\leq |v(\cdot,t)|_\infty \sup\nolimits_{x\in\mathbb{R}} |xv_x(x,t)| \\ &\leq C(1+t)^{-1}. \end{aligned} \tag{2.68}$$

Note also that

$$v_{xx} = \frac{w_{xxx}w^2 - 3w_{xx}w_xw + 2w_x^3}{w^3}.$$

Using Lemma 4 and (2.65) we find

$$\begin{aligned} &\sup\nolimits_{x\in\mathbb{R}} |xv_{xx}(x,t)| \\ &\leq C\sup\nolimits_{x\in\mathbb{R}} \left\{|xw_{xxx}(x,t)| + |xw_{xx}(x,t)||w_x(x,t)| + |xw_x(x,t)||w_x^2(x,t)|\right\} \\ &\leq C((1+t)^{-1} + (1+t)^{-1} + (1+t)^{-1}) \leq C(1+t)^{-1}. \end{aligned} \tag{2.69}$$

The definition of u, and the estimates (2.8), (2.68) and (2.69), imply

$$|u(\cdot,t)|_\infty \leq C_1(1+t)^{-\frac{1}{2}} + \varepsilon C_2(1+t)^{-1}, \tag{2.70}$$

where C_1 depends on $||v_0||_{L_1(\mathbb{R})\cap L_\infty(\mathbb{R})}$ while C_2 depends on $||v_0||_{\mathcal{H}_2^3(\mathbb{R})}$. $\diamondsuit$

Theorem 1. *Let $u_0(x)$ and $u_1(x)$ be in $\mathcal{H}_4^5(\mathbb{R})$ and $\mathcal{H}_2^3(\mathbb{R})$, respectively. Define $\tilde{v}(x,0,\varepsilon)$ by (2.1). Let $\tilde{v}(x,t,\varepsilon)$ be the solution of the Burgers equation (2.6) with the initial condition $\tilde{v}(x,0,\varepsilon)$. Define $\tilde{u}(x,t,\varepsilon)$ by*

$$\tilde{u}(x,t,\varepsilon) = \tilde{v} + \varepsilon\left\{\frac{1}{2}(2A-\Gamma)\tilde{v}^2 + \frac{3A-B}{2}\tilde{v}_x\partial_x^{-1}\tilde{v} - \frac{A}{2}x(\tilde{v}_{xx} + 2\tilde{v}\tilde{v}_x)\right\}. \quad (2.71)$$

Then $\tilde{u}$ satisfies the generalized Burgers equation

$$\tilde{u}_t = \tilde{u}_{xx} + 2\tilde{u}\tilde{u}_x + \varepsilon\left[A\tilde{u}_{xxx} + B\tilde{u}\tilde{u}_{xx} + \Gamma\tilde{u}_x^2 + (\Gamma + \frac{B}{2} - \frac{3A}{2})\tilde{u}^2\tilde{u}_x\right] + O(\varepsilon^2). \quad (2.72)$$

There exist constants C_1 and C_2 depending on $||u_0||_{\mathcal{H}_4^5(\mathbb{R})}$ and $||u_1||_{\mathcal{H}_2^3(\mathbb{R})}$, such that

$$|\tilde{u}(\cdot,t)|_\infty \leq C_1(1+t)^{-\frac{1}{2}} + C_2\varepsilon(1+t)^{-1}, \quad \text{for all} \quad t \geq 0. \quad (2.73)$$

Furthermore, if u_0 and u_1 lie in $\mathcal{H}_6^7(\mathbb{R})$ and $\mathcal{H}_4^5(\mathbb{R})$, respecticely, one has

$$|(u-\tilde{u})(\cdot,t,\varepsilon)|_\infty \leq C\varepsilon^2, \quad \text{for all} \quad t > 0, \quad (2.74)$$

where u satisfies (2.72) with initial data

$$u(x,0,\varepsilon) = u_0(x) + \varepsilon u_1(x), \quad (2.75)$$

and C depends on $||u_0||_{\mathcal{H}_6^7(\mathbb{R})}$ and $||u_1||_{\mathcal{H}_4^5(\mathbb{R})}$.

Proof: By Lemma 3 and Corollary 1, $\tilde{u}$ is well-defined and has the decay property (2.73). Let $v(x,t,\varepsilon)$ be the solution of the Burgers equation (2.6) with the initial condition $v(x,0,\varepsilon)$ defined in (2.3). Applying Lemma 2, one shows that

$$\begin{aligned} &|(v_x\partial_x^{-1}v - \tilde{v}_x\partial_x^{-1}\tilde{v})(\cdot,t,\varepsilon)|_\infty \\ &\leq |(v_x-\tilde{v}_x)\partial_x^{-1}v(\cdot,t,\varepsilon)|_\infty + |\tilde{v}_x\partial_x^{-1}(v-\tilde{v})(\cdot,t,\varepsilon)|_\infty \\ &\leq C[|(v_x-\tilde{v}_x)(\cdot,t,\varepsilon)|_\infty + |(v-\tilde{v})(\cdot,t,\varepsilon)|_1] \\ &\leq \varepsilon^2 C(||u_0||_{6,4}, ||u_1||_{4,2}). \end{aligned} \quad (2.76)$$

Following the line in proving Lemma 3, one can easily show that

$$||(v_x-\tilde{v}_x)(\cdot,t,\varepsilon)||_{2,2} \leq \varepsilon^2 C(||u_0||_{7,6}, ||u_1||_{5,4}), \quad (2.77)$$

because of the choice of $\tilde{v}(x,0,\varepsilon)$. By Lemma 2, (2.51) and (2.77) one shows that

$$\begin{aligned} &\sup_{x\in\mathbb{R}} |(xvv_x - x\tilde{v}\tilde{v}_x)(x,t,\varepsilon)| \\ &\leq \sup_{x\in\mathbb{R}} |x(v_x-\tilde{v}_x)v(x,t,\varepsilon)| + \sup_{x\in\mathbb{R}} |(v-\tilde{v})(x\tilde{v}_x)(x,t,\varepsilon)| \\ &\leq C[\sup_{x\in\mathbb{R}} |x(v_x-\tilde{v}_x)(x,t,\varepsilon)| + |(v-\tilde{v})(\cdot,t,\varepsilon)|_\infty] \\ &\leq C[||(v_x-\tilde{v}_x)(\cdot,t,\varepsilon)||_{1,2} + |(v-\tilde{v})(\cdot,t,\varepsilon)|_\infty] \\ &\leq \varepsilon^2 C(||u_0||_{7,6}, ||u_1||_{5,4}). \end{aligned} \quad (2.78)$$

Applying Lemma 2, (2.76), (2.77) and (2.78) to the difference $u(x,t,\varepsilon)-\tilde{u}(x,t,\varepsilon)$ in L_∞-norm shows that

$$\begin{aligned}
&|(u-\tilde{u})(\cdot,t,\varepsilon)|_\infty \\
&\le |(v-\tilde{v})(\cdot,t,\varepsilon)|_\infty + C\varepsilon\Big[|(v^2-\tilde{v}^2)(\cdot,t,\varepsilon)|_\infty + |(v_x\partial_x^{-1}v-\tilde{v}_x\partial_x^{-1}\tilde{v})(\cdot,t,\varepsilon)|_\infty \\
&\qquad + \sup\nolimits_{x\in\mathbb{R}}|(xv_{xx}-x\tilde{v}_{xx})(x,t,\varepsilon)| + \sup\nolimits_{x\in\mathbb{R}}|(xvv_x-x\tilde{v}\tilde{v}_x)(x,t,\varepsilon)|\Big] \\
&\le |(v-\tilde{v})(\cdot,t,\varepsilon)|_\infty \\
&\qquad + C\varepsilon\Big[|(v-\tilde{v})(\cdot,t,\varepsilon)|_\infty + |(v-\tilde{v})(\cdot,t,\varepsilon)|_1 + ||(v_x-\tilde{v}_x)(\cdot,t,\varepsilon)||_{2,2}\Big] \\
&\le \varepsilon^2 C(||u_0||_{7,6}, ||u_1||_{5,4}). \qquad \diamondsuit
\end{aligned}$$

References

[1] F. Calogero, *What is Integrability,* V.E. Zakharov, ed. Springer-Verlag, 1992.
[2] Y. Kodama, Phys. Lett. A **107**, 245 (1985); Physica D **16**, 14 (1985).
[3] A.S. Fokas and Q.M. Liu, *Asymptotic Integrability of Water Waves,* preprint, 1995.
[4] A.S. Fokas and B. Fuchssteiner, Phys. Lett. A **86**, 341 (1981).
[5] Q.M. Liu, Ph.D. Dissertation, Clarkson University, 1995.
[6] A.S. Fokas, R. Grimshaw and D. Pelinofsky, *On the Asymptotic Integrability of a Higher-Order Evolution Equation Describing Internal Waves in a Deep Fluid,* preprint, 1995.
[7] M.E. Schonbek, *Uniform Decay Rates for Parabolic Conservation Laws,* Nonlinear Anal. T.M.A., **10** (9), 943-953 (1986).
[8] J.L. Bona and L. Luo, *More Results on the Decay of Solutions to Nonlinear, Dispersive Wave Equations, Discrete and Continuous Dynamical Systems.* **1**, pp. 151-193 (1995)

Department of Mathematics, Imperial College of Science, Technology & Medicine, London, SW7 2BZ, U.K.

E-mail address: `asf@ic.ac.uk AND l.luo@ic.ac.uk`

Contemporary Mathematics
Volume **200**, 1996

SPATIAL INSTABILITIES AND CHAOS IN HIGH-ORDER HAMILTONIAN STANDING WATER WAVES

by M. Glozman and Y. Agnon

ABSTRACT. The spatial instabilities of standing water waves are thoroughly studied using a high-order (up to 18) Hamiltonian formulation. The interaction of the harmonics of the standing wave with other, three-dimensional modes, forms a dynamical system, in which the standing waves are periodic solutions. These solutions are unstable to certain perturbations. The long time evolution of the instabilities reveals four distinct classes of chaotic behavior including a new type of thermalization in which no resonance overlap was identified.

1. Introduction

The evolution of water waves has been studied by deterministic models on one hand and by stochastic models on the other hand. Low order models have been shown to lead to chaos which is often fictitious [6]. The development of stochastic motion in a deterministic model is of great interest, and forms the basis for using stochastic models. Standing waves are a natural choice for a study which uses discrete modes, which imply spatial periodicity. Instability of standing waves has been studied by Mercer and Roberts [10], who studied a two dimensional problem using a semi-Lagrangian approach. They found primarily side-band instabilities. Bryant and Stiassnie [2] and Agnon and Glozman [1] have also studied two-dimensional instabilities. Okamura [11] has studied the instability of a low order standing wave to travelling wave disturbances using a third order Zakharov equation.

In the present study we use a most accurate method. In [6] we gave the derivation of an expansion of the Hamiltonian to arbitrary order for an arbitrary number of modes. Using this formulation, we focus in the present work on the stability of the standing Stokes wave to spatial perturbations. This is done using Floquet theory.

1991 *Mathematics Subject Classification.* Primary 76B15; Secondary 58F05.

The research was supported by the US Office of Naval Research (Grant N0014-91-J-1449).

Section 2 formulates the problem. In Section 3 the regions of instability are mapped and in Section 4 the long time evolution in different regions is studied. Various types of chaos, and a new type of thermalization are discovered. Section 5 presents a second order instability. Appendix A addresses the nontrivial issue of effectively and consistently truncating the Hamiltonian. Some important properties of complex eigenvalues and eigenvectors, relevant to the instabilities discovered, are given in Appendix B. Special emphasis was placed on verifying the instabilities that were found by checking their existence and range in several independent truncations.

2. Formulation of the problem

We consider irrotational incompressible and inviscid standing gravity waves in a rectangular basin of infinite depth, with a free surface at $z = 0$. The surface of the basin, Ω, is bounded at

$x = 0, L_1$;

$y = 0, L_2$.

Denoting the free surface elevation by η, and the velocity potential at the surface by ψ, there is a Hamiltonian H, which is the total energy of the system per unit density

$$H = T + V \tag{2.1}$$

where

$$T = \frac{1}{2}\int_\Omega \psi\eta_t d\vec{x} \tag{2.2}$$

and

$$V_g = \frac{1}{2} g \int_\Omega \eta^2 d\vec{x} \tag{2.3}$$

are the kinetic and gravitational energies, respectively. Capillary effects, which were included in GAS, are neglected. The evolution equations for the the wave field, were given by Zakharov [13]:

$$\eta_t = \frac{\delta H}{\delta \psi}; \qquad \psi_t = -\frac{\delta H}{\delta \eta} \tag{2.4}$$

where δ is a functional derivative (e.g. [8]).

The appearance of η_t in (2.2) makes it necessary to solve the boundary value problem for the potential in the (time dependent) fluid domain. This makes the evolution equations depend on η nonlinearily (to infinite order). The expansion of the evolution equations, as well as the Hamiltonian, (to arbitrary order) is given in [6].

For standing waves, the functions ψ and η can be expanded into double Fourier series:

$$\eta(x,y) = \frac{1}{\sqrt{L_1 L_2}} \sum_m \sum_n \hat{\eta}(k_m, \ell_n) e^{-i(k_m x + \ell_n y)} \tag{2.5}$$

(m, n take all integer values), and a similar expression for $\psi(x,y)$.

The boundary conditions at $x = 0, L_1$; $y = 0, L_2$; dictate:

$$\begin{aligned} k_m &= \frac{\pi m}{L_1} \qquad m = \ldots, -1, 0, 1, \ldots \\ \ell_n &= \frac{\pi n}{L_2} \qquad n = \ldots, -1, 0, 1, \ldots \end{aligned} \tag{2.6}$$

and

$$\hat{\eta}(k_m, \ell_n) = \hat{\eta}(-k_m, \ell_n) = \hat{\eta}(k_m, -\ell_n) = \hat{\eta}(-k_m, -\ell_n) \tag{2.7}$$

and the same symmetry requirements for $\hat{\psi}$.

$\hat{\eta}(0) = \hat{\psi}(0) = 0$ due to mass conservation.

The evolution equations are then written as an autonomous dynamical system in the form of polynomials of infinite order in $\hat{\eta}(k_m, \ell_n)$ and $\hat{\psi}(k_m, \ell_n)$(cf. [6]).

In practice, the number of modes [M] and the order of the system need to be truncated in order to perform computations. We thus obtain an energy surface in the phase space of the form $(\hat{\eta}_1, \ldots, \hat{\eta}_M, \hat{\psi}_1, \ldots, \hat{\psi}_M; H = const.)$ which has 2M-1 dimensions. The subscripts stand for a numbering of the selected modes.

In Appendix A we present a condensed discussion regarding the considerations involved in different choices of truncation schemes that lead to accurate and reliable solutions.

To study the dynamical system, it is most helpful to find periodic orbits, which are the most ordered form possible for standing waves. The periodic orbits correspond to periodic points on the return map obtained on a Poincare section of the system with the hyperplane $\hat{\psi}_1 = 0$. To find such periodic points, we define $\vec{D}_j(\vec{X})$ to be the Euclidian distance in the phase space between some initial point, $\vec{X}$, on the Poincare section and its j-th return $\vec{X}_j$. The function D_j found by numerical integration of the evolution equations, is regular enough for conducting a search to find the periodic orbits (the first return $\vec{X}_1 = F(\vec{X})$ defines a symplectic map). Solving the nonlinear equation $\vec{D}_j(\vec{X}) = 0$ with some additional conditions, yields the desired periodic solutions. Focusing on two dimensional ($\ell_n = 0$) periodic solutions, we have previously found high order expansions for the Stokes standing waves, which converge for high wave steepnesses, and for additional new standing waves, in which resonance occurs among modes of the type [k_m; m = $1^2, 2^2, 3^2$,...] (cf.[1]).

3. Stability of standing Stokes waves to spatial perturbations

Many methods that were used for studying water waves were limited to two dimensional problems. The present method is equally capable of studying fully three dimensional waves. Here we study the stability of a standing Stokes wave to three dimensional perturbations.

We denote by (p,q) the mode with wave vector $\vec{K}_{pq} = (k_p, \ell_q) = (\pi p / L_1, \pi q / L_2)$. The Stokes wave was represented at the different stages of the study by an increasing number of modes and the order of the model was also increased correspondingly.

Once a set of modes is chosen, the period T of the periodic Stokes wave is determined. To perform the Floquet analysis, we define a time-shift mapping F(X) by integration of the evolution equations over the period T. By numerical differentiation we get the Jacobian of F(X). The eigenvalues and eigenvector of this Jacobian determine the linear stability of the Stokes wave in this system. An eigenvalue λ_i for which $|\lambda_i| > 1$ indicates instability.

In each of the instabilities that we found, were involved mainly a group of one or more modes of the form $(k_p, \ell_q)$with a specific value of ℓ_q. The value of $\ell_q$can vary continuously by varying L_2. We denote $\alpha = L_1/L_2$, so that $\ell_q/k_1 = \alpha q$. By uniformly scaling by k_1, and collecting all the results, we get on the plane (ak_1,αq) many regions of instability. Here ak_1 is the steepness of the Stokes wave. The picture obtained is typical of pictures of parametric resonance. The computations used up to 39 modes (9 for the Stokes wave and 30 spatial modes) using a 9th order balanced model (after comparing with a 7th order uniformly truncated model and checking the Stokes wave against a 17th order balanced model).

The basins studied had many different aspect ratios $\alpha = L_1{:}L_2$ (see Table 1).

The perturbation modes had $\vec{k}_{pq} = (\pi p / L_1, \pi q / L_2)$ p=0,...5, q = 1,...,5.

The region considered here included values of ak_1, up to 0.48 and ℓ_q/k_1 in the range [0,4]. Instabilities outside this range are rare and weak. The regions of instability are plotted in Figure 1. Table 1 summarized the results of the stability analysis.

We define the order of instability as the order n in ε of the terms which excite the unstable modes. It may be the nth harmonic of the Stokes wave or some combination of harmonics which is of order n.

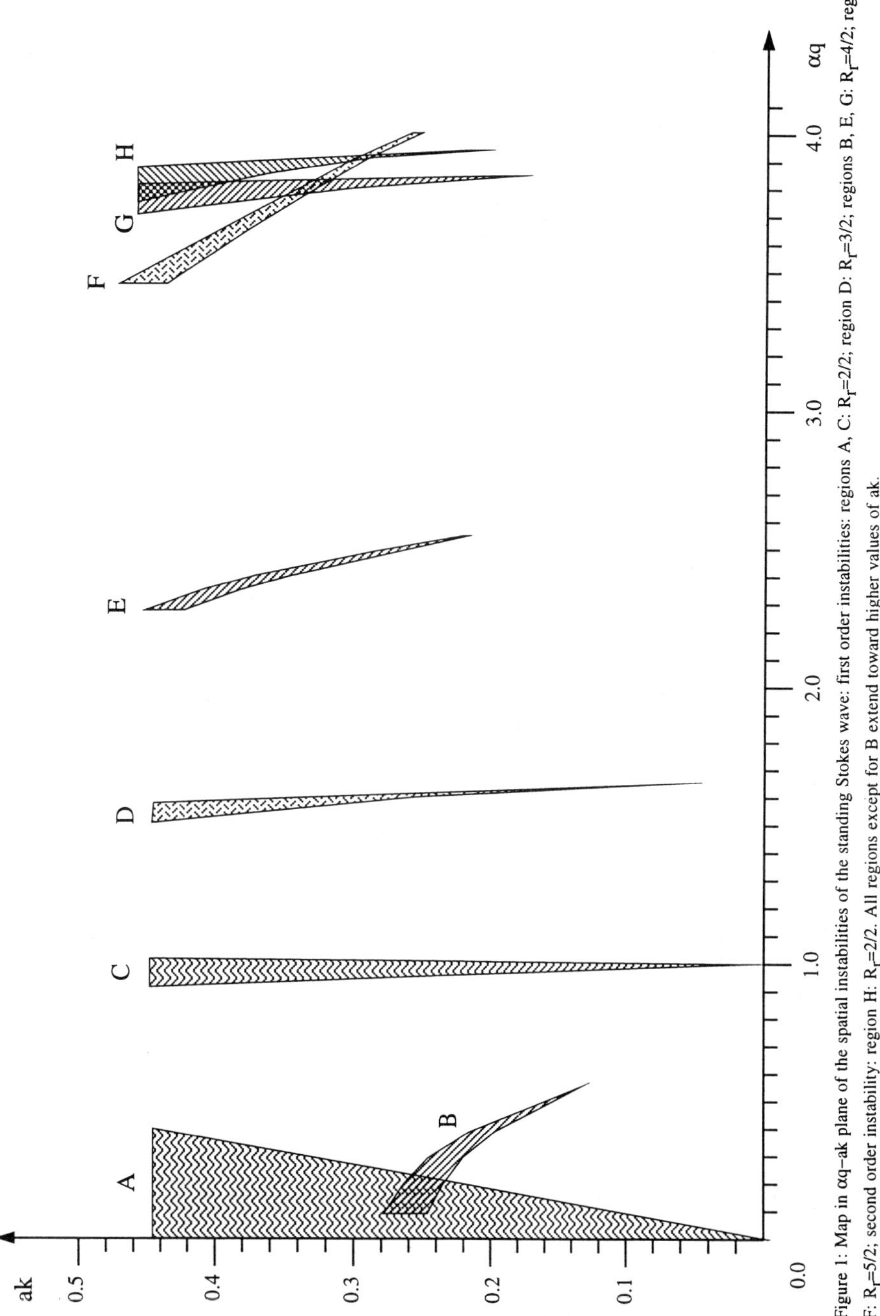

Figure 1: Map in αq–ak plane of the spatial instabilities of the standing Stokes wave: first order instabilities: regions A, C: R_r=2/2; region D: R_r=3/2; regions B, E, G: R_r=4/2; region F: R_r=5/2; second order instability: region H: R_r=2/2. All regions except for B extend toward higher values of ak.

Table 1: Spatial instabilities of the standing Stokes wave

Order of instability	Resonant ratio R_r	Main perturbations involved	Region in Fig.7.3	αq	ak range	ak for maximal instability	Maximal Floquet multiplier: Value	Maximal Floquet multiplier: Modulus	Long-time evolution: Class	Long-time evolution: Fig. #	Remarks
1	2/2	(1,q)(2,q)	A	0.10	0.110-0.358*	0.358	1.078347	1.078347	2	7.9	
				0.20	0.221-0.358*	0.358	1.108987	1.108987	2	7.9	
				0.25	0.274-0.358*	0.358	1.103724	1.103724	2	7.9	
				0.30	0.332-0.358*	0.358	1.071367	1.071367			
1	4/2	(3,q)(4,q)(5,q)	B	0.10	0.247-0.283	0.266	-0.023437+1.013227i	1.013498	3	7.10	overlap AB
				0.20	0.237-0.268	0.258	-0.020545+1.011433i	1.011641	3	7.10	overlap AB
				0.25	0.230-0.260	0.249	-0.018234+1.010145i	1.010309	1	7.4	
				0.30	0.222-0.250	0.241	-0.015781+1.008221i	1.008345			
				0.40	0.198-0.218	0.213	-0.008303+1.005068i	1.005102			
				0.50	0.164-0.173	0.169	0.001581+1.002592i	1.002593			
				0.577	0.127	0.127	0.010364+1.000232i	1.000286			
1	2/2	(0,q)	C	0.95	0.292-0.358*	0.358	1.126733	1.126733			
				1.00	0.09**-0.358*	0.358	1.096127	1.096127	2	7.8	
				1.01	0.335-0.388*	0.388	1.055186	1.055186			
1	3/2	(1,q)(2,q)(3,q)	D	1.55	0.379-0.470*	0.470	-0.715855+0.797448i	1.071621			
				1.60	0.263-0.384	0.325	-0.735971+0.713322i	1.024929	4	7.15	simplified-Fig. 7.16
				1.65	0.049	0.049	-0.767238+0.641532i	1.000108			
1	4/2	(2,q)(3,q)(4,q)(5,q)	E	2.28	0.427-0.456	0.444	1.025019	1.025019			
				2.35	0.385-0.412	0.400	1.022513	1.022513	1	7.5	
				2.40	0.350-0.377	0.364	1.018927	1.018927			
				2.50	0.267-0.288	0.282	1.008461	1.008461			
				2.55	0.217-0.227	0.222	1.004481	1.004481			

Table 1 (cont.)

Order of instability	Resonant ratio R_r	Main perturbations involved	Region in Fig.7.3	αq	ak range	ak for maximal instability	Maximal Floquet multiplier: Value	Maximal Floquet multiplier: Modulus	Long-time evolution: Class	Long-time evolution: Fig. #	Remarks
1	5/2	(3,q)(4,q)(5,q)	F	3.464	0.441-0.477	0.460	-0.893976+0.564234i	1.057143			
				3.65	0.383-0.406	0.394	-0.861985+0.566375i	1.031406	4	7.17	
				3.75	0.352-0.369	0.358	-0.852545+0.560906i	1.020514	4	7.17	
				3.80	0.334-0.349	0.345	-0.848767+0.556279i	1.014816	4	7.17	overlap FG
				3.85	0.318-0.330	0.325	-0.848481+0.551905i	1.012185			
				3.90	0.299-0.309	0.304	-0.849720+0.546263i	1.010162	3	7.12	overlap FH
				4.00	0.261-0.266	0.263	-0.853846+0.530451i	1.005202	1	7.17	
1	4/2	(0,q)(1,q)(2,q)	G	3.75	0.395-0.470*	0.470	1.116942	1.116942	1	7.6	
				3.80	0.309-0.470*	0.470	1.115813	1.115813	3	7.11	overlap GH
				3.85	0.175-0.203	0.192	1.004368	1.004368			
2	2/2	(0,q)(2,q)	H	3.80	0.425-0.470*	0.470	1.104620	1.104620			
				3.85	0.363-0.470*	0.470	1.108954	1.108954		7.18	
				3.90	0.293-0.392	0.339	1.031298	1.031298			
				3.95	0.204-0.225	0.213	1.005869	1.005869			

* The instability region extends toward higher values of ak; values in columns 7, 8 and 9 refer to the maximal value of Floquet multiplier found so far.
** The instability region extends towards lower values of ak.

The resonance ratio, R_r = m/2 is taken to be the closest expression of this form to $\omega/(n\omega_1)$, where ω is an average which represents the frequencies of the most unstable modes, n is the order of the instability and ω_1 is the frequency of the Stokes wave. Note that in some regions (A and B, F, G and H) resonance overlap occurs. This is very significant to the development of chaos which will be discussed in the next section on long time evolution.

4. Long time evolution of first order instabilities

The long time evolution of the spatial instabilities is studied in a 5th order uniformly truncated model with 12 modes. These are the first 6 harmonics of the Stokes standing wave and 6 perturbation modes with a fixed transverse wave number q. This model is suitable for steepness in the range considered.

Since one value of q is specified, resonance overlap will be described only for resonances with the same q value. Overlaps of resonances with different q occur more often as L_1/L_2 decreases. This case will be studied separately. Several types of chaotic behavior were encountered. In the figures, we plot the following, versus time in periods:

a. The distance (in the phase space, in logscale) of the trajectory from the periodic standing wave. This indicates the divergence of the trajectories. The linear part of the curve exhibits the exponential growth of the linear instability.

b. The relative error in the Hamiltonian, which indicates the extreme accuracy and numerical stability of the model.

c. Moduli of certain modes, showing the transfer of energy among the modes.

d. The phases of certain modes - the loss of coherence indicates transition to chaos.

In some cases we plot, in addition, in the phase plane $(\hat{\eta}_{pq}, \hat{\psi}_{pq})$ trajectories of certain modes and the correlation matrices among the variables, which also indicate transition to chaos. The scaling is such that $\hat{\eta}_1$, the first harmonic of the Stokes wave, equals 1.

The modes shown are: Mode (1,0) of the Stokes wave, one of the leading unstable modes, a higher harmonic of the Stokes wave and a nonleading perturbation.

We found four major classes of chaotic regimes for first order spatial instabilities of Stokes waves. They are:

4.1 Class 1: Encapsulated instabilities. These are instabilities in which the perturbed trajectory remains close to the initial periodic trajectory [9].

This type of instabilities seems generic to instabilities with resonance ratio R_r = 4/2 (regions B, E and G in Fig. 1). The long time evolution for regions B and G is given in Figs. 2 and 3, respectively.

For the encapsulated instabilities the energy transfer is small (no more than a few percent of the initial energy of the mode (1,0)). This is true even for $ak_1 \geq 0.4$ (see Fig. 3). Region B does not reach high values of ak.

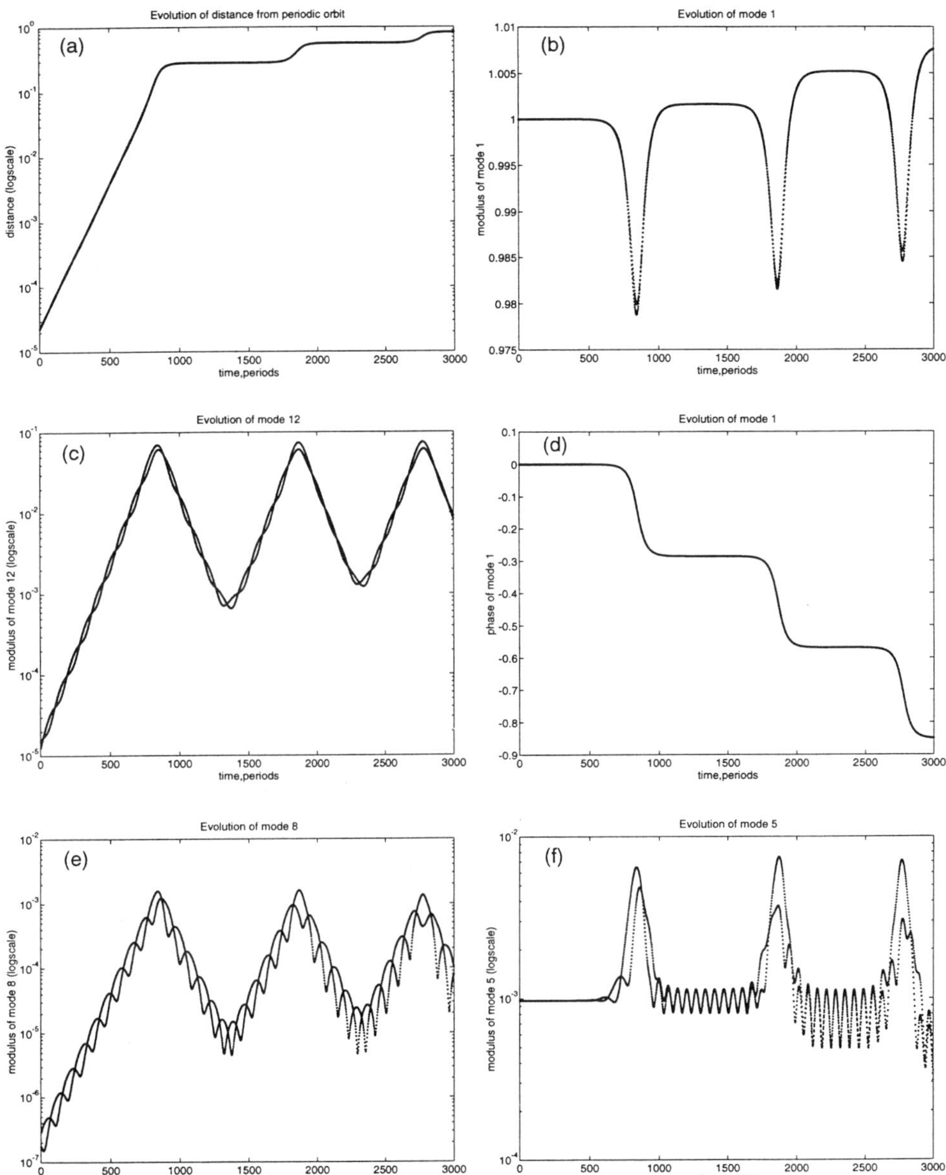

Figure 2: First order spatial instability of standing Stokes wave: $\alpha q=0.25$; ak=0.2494; $R_r=4/2$: long–time evolution of perturbed solution: class 1 of chaotic behavior–encapsulated instability

a) distance from unperturbed periodic solution (logscale); b) modulus of mode 1 (1,0);

c) modulus of mode 12 (5,q) (logscale); d) phase of mode 1 (1,0);

e) modulus of mode 8 (1,q) (logscale); f) modulus of mode 5 (5,0) (logscale);

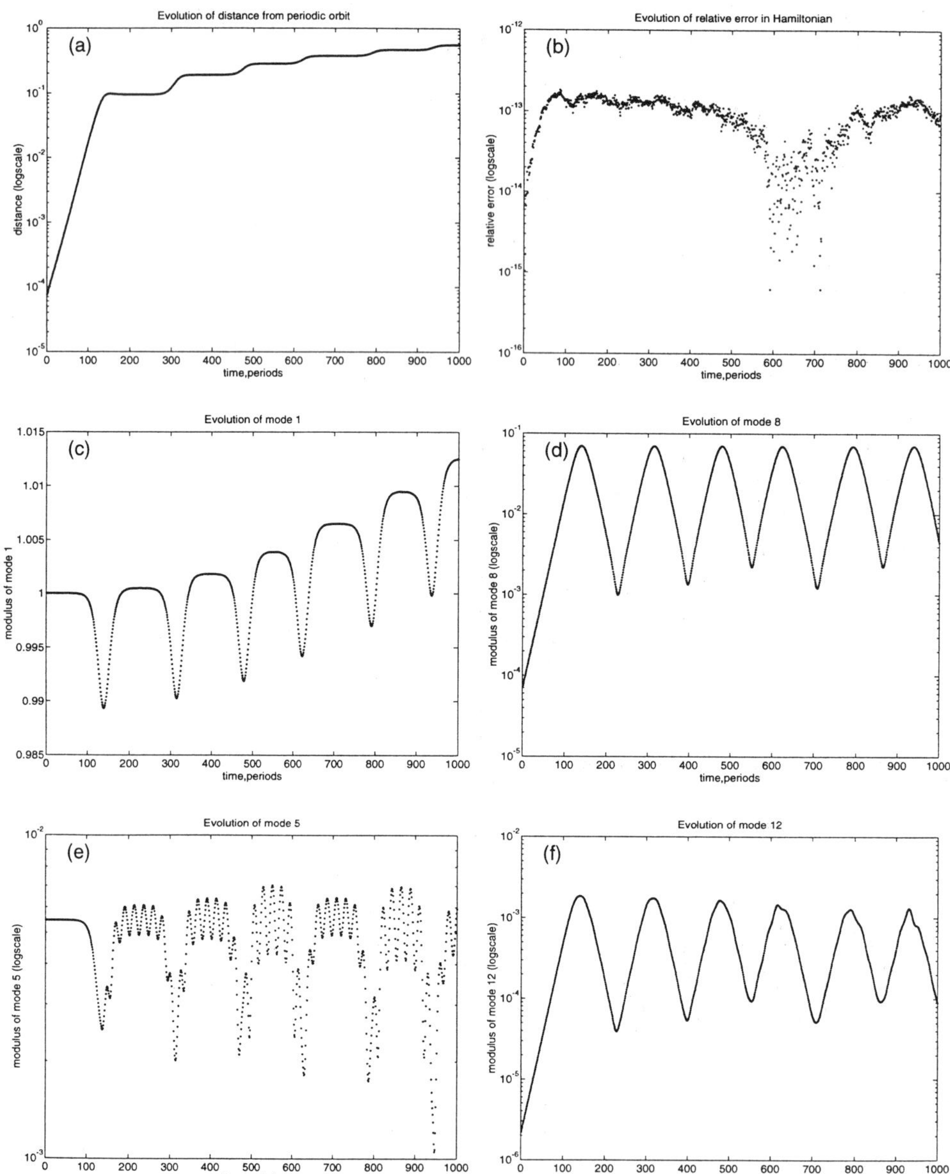

Figure 3: First order spatial instability of standing Stokes wave: αq=3.75; ak=0.4169; R_r=4/2: long–time evolution of perturbed solution: class 1 of chaotic behavior–encapsulated instability
a) distance from unperturbed periodic solution (logscale); b) relative error in Hamiltonian (logscale);
c) modulus of mode 1 (1,0); d) modulus of mode 8 (1,q) (logscale);
e) modulus of mode 5 (5,0) (logscale); f) modulus of mode 12 (5,q) (logscale).

The correlations existing in the initial periodic solution are preserved over a very long time (see Fig. 13a). The changes in amplitudes and phases of different modes are well synchronized with the leading mode (1,0). The evolution is characterized by short synchronous bursts of perturbations. The main effect of these bursts is a phase shift in the leading mode and consequently in the solution as a whole (see Fig. 2). As can be observed in Fig. 2 and Fig. 3c,d such bursts have no perfect period and cannot be predicted for long times. This is because at the recurrence points, when the perturbed solution returns to the close vicinity of the periodic orbit, the behavior of the perturbations is unpredictable.

The internal structure of the Stokes wave itself is also disturbed during evolution, but rather slightly. When the perturbations burst, all modes of the Stokes wave change their amplitudes. Near the recurrence points the higher modes of the Stokes wave remain disturbed, as shown in Fig. 2f and Fig. 3e.This unpredictable disturbance is caused by leading perturbations which even near the recurrence points cannot be considered small compared with higher modes of the Stokes wave.

From the global point of view the major effect of encapsulated instabilities is loss of phase information after a few bursts. However, only the phase of the solution as a whole becomes fully unpredictable. The phases of individual modes within the Stokes wave remain correlated to some extent, although in higher modes some quasiperiodic disturbances exist. The energy spectrum remains almost unchanged.

Note that instabilities in regions D and F with resonant ratios 3/2 and 5/2 respectively (complex Floquet multipliers) for small ak look like encapsulated (see Fig. 9a). However, comparison with Fig. 2 (region B, also complex Floquet multipliers) reveals the significant difference: for region B there are no signs of chaotization for nonleading perturbations or higher modes of the Stokes wave, whereas for regions D and F such chaotization exists even for small ak_1 (Fig. 9b).

4.2 **Class 2: Global instabilities with recurrence.** This type of behavior seems to be generic to instabilities with resonance ratio $R_r = 2/2$ (regions A and C in Fig. 1). Long-time evolution for these regions is given in Figures 5 and 4 respectively.

The evolution in this case is characterized by short, strong bursts of the leading perturbation, which at its maximum concentrates half or more of the total energy of the system (Fig. 4c,d, Fig. 5e,b). These bursts are not perfectly periodic and thus cannot be predicted for a long time. This is because at the recurrence points, when the perturbed solution returns to some vicinity of the periodic orbit (with a phase shift) the behavior of the perturbations is unpredictable. Only the behavior of the leading modes is identical in both areas (A and C). The behavior of higher modes is completely different (compare correlation matrices in Fig. 11b,c).

For region C all the perturbations are well synchronized - they simultaneously reach their maxima as well as recurrence points (Fig. 4f). The internal structure of the Stokes wave is only slightly disturbed during the evolution: when the perturbations burst, all modes of the Stokes wave decrease in amplitude. Near the recurrence points the higher modes of the Stokes wave remain slightly disturbed (Fig. 4e). This unpredictable disturbance is caused by leading perturbations which even near recurrence points cannot be considered small compared with higher modes of the Stokes wave. The recurrence for region C is to close vicinity of the periodic

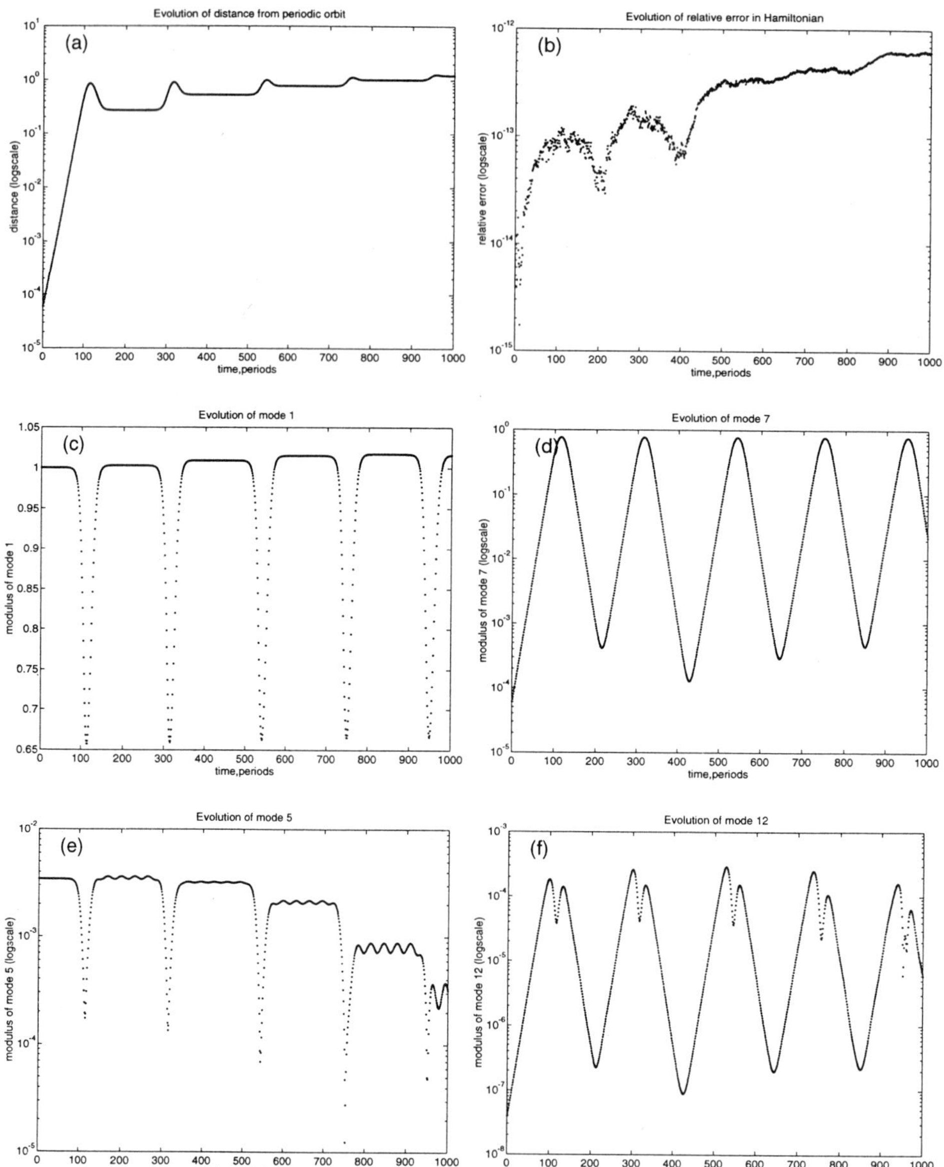

Figure 4: First order spatial instability of standing Stokes wave: αq=1.00; ak=0.3579; R_r=2/2: long–time evolution of perturbed solution: class 2 of chaotic behavior–global instability with recurrence
a) distance from unperturbed periodic solution (logscale); b) relative error in Hamiltonian (logscale);
c) modulus of mode 1 (1,0); d) modulus of mode 7 (0,q) (logscale);
e) modulus of mode 5 (5,0) (logscale); f) modulus of mode 12 (5,q) (logscale).

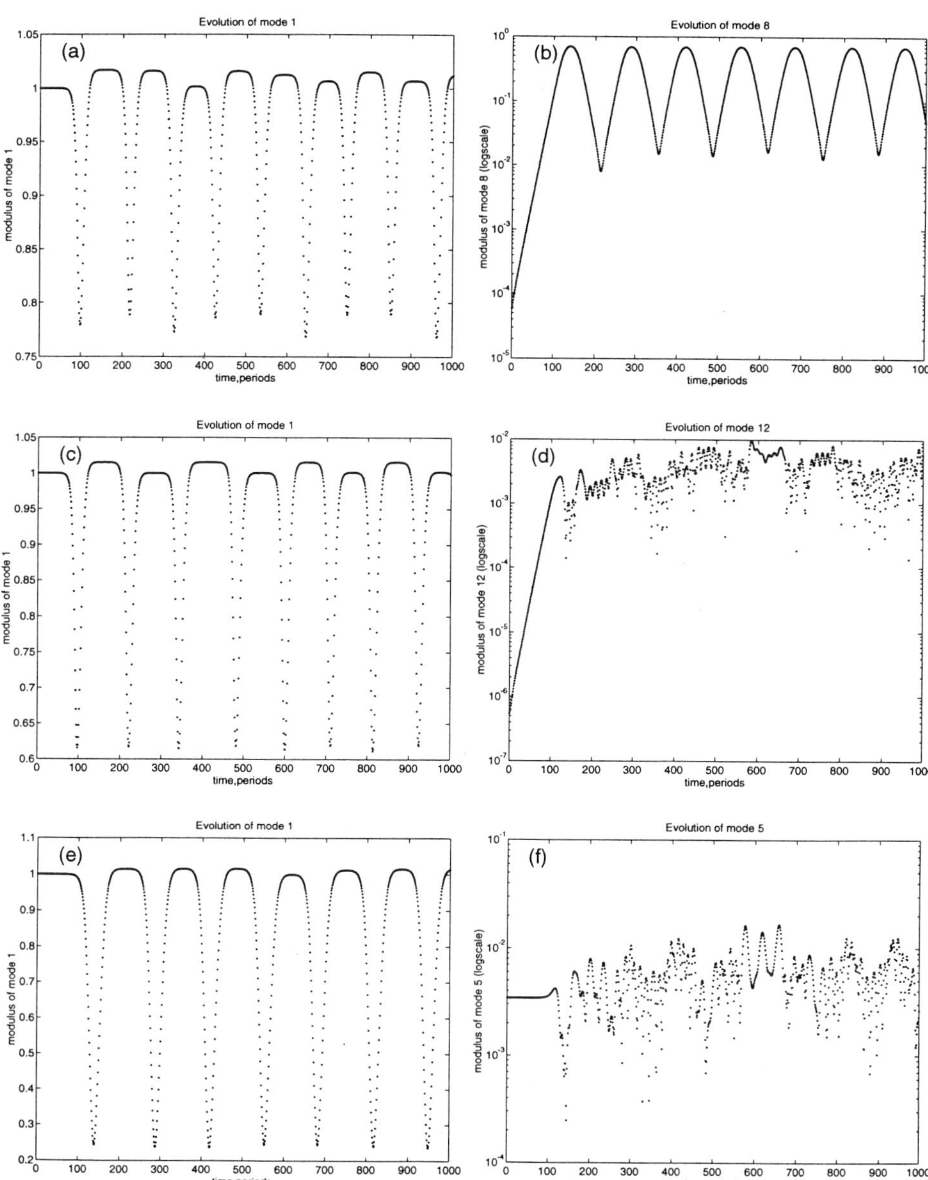

Figure 5: First order spatial instability of standing Stokes wave: ak=0.3579; R_r=2/2; αq<<1:
long–time evolution of perturbed solutions: class 2 of chaotic behavior–global instability with recurrence
a) modulus of mode 1 (1,0) for αq=0.25;
b) modulus of mode 8 (1,q) for αq=0.10 (logscale);
c) modulus of mode 1 (1,0) for αq=0.20;
d) modulus of mode 12 (5,q) for αq=0.10 (logscale);
e) modulus of mode 1 (1,0) for αq=0.10;
f) modulus of mode 5 (5,0) for αq=0.10 (logscale).

orbit.

For region A the energy transfer from mode (1,0) to the leading perturbation increases when $\ell_q / k_1 \to 0$ (Fig. 5a,c,e). Here the leading perturbation even at recurrence points is not very small (compare Fig. 5b and Fig. 4d). In higher modes perturbations there is no recurrence at all: they reach their maximal values and afterwards oscillate chaotically (Fig. 5d). Higher modes of the Stokes wave experience unpredictable oscillations around their mean level (Fig. 5f). It must be mentioned that in all the cases given in Fig. 5 the relative error in the Hamiltonian does not exceed $3 \cdot 10^{-13}$. For region A at the global level we still have recurrence to some (not close) vicinity of the periodic orbit. At the local level, the motion is quite chaotic, and phases of higher modes become actually random. However, there is no thermalization here - near the recurrence points the energy spectrum is quite close to its initial shape.

4.3 Class 3: Resonance overlap. Resonance overlap is known to be a mechanism that leads to chaos and thermalization. In Fig. 1. we can see several regions of resonance overlap, we shall consider all of them here.

Region AB (Fig. 6) has an overlap of instabilities with resonant ratios $R_{r_1} = 2/2$, $R_{r_2} = 4/2$. The individual instabilities were described in the two previous sub-sections. In Fig. 6 two cases are given, with $\ell_q / k_1 = 0.2$ and 0.1 respectively. In both cases the initial perturbation was in the direction of the weak instability (with $R_r = 4/2$). As can be seen from the change of slope in Fig. 6a, this instability is dominant at the initial stage; afterwards the strong instability becomes dominant. As shown in Fig. 6b up to about 2000 periods the global behavior (leading modes) in case 1 seems to be similar to that given in Fig.5 for the strong instability only. However, the energy transfer is now weaker, and the distance from the periodic orbit at recurrence points larger. The higher modes in the same interval are chaotic and gradually increasing (Fig. 6d). After about period 2000 the recurrence pattern is rapidly destroyed: the mode (1,0) loses its energy, whereas all higher modes grow. Near period 3000 the system is actually thermalized - all modes are of the same order of magnitude.

In case 2 (Fig. 6c) energy transfer is much stronger and partial thermalization occurs relatively soon (Fig. 6e,f). However, complete destruction of the recurrence pattern and full thermalization occurs only near period 3000.

Region FG has a resonance overlap which does not impact the motion. This is because in this region exists another, more powerful mechanism which leads to thermalization (see next section).

Region GH has an overlap of second order instability with $R_r = 2/2$ (see §5) and first order instability with $R_r = 4/2$ (see §4.1). The initial perturbation was given in the direction of the second order instability. Unfortunately, this overlap occurs at quite high values of ak_1, and the solver has failed after 415 periods. However, the initial stage of chaotization and thermalization can be seen very clearly. Note, that even under such conditions the relative error in the Hamiltonian at the end of integration does not exceed $3 \cdot 10^{-13}$.

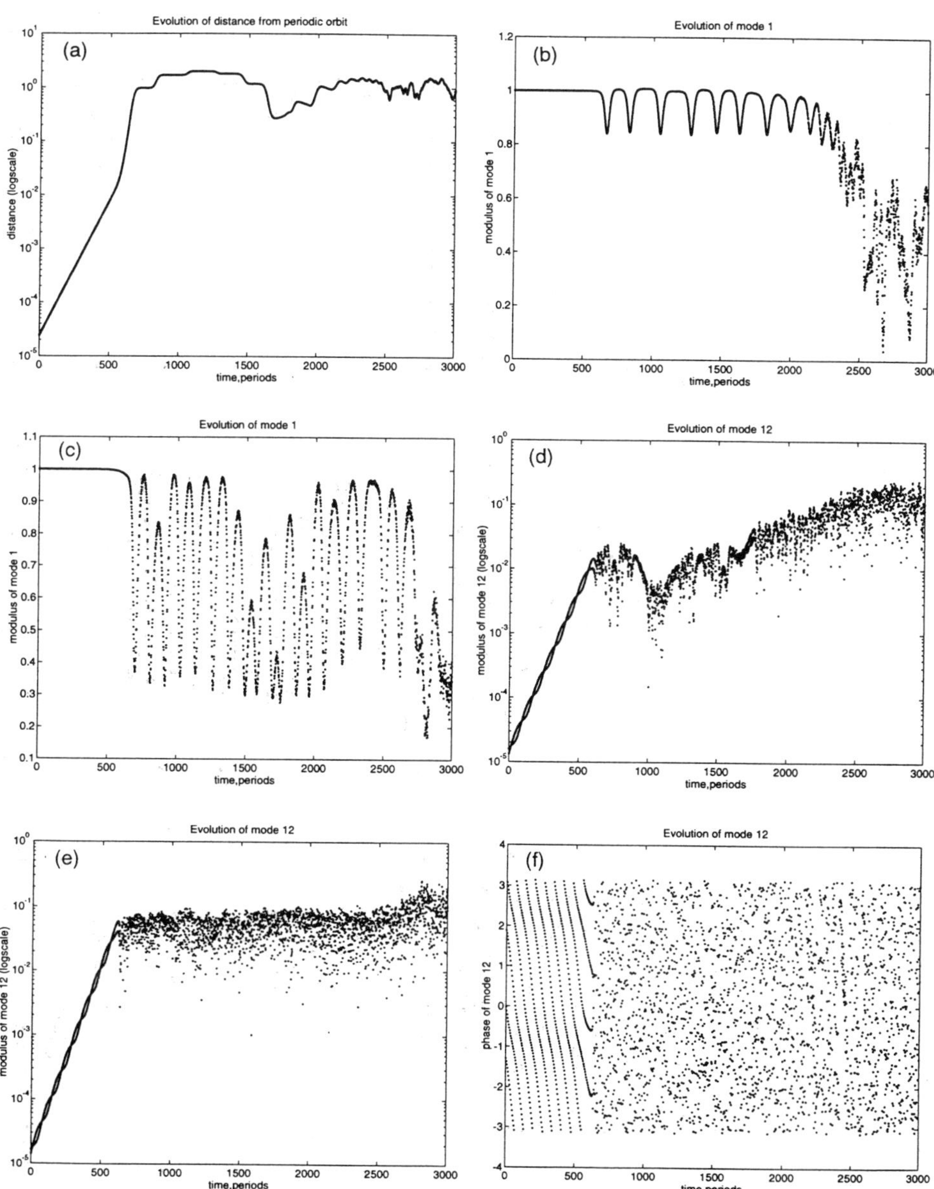

Figure 6: First order spatial instability of standing Stokes wave: case1: $\alpha q=0.2$, ak=0.258; case2: $\alpha q=0.1$, ak=0.266: long–time evolution of perturbed solutions: class 3 of chaotic behavior–resonance overlap: $R_{r1}=2/2$, $R_{r2}=4/2$

a) distance from unperturbed periodic solution for case1 (logscale); b) modulus of mode 1 (1,0) for case1;
c) modulus of mode 1 (1,0) for case2; d) modulus of mode 12 (5,q) for case1 (logscale);
e) modulus of mode 12 (5,q) for case2 (logscale); f) phase of mode 12 (5,q) for case2;

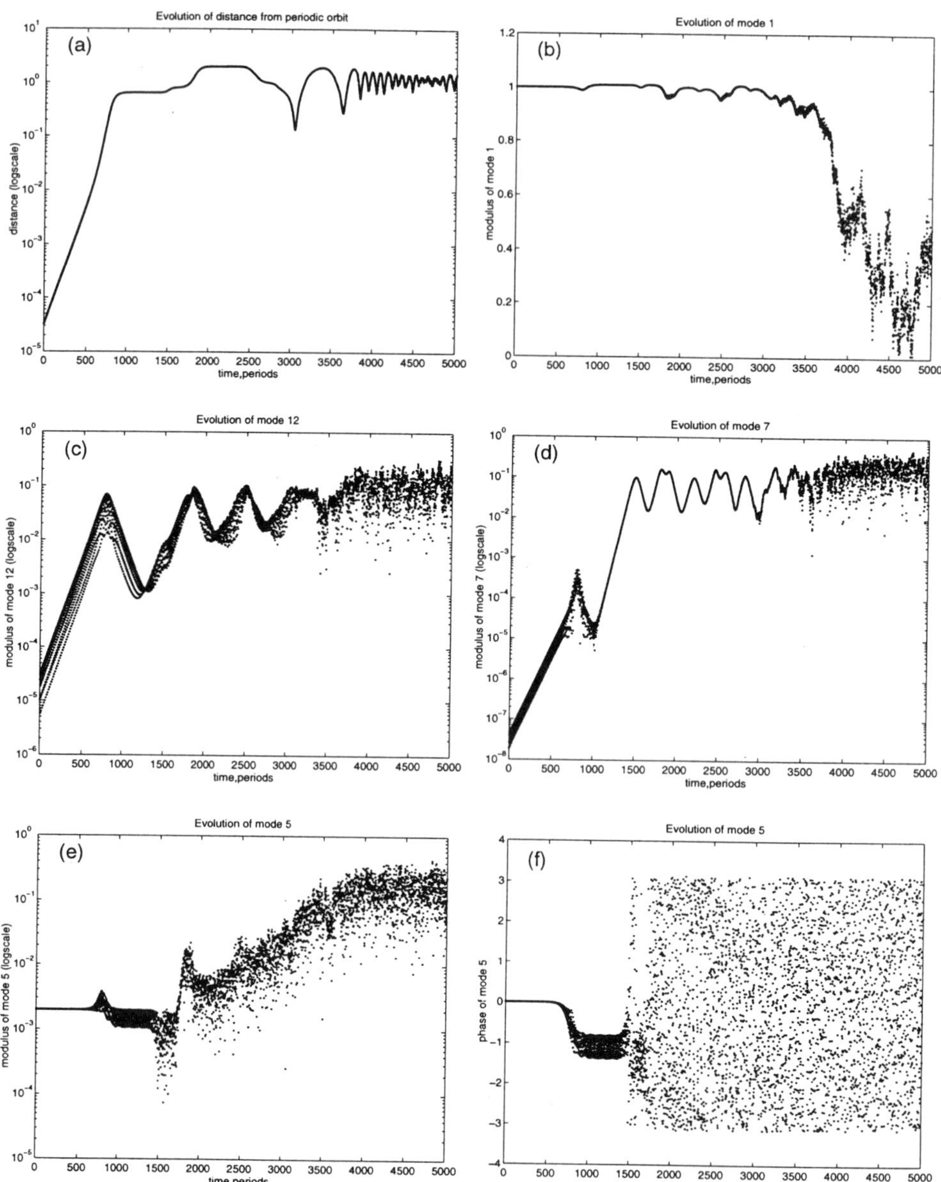

Figure 7: Spatial instability of standing Stokes wave: αq=3.90; ak=0.304; R_{r1}=2/2 (second order instability): long–time evolution of perturbed solution: class 3 of chaotic behavior–resonance overlap: R_{r2}=5/2 (1st order instability)

a) distance from unperturbed periodic solution (logscale); b) modulus of mode 1 (1,0);
c) modulus of mode 12 (5,q) (logscale). d) modulus of mode 7 (0,q) (logscale);
e) modulus of mode 5 (5,0) (logscale); f) phase of mode 5 (5,0);

The long-time evolution for the overlap region FH is shown in Figure 7. Here the instability with the resonant ratio $R_r = 5/2$ is still beneath its thermalization threshold (a situation like that shown in Fig. 10a,b). The initial perturbation was given in the direction of this instability, which is the weakest in the overlapping pair. The strongest instability in this pair is the second order instability with $R_r = 2/2$ (see §5). As seen in Fig. 7d, up to the first recurrence point the instability behaves like a single instability with $R_r = 3/2$ (compare with Fig. 11c). Afterwards the strongest instability becomes dominant in the mode (0,q) - it is its leading mode. At this stage (Fig. 7b-d) the global picture looks like a simple superposition of two instabilities. However, at the local level (Fig. 7e) there is clear energy transfer to the higher modes of the Stokes wave and significant chaotization of these modes even at this early stage. This energy transfer (higher modes of the Stokes wave grow exponentially) leads to gradual erosion and chaotization even of leading perturbations. After few recurrences the internal structure of the Stokes wave becomes completely destroyed and so the recurrence itself - the thermalization process is completed and the system reaches an equilibrium stage of stochastic motion.

All cases of resonance overlap described above are overlaps within a single group of perturbations. Let us consider now what will happen when the system has simultaneously instabilities in two different groups of perturbations. As we have mentioned previously, we cannot work with our regular precision with two groups perturbations (due to computer limitations). Thus, a simplified 5th order 9 modes model was used with 3 modes in the standing Stokes wave and 3 modes in each group of perturbations. Two cases were considered: when the groups of perturbations are distant ($q_1 = 1, q_2 = 5$) and close ($q_1 = 1, q_2 = 2$). The results of the second case are given in Fig. 8. The leading modes of the Stokes wave and of both groups of perturbations are shown in the left column, whereas in the right column are given the last mode of the Stokes wave and the weakest perturbations of both groups. There are no signs of thermalization - the pictures look like simple superposition of instabilities. We can conclude that such multigroup resonance overlaps do not lead to thermalization even for two close groups of instabilities (at least up to $ak_1 = 0.36$).

4.4 Class 4 - thermalization without resonance overlap. This seems to be a new type of chaotic behavior which, as far as we know, has not been previously reported. It is generic for instabilities with resonant ratios $R_r = 3/2$ and 5/2 (regions D and F respectively in Fig.1). Full thermalization occurs for sufficiently large values of ak_1; However, even for small steepnesses some chaotization can be observed.

We start from region D. As can be seen in Fig. 1, this region has no resonance overlaps. Evolution of perturbed solution for sufficiently large ak_1=0.332 is given in Fig. 9. As can be seen from the figure, in this case there is no recurrence at all. All modes, including the leading ones, become chaotic very soon (Fig. 9b-d). The thermalization process is very fast (about 400 periods, Fig. 9e,f). The correlation matrices (Fig. 13) show how rapidly the correlations in such a system are destroyed. Note that here, as in all systems with complex Floquet

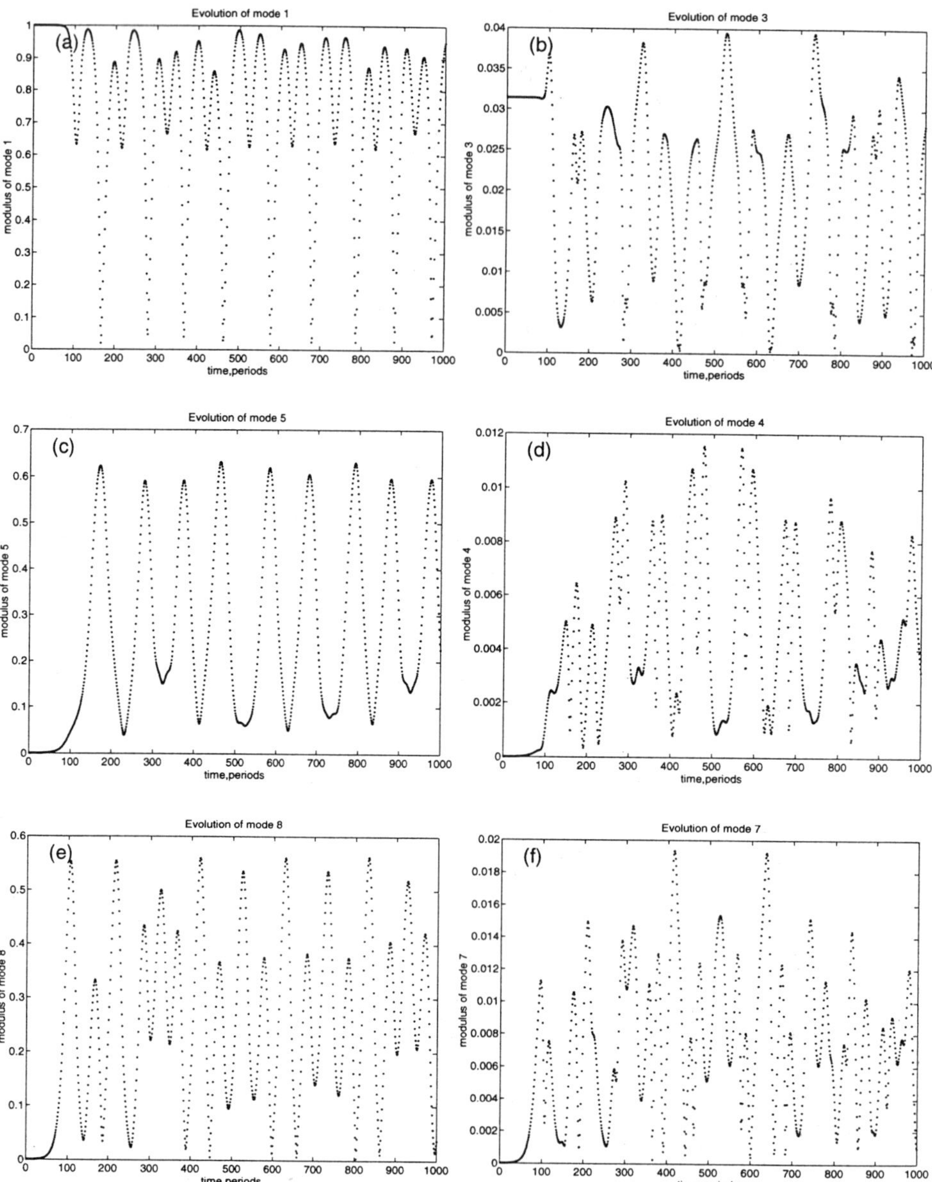

Figure 8: Multigroup spatial instability of simplified standing Stokes wave (three modes only): α=0.1; ak=0.357; $R_{r1}=R_{r2}$=2/2; q_1=1; q_2=2: long–time evolution of perturbed solution: superposition of instabilities

a) modulus of mode 1 (1,0); b) modulus of mode 3 (3,0);

c) modulus of mode 5 (1,1); d) modulus of mode 4 (0,1);

e) modulus of mode 8 (1,2); f) modulus of mode 7 (0,2).

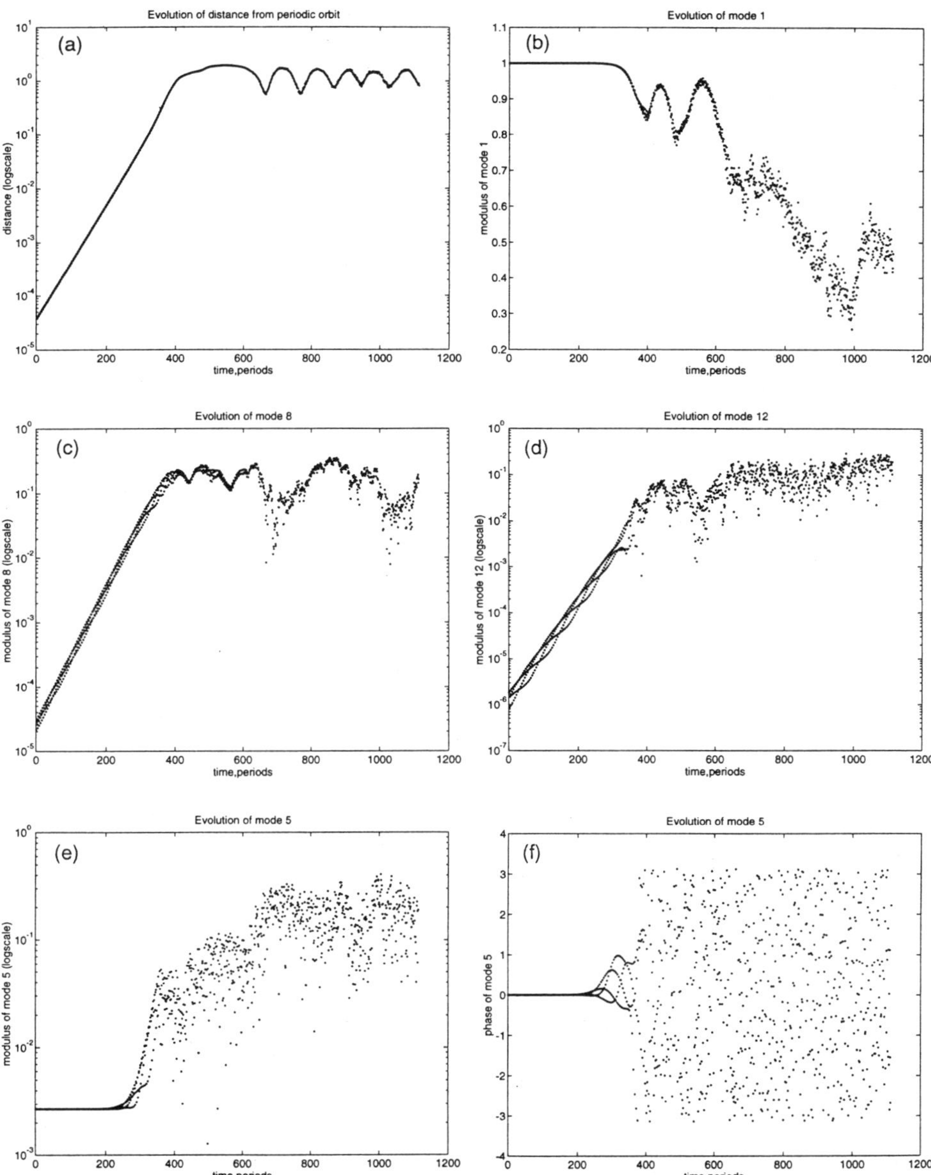

Figure 9: First order spatial instability of standing Stokes wave: $\alpha q=8/5$; ak=0.3319; $R_r=3/2$:
long–time evolution of perturbed solution: class 4 of chaotic behavior–thermalization without resonance overlap
a) distance from unperturbed periodic solution (logscale); b) modulus of mode 1 (1,0);
c) modulus of mode 8 (1,q) (logscale); d) modulus of mode 12 (5,q) (logscale);
e) modulus of mode 5 (5,0) (logscale); f) phase of mode 5 (5,0);

multipliers, the evolution is quite complicated even at the initial stage where the Floquet theory is applicable (see Appendix B).

As we have mentioned above, in this case there is no local resonance overlap (i.e. our periodic solution has only one instability). However, there exists some possibility of nonlocal resonance overlap with another, sufficiently close periodic solution. As we know [1] in the system of 6 modes there are four additional primary periodic solutions with periods close to that of the standing Stokes wave, and two of them are unstable. In order to eliminate any possibility of non-local resonance overlap, we have considered a simplified system, in which the Stokes wave was represented by only three modes. The above mentioned additional periodic solutions do not exist in this simplified system. Also, one of the nonleading perturbations (5,q) was omitted. Hence, the simplified system has only 8 modes. The results for such a system are given in Fig. 10. To estimate the reliability of the simplified system, we have compared the extent of the instability region in this model with our regular 12 modes model (Fig. 10f, dashed and solid line, respectively). As we could expect, the difference between the two models is small for relatively low ak_1, but rapidly increases when ak_1 grows. We can conclude that for $ak_1 \geq 0.3$, three-modes representation of the standing Stokes wave is not good enough. However, we shall use it here to get the qualitative picture.

Figure 8 represents long-time evolution of the disturbed simplified system for two different steepnesses: $ak_1 = 0.274$; 0.331. We can see (Fig. 9a) that for the first case the behavior in global scale is recurrence. However, the system does not return to very close vicinity of the periodic orbit, and there are clear signs of chaotization in the high perturbations (Fig. 10b) and in the last mode (3,0) of the Stokes wave. The second case refers actually to the same energy level as in Fig. 9. Although now the maximal Floquet multiplier is significantly smaller (see Fig. 10f) the long-time behavior is qualitatively the same as in Fig. 9 - no recurrence and fast thermalization (Fig. 10c,d). It seems that there exists some critical level of energy transfer, after which the thermalization process begins.

Let us consider now region F (Fig. 11). This region has two resonance overlaps - with regions G and H (see Fig. 1). One of these overlaps was considered in the previous section. The second has no actual influence - the new thermalization mechanism is clearly dominant.

For relatively small $ak_1 = 0.263$, this instability looks like encapsulated; however, even for this steepness there are clear signs of chaotization in non-leading perturbations (Fig. 11c) and in higher modes of the Stokes wave (Fig. 10b,d). The next two cases have $ak_1 = 0.345$ and 0.358; the first of them lies in the FG overlap region, the second lies out of any overlap region. Comparing the two series of pictures we see actually the same behavior - fast thermalization without any signs of recurrence. This means that for the case with $ak_1 = 0.345$ the resonance overlap has no influence on the motion. For the case with $ak_1 = 0.358$ special tests were made to check the possible mechanisms of thermalization. For different points on the evolution trajectory the deformations of a very narrow trajectory pipe surrounding the original trajectory were computed. We call the eigenvalues of the resulting Jacobian matrix pseudo-Floquet multipliers. These computations indicated that when the perturbations reach some critical value, an energy transfer channel opens from the perturbations to the higher modes of the Stokes wave. Shortly

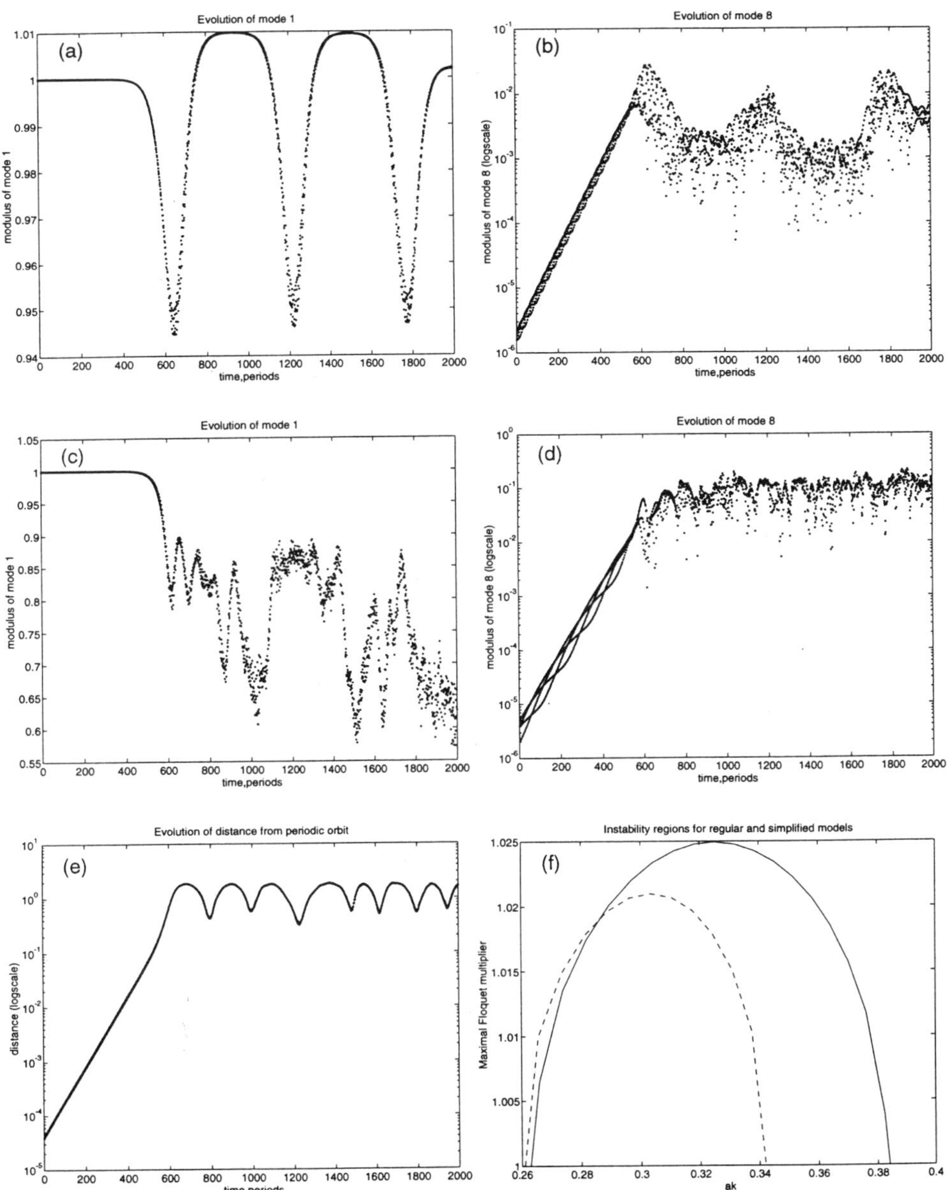

Figure 10: Spatial instability of simplified standing Stokes wave (three modes only): $\alpha q=8/5$; $R_\Gamma=3/2$:

long–time evolution of perturbed solutions for different values of ak:

a) modulus of mode 1 (1,0) for ak=0.2737; b) modulus of mode 8 (4,q) for ak=0.2737 (logscale);

c) modulus of mode 1 (1,0) for ak=0.3310; d) modulus of mode 8 (4,q) for ak=0.3310 (logscale);

e) distance from unperturbed periodic solution for ak=0.3310 (logscale);

f) instability regions for regular (solid line) and simplified (dashed line) models.

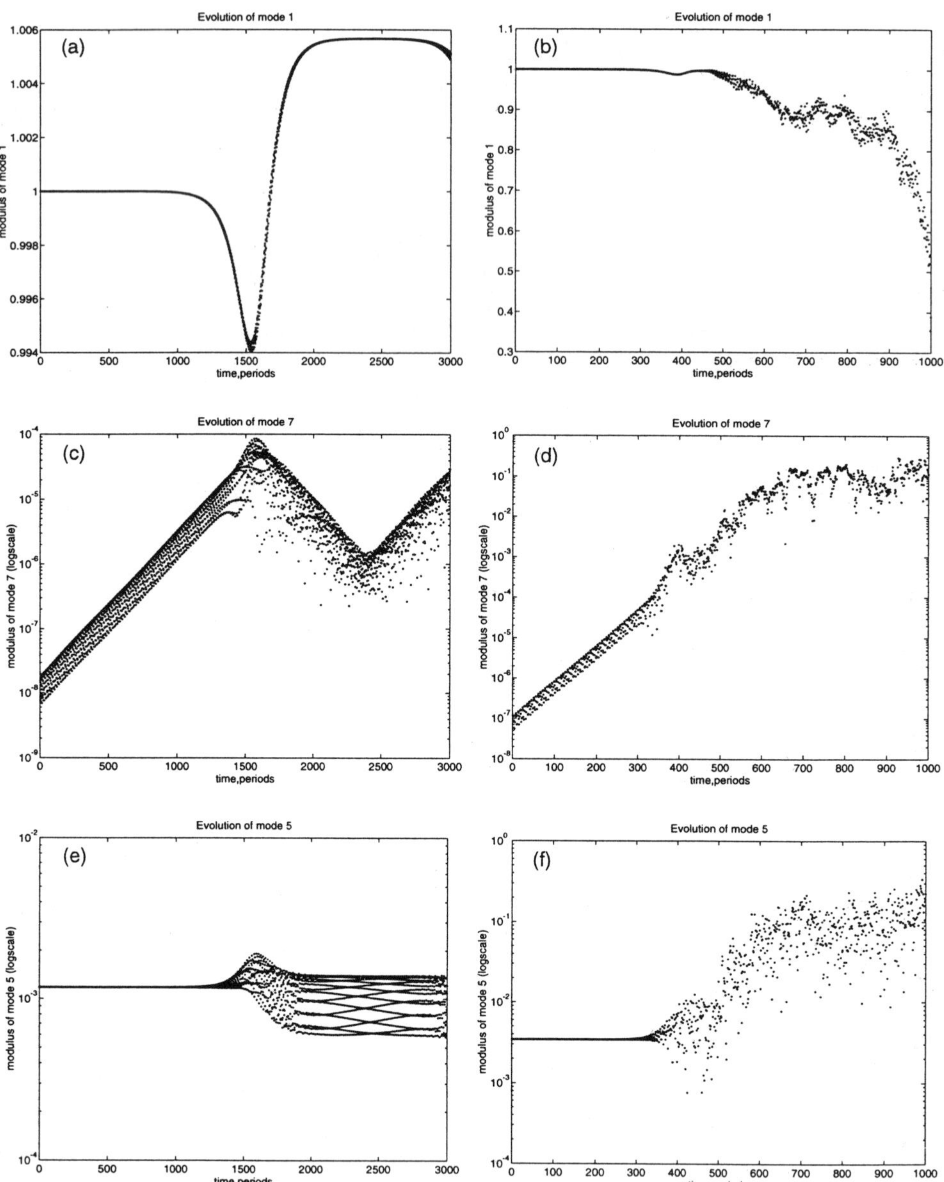

Figure 11: First order spatial instability of standing Stokes wave: R_r=5/2:

long–time evolution of perturbed solutions: class 4 of chaotic behavior–thermalization without resonance overlap

a) modulus of mode 1 (1,0) for αq=4.00, ak=0.2628; b) modulus of mode 1 (1,0) for αq=3.75, ak=0.3579;

c) modulus of mode 7 (0,q) for αq=4.00, ak=0.2628; d) modulus of mode 7 (0,q) for αq=3.75, ak=0.3579.

e) modulus of mode 5 (5,0) for αq=4.00, ak=0.2628 (logscale);

f) modulus of mode 5 (5,0) for αq=3.75, ak=0.3579 (logscale).

afterwards the motion becomes extremely complicated. As could be expected, when ak_1 grows the thermalization process becomes faster and faster.

The above mentioned results regarding class 4 of chaotic behavior can be considered only as preliminary ones. To understand better this phenomenon, more extensive investigation is required.

5. An example of second order spatial instability of the standing Stokes wave

Region H in Fig. 1 is a region of second order instability with $R_r = 2/2$. An important feature of the second order instability is that there exist two kinds of periodic coefficients of second order in ε: one involving squares of mode (1,0) and another involving the mode (2,0). As a result, within the same second order instability two components may be present, corresponding to interactions of perturbations with modes (1,0) and (2,0). An example is given in Fig. 12.

The global evolution of modes (1,0) and (2,0) as shown in Fig. 12a,b is typical for Stokes wave. However, usually when mode (2,0) decreases, the energy is transferred to mode (1,0); here it is transferred to perturbations and to mode (4,0) (Fig. 12c). Here modes (4,0) and (1,q) show some signs of recurrence; in the leading perturbation (0,q) such signs are very weak. Figure 13 displaying the correlations among the variables illustrates the loss of coherence when instabilities or chaos develop.

6. Conclusions

The Hamiltonian approach to the water wave problem, especially in its generalized fashion developed here, is a very powerful, flexible and reliable method. It has many advantages over other existing methods: a broad variety of truncation techniques, the possibility to work with three-dimensional systems, the possibility to pick out any desirable set of modes, and many more. As an independent check on the stability of the computations, we have calculated the value of the Hamiltonian throughout the evolution. The relative error after up to 8000 periods was typically well under 10^{-11}. This indicates the absence of numerical instabilities.

In this research we have studied standing water waves in a rectangular basin, which form a system with natural discretization and are particularly convenient for investigation. At the preliminary stage (two-modes models) we have shown that high-order models are necessary to get reliable results. However, only models with many degrees of freedom can provide a realistic simulation of the physical system.

We have computed the standing Stokes wave up to order 18 in ε for $ak \leq 0.42$ in the region where the traditional perturbation methods require Pade approximants to sum the series. Our solutions are fully consistent and show good convergence - an additional numerical evidence to existence of the standing Stokes wave. In the present work we have thoroughly studied the stability of this wave to spatial (three-dimensional) perturbations. This investigation was done for the case of infinite depth. This was really terra incognita and now we have its map (Fig. 1). Note that we have studied stability of the standing Stokes wave to a "continuous" **group** of six perturbations.

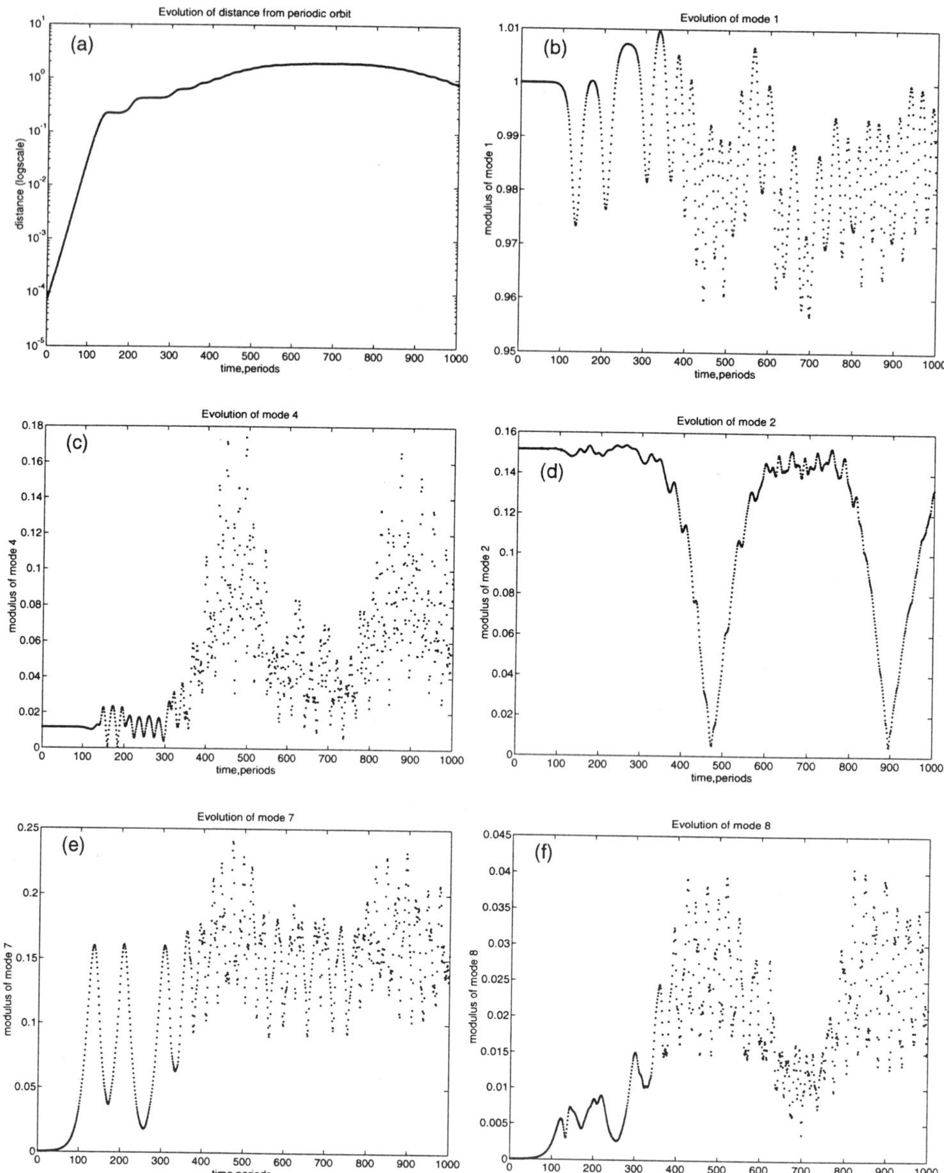

Figure 12: Second order spatial instability of standing Stokes wave: $\alpha q=3.85$; $ak=0.400$; $R_r=2/2$:
long–time evolution of perturbed solution: energy exchange between perturbations and two leading modes
a) distance from unperturbed periodic solution (logscale); b) modulus of mode 1 (1,0);
c) modulus of mode 4 (4,0); d) modulus of mode 2 (2,0);
e) modulus of mode 7 (0,q); f) modulus of mode 8 (1,q).

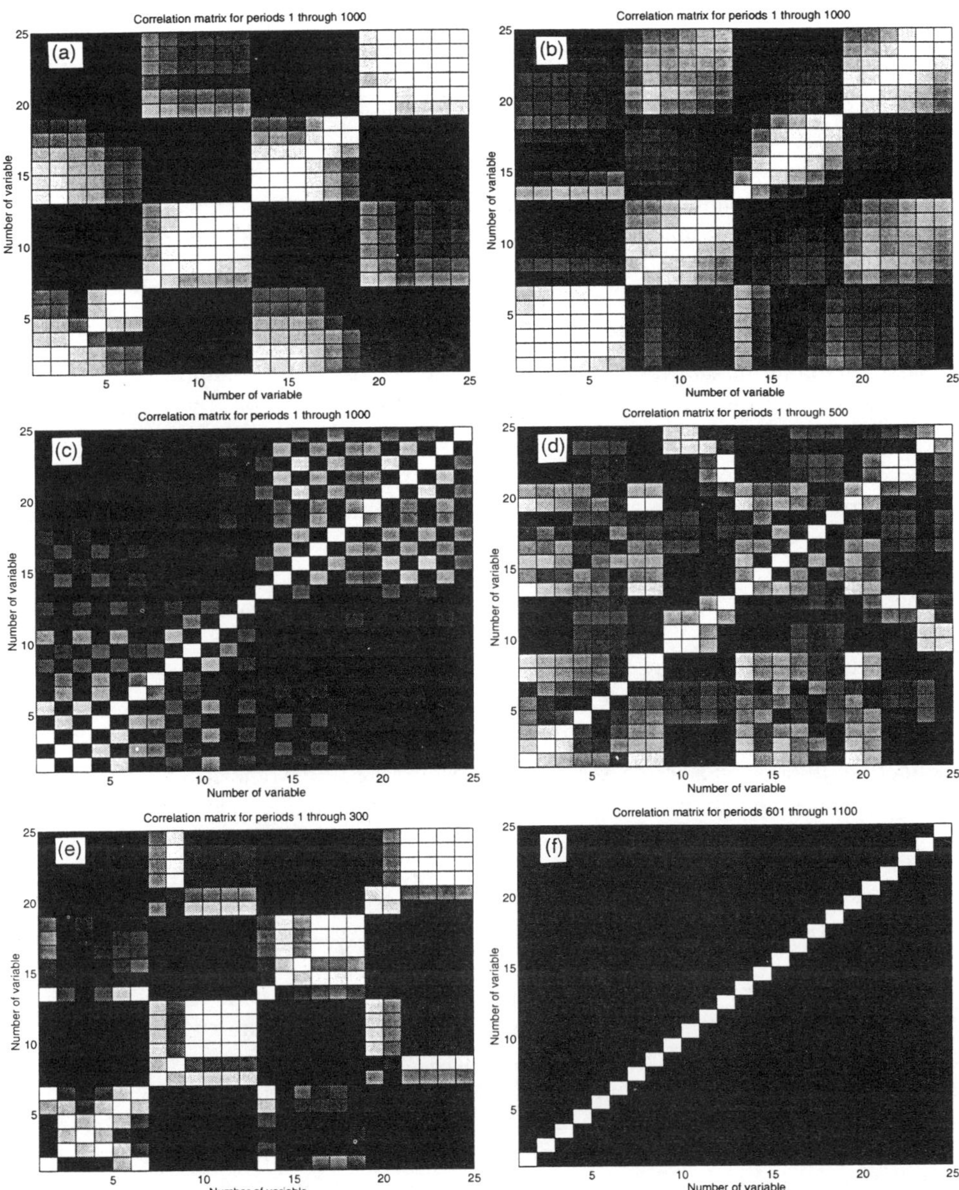

Figure 13: Spatial instabilities of standing Stokes wave: correlation matrices for different classes of chaotic behavior: (dark areas indicate lack of correlations)

class 1: encapsulated instabilities: a) region G, ak=0.417;

class 2: global instabilities with recurrence: b) region C, ak=0.358; c) region A, ak=0.358;

class 3: resonance overlap: region AB, ak=0.258: d) initial stage of evolution;

class 4: thermalization without resonance overlap: region D, ak=0.332:

e) initial stage of evolution; f) fully randomized motion.

Our main goal in this research was to study the route from order to chaos in standing water waves. Our approach was similar to that of Fermi, Pasta and Ulam [5]: we studied the long-time evolution of perturbed, unstable periodic solutions. Indeed, in a system of standing water waves periodic solutions are the highest possible level of organization. Studying the numerous spatial instabilities of the standing Stokes wave we found that the different scenarios of chaotic behavior can be grouped into four major classes:

Class 1: Encapsulated instabilities. Generic for instabilities with resnonant ratio $R_r = 4/2$. Energy transfer in this case is weak; the main impact of these instabilities is a phase shift, which after a few bursts of perturbations leads to loss of phase information.

Class 2: Global instabilities with recurrence. Generic for instabilities with $R_r = 2/2$. Energy transfer between the leading mode of the Stokes wave and the leading perturbations is strong. Bursts of perturbations are followed by recurrences to the vicinity of the periodic orbit. Higher modes are significantly disturbed even at the recurrence points. The main impact of these instabilities is in loss of phase information.

Class 3: Resonance overlap. Resonance overlap between two different instabilities in the same group of perturbations leads to gradual thermalization. Resonance overlap between two different groups of perturbations does not lead to thermalization (at least for $ak \leq 0.36$).

Class 4: Thermalization without resonance overlap. Generic for instabilities with resonant ratios $R_r = 3/2$ and $5/2$. In this case a single instability with complex Floquet multipliers leads to fast thermalization, starting from some critical value of ak. There are no visible resonance overlaps (both local or nonlocal) in this case.

To explain the different types of evolution and the impact of many degrees of freedom, let us consider firstly the limitations, which are generic for Hamiltonian systems with two degrees of freedom. For such a system stability of a periodic solution (periodic point on Poincare section) is determined by two eigenvalues only. Due to Liouville's theorem they are either a complex-conjugated pair, lying on a unit circle, or two real eigenvalues with product equal to one. As a consequence, a periodic point (orbit) cannot have instabilities, caused by complex eigenvalues; no more than one instability, caused by a real eigenvalue, may occur. The periodic point is either elliptic or hyperbolic (moduli of both eigenvalues are simultaneously equal or not equal to one). Resonance overlap which produces global chaos and thermalization can only be nonlocal in a system with two degrees of freedom: at least two different unstable periodic solutions are needed to get resonance overlap [12].

In a system with many degrees of freedom all these limitations are removed. The first important new phenomenon is instability, caused by a pair of complex-conjugated eigenvalues. The behaviour of the system with a single instability of this kind is much more complicated than that for the case of a real eigenvalue (see Appendix B). The reason is that here the unstable manifold is a plane, and at each iteration of the map the perturbation vector is simultaneously stretched and rotated. In the case of resonance overlap this can provide the mixing mechanism, needed for thermalization.

Another new feature is, that in a system with many degrees of freedom, an unstable periodic point is not necessarily hyperbolic. This means that some of the eigenvalues can still have a modulus equal to one, and additional bifurcations are possible.

Perhaps most important is that for a system with many degrees of freedom local resonance overlap is possible. This means that a **single** periodic solution with two or more different instabilities can produce resonance overlap and thermalization (see § 4.3).

Another new and very important phenomenon is that instabilities with complex Floquet multipliers and resonant ratios 3/2 or 5/2 can lead to fast thermalization for sufficiently high steepnesses without any visible resonance overlap. This is a new mechanism of thermalization and it requires further investigation.

Appendix A: Truncation methods for the Hamiltonian approach

In our numerical investigations we have two main tasks: 1. Exploration of long-time evolution and 2. Search for periodic solutions and investigation of their properties. Each of these tasks has its own optimal truncation methods. The truncation determines the subset of modes, included in the truncated system, the order of nonlinearity, and the expected order of ε (for periodic solutions).

All numerical models are truncated ones. In most of the existing numerical models (namely in models working in the physical space or in both the physical and Fourier spaces) the truncation is mainly implicit. Although we usually know exactly the number of collocation points, there is no straightforward relation between the number of these points and the number of modes which can be reliably represented. For example, Mercer and Roberts [10] have found that in their model for standing Stokes waves with 8 mesh points per wavelength only the two first modes are reliably represented, and for the system with 64 points per wavelength only modes 1 through 16 can be considered well represented for all steepnesses, up to ak = 0.53. For higher steepnesses the number of well represented modes decreases. It must be mentioned that Mercer and Roberts have not used smoothing in their model. In other models, which use smoothing in order to suppress the high-mode instability, e.g. [3,4], the ratio between the number of mesh points and number of well represented modes can be significantly higher. In a comparison with a finite differences model, we found that 128 mesh points per wavelength represented reliably about 16 modes.

The various truncation methods introduced are the following:

A.1 Uniform truncation. This truncation method is intended for exploration of long-time evolution. For the desired group of modes all terms up to order n+1 are retained in the Hamiltonian, where n is the desired order of nonlinearity. Such a uniformly truncated model is often considered as "universal". However, the real precision of the model, i.e. its order in ε, depends on the initial conditions. For example, the 16-modes 5-th order model can be precise up to ε^5 if the energy is initially concentrated in the first few modes. On the other hand, if initially modes 1 and 9 are of order ε, the same model is precise only up to ε^1.

In [6] we have found that models of order less than 5 can produce fictitious phenomena. Investigating systems with many degrees of freedom we have found additional evidence of this. Thus we do not recommend using models of order less than 5.

Notes:

1.Due to the uniform nature of this truncation method its precision in ε is relatively low. It is thus not suitable for precise search for periodic solutions.

2.Uniformly truncated models do not provide any benefit in resolution. The main benefit is in their excellent numerical stability, which permits investigation of evolution for very long times (many thousands of periods). Other benefits are the ability to choose any subset of modes including high harmonics and to investigate spatial interactions.

A.2 ε-truncation. This truncation method is intended for search and investigation of periodic solutions. With this method, firstly the subset of modes is constructed and the expected order in ε is assigned to each mode. This is done using some heuristic approach. We will describe it here for the example of the standing Stokes wave. For the standing Stokes wave mode 1 is the leading mode and thus we assign to it order 1 in ε. It is also the only free mode. Mode 2 (the second harmonic) is locked and is induced by mode 1; the appropriate nonlinear term, considered as a forcing term, is of order ε^2; thus we assign to mode 2 order 2 in ε. Considering analogously other modes, we found that for mode n its expected order in ε is n. For three-dimensional periodic solutions and for two-dimensional periodic solutions with more than one free modes the consideration is very similar and the only difference is that for each order of ε we have a group of modes instead of a single mode.

For the ε-truncation, in the evolution equations are retained all terms up to order m in ε and order n of nonlinearity. Usually $m = n$ is a good choice. Of course, the evolution equations will contain also terms for which the order in ε is larger than m; these terms are needed in order to preserve the Hamiltonian structure of the problem.

Notes:

1.Due to its nature, the ε-truncated model can be used only to find the periodic solution or to investigate the behavior of the system in close vicinity of this periodic solution (for example, to check its linear stability).

2.The main benefit of ε-truncation is in its precision. At the price of the same computational effort as in 7th order 7 modes uniformly truncated model, using ε-truncated model we can get solution of 14th or 15th order in ε.

A.3 The balanced truncation. The well known idea of the ε-truncation is, that the leading modes are the most important ones, and for them a very large number of interactions must be taken into account; the high-order modes are less important, and only a few of the possible interactions between such modes are taken into account. For example, in the n-th order, ε-truncated model for the standing Stokes wave the evolution equations for mode 1 contain terms of orders 1 through 2n-1 in ε; the evolution equations for mode n contain terms only for order n in ε.

The high order modes, given by the ε-truncated model, are thus less reliable, not only due to truncation in the number of modes, but also because many important interactions between existing modes are lost. When the order of model increases the gap in reliability between leading modes and high-order modes increases as well.

This can lead to two kinds of artifacts: artifacts in modal amplitudes and artifacts in stability. In the distribution of modal amplitudes the amplitudes of the last few modes can become inconsistent. In the stability analysis starting from a certain value of ε a fictitious instability appears in which a group of last modes are involved. When ε grows this instability grows very rapidly, and the Floquet multipliers can easily reach 100 and more. Because the high-order modes are very small, we can expect, that in order to describe correctly the nonlinear interactions among them, it is enough to take into account only terms of few first orders in nonlinearity. For the standing Stokes wave, the ε-truncated model contain all interactions to second order in nonlinearity. However, due to the δ-filter, not all interactions are possible in this order. To balance the ε-truncated model we have added to it the missing terms of third order in nonlinearity. We have expected that the added terms will permit energy redistribution in order to correct the amplitudes of the last modes, as well as necessary detuning in order to remove or reduce the fictitious instability. Indeed, the computations show, that at the price of negligible additional computational effort we achieve a significant improvement: the modal amplitudes become self-consistent for much larger values of ε, and the fictitious instability was largely removed (in some cases it was removed totally). In the remaining instability, the maximum Floquet multiplier is about 2 (in our computations), and it grows slowly. This remaining instability, can be totally removed when adding terms of fourth order in nonlinearity; however, we do not advise to do this, because the additional computational effort is very large.

Resuming, we define the balanced truncation as a combination of uniform and ε-truncations, where the terms of orders 2 and 3 in nonlinearity, are taken from the third order uniformly truncated model, and the terms of orders 4 through n in nonlinearity from the n-th order ε-truncated model.

Notes:

1.Balanced truncation should be used only if the appropriate ε-truncation model generates solutions with the above mentioned artifacts.

2.For periodic solutions with large number of modes (such as the spatial standing wave or two-dimensional standing wave with two or more free modes), the additional computational effort associated with balanced truncation is too heavy, and for these solutions it is preferable to increase the order of ε-truncated model.

A.4 The combined truncation. This kind of truncation is useful when checking the linear stability of certain periodic solutions to perturbations in modes, which do not belong to the solution itself. For the unperturbed solution, the amplitudes of such modes are equal to zero.

For the perturbed solution, they are very small quantities. When the number of such "perturbation" modes is small, uniform truncation of 5-th order seems to be appropriate. For a large number of modes, however, the computational effort for a model of this kind is too large. This situation is typical, for example, for the case of spatial perturbations. The idea of the combined truncation is to combine the original Hamiltonian with which the periodic solution was otbained, with additional terms, which describe interactions with "perturbation" modes and among such modes. The most straightforward approach is simply to take the terms of orders 2 and 3 from the uniformly truncated third order model, which contains all modes, and the terms of higher orders - from the model used to get the original periodic solution. This method will work if the original model was uniformly truncated or balanced. This simple method is similar to the approach in the original third-order Zakharov equation. It has two defects: firstly, it takes into account all interactions of third order, including interactions between perturbations only. This results in significant computational effort. The second defect is, that interactions in the 4th and higher orders in nonlinearity between leading modes and perturbations are lost. As a result, part of the instabilities obtained by this method, are fictitious, whereas some true instabilities are lost. The precision of bounds of the instability regions is for part of the cases very bad.

To overcome these difficulties, we use combination of two ε-truncated models. Each of them contains all modes from the original periodic solution plus all perturbations. The first model is of order n in nonlinearity and order n in ε, where n is the order in ε of the original periodic solution. Orders in ε, assigned to the original modes, are the same as in the periodic solution; to all perturbations we assign the same order in ε, $[n/2]$, where square brackets denote the integer part.

The second model is of order 3 in nonlinearity and order 3n in ε. To the original modes we assign the same orders in ε as in the first model; to all of perturbations we assign order $[3n/2]$ in ε. The combined model consists of terms of orders 2 and 3 in nonlinearity, taken from the second model, and terms of 4-th and higher orders in nonlinearity, taken from the first model. This combined model is compatible with periodic solutions, obtained using balanced truncation.

From comparison between these approaches it turns out that the second approach is much faster and much more reliable.

Appendix B: Some important properties of complex eigenvalues and eigenvectors

Let us consider the behavior of dynamical system with m degrees of freedom in some near vicinity of periodic orbit with period T. If F is a time-shift operator, shifting over T, and x_0 is some point on the periodic orbit in the 2m dimensional phase space $P = R^{2m}$, then $F(x_0) = x_0$, and x_0 is a stationary point of F. For an arbitrary point y we denote the shifted value by y^+ : $y^+ = F(y)$, and value, shifted by n periods by y^{n+} : $y^{n+} = F(y^{(n-1)+})$. If x_p is a small perturbation, added to x_0, then its shifted values are given by:

$$x_p^+ = DFx_p = Ax_p \tag{B.1}$$

$$x_p^{n+} = A^n x_p \tag{B.2}$$

where DF = A is the Jacobian matrix for the operator F at the point x_0.

Eigenvalues and eigenvectors of A determine the stable and unstable manifolds. If some eigenvalue λ and the corresponding eigenvector x are real, the appropriate manifold M is one-dimensional, and the behavior in phase space is quite trivial:

(B.3) $$x^{+} = \lambda x; \qquad x^{n+} = \lambda^{n} x$$

This is because $x \in M$

For the complex case

(B.4) $$\left.\begin{aligned} \lambda &= \lambda_r + i\lambda_i = \beta e^{i\varphi} \\ x &= x_r + ix_i \end{aligned}\right\}$$

the behavior of perturbation in the phase space is more complicated. In this case the appropriate manifold M is a plain containing x_r and x_i; each vector $q \in M$ can be written as

(B.5) $$q = \alpha_r x_r + \alpha_i x_i = \alpha^T X$$

where

(B.6) $$X = \begin{bmatrix} x_r \\ x_i \end{bmatrix}, \qquad \alpha = \begin{bmatrix} \alpha_r \\ \alpha_i \end{bmatrix}, \qquad \alpha_r, \alpha_i \in R$$

For this case we have

(B.7) $$\left.\begin{aligned} x^{+} &= \lambda x \\ x_r^{+} &= \lambda_r x_r - \lambda_i x_i \\ x_i^{+} &= \lambda_r x_i + \lambda_i x_r \end{aligned}\right\}$$

(B.8) $$q^{+} = (\alpha_r \lambda_r + \alpha_i \lambda_i) x_r + (\alpha_i \lambda_r - \alpha_r \lambda_i) x_i$$

(B.9) $$\left.\begin{aligned} \alpha_r^{+} &= \alpha_r \lambda_r + \alpha_i \lambda_i \\ \alpha_i^{+} &= \alpha_i \lambda_r - \alpha_r \lambda_i \end{aligned}\right\}$$

Using (B.4) we can rewrite (B.9) as

(B.9') $$(\alpha^T)^{+} = \beta \alpha^T \begin{vmatrix} \cos\varphi & -\sin\varphi \\ \sin\varphi & \cos\varphi \end{vmatrix}$$

and

(B.10) $$(\alpha^T)^{n+} = \beta^n \alpha^T \begin{vmatrix} \cos n\varphi & -\sin n\varphi \\ \sin n\varphi & \cos n\varphi \end{vmatrix}$$

(B.11) $$q^{n+} = (\alpha^T)^{n+} X$$

The evolution of the vector α in the coefficients plane R^2 produces an exponential spiral: in polar coordinates r,θ

(B.12) $$\left.\begin{aligned} r^{n+} &= \beta^n r \\ \theta^{n+} &= (\theta - n\varphi)\mathrm{mod}2\pi \end{aligned}\right\}$$

However, the evolution of the perturbation q in the phase space P is more complicated; this is because in the general case vectors x_r and x_i cannot be made orthonormal.

If $x = x_r + ix_i$ is an arbitrary eigenvector corresponding to a complex eigenvalue λ, then all possible eigenvector corresponding to the same eigenvalue can be obtained as

$$x' = \mu x$$

where $\mu = se^{i\gamma}$ is a complex constant.

Choosing s properly, we can uniformly scale x_r' and x_i' (usually to get $|x_r'|^2 + |x_i'|^2 = 1$). Choosing γ properly, we can satisfy one additional condition. With some simple algebra one can show that, if we demand that $x_r' \cdot x_i' = 0$, γ must satisfy the condition

(B.13) $$\mathrm{ctg}\ 2\gamma = \frac{|x_r'|^2 - |x_i'|^2}{2x_r \cdot x_i}$$

Analogously, if we demand that $|x_r'| = |x_i'|$, γ must satisfy another condition

(B.14) $$(b/\mathrm{tg}\ 2\gamma = \frac{|x_r'|^2 - |x_i'|^2}{2x_r \cdot x_i}$$

These conditions are totally incompatible ($\mathrm{tg}\ \varepsilon + \mathrm{ctg}\ \varepsilon \neq 0$ for any ε).

It must be mentioned, that standard subroutines for eigenproblems from IMSL and NAG perform only uniform scaling. It is always desirable to get the complex eigenvectors in some standard (and unique) form. Orthogonalization according to (B.13) seems to be appropriate for this purpose.

Let us consider now the evolution of perturbation in R-I plane of mode j.

From (B.11) we have

(B.15) $$W_j^{n+} = (\alpha^T)^{n+} V_j,$$

where

(B.16)
$$W_j = \begin{bmatrix} q_j \\ q_{j+m} \end{bmatrix};\ V_j = \begin{bmatrix} y_j \\ z_j \end{bmatrix};\ y_j = \begin{bmatrix} x_{r_j} \\ x_{r_{j+m}} \end{bmatrix};\ z_j = \begin{bmatrix} x_{i_j} \\ x_{i_{j+m}} \end{bmatrix}$$
)

Note that even when $x_r \cdot x_i = 0$ usually $y_j \cdot z_j \neq 0$.

For $|W_j^{n+}|$ we have

(B.17)
$$|W_j^{n+}|^2 = \beta^{2n}\left\{\frac{1}{2}(\alpha_r^2+\alpha_i^2)(|y_i|^2+|z_j|^2)+\right.$$
$$+\cos 2n\varphi\left[2\alpha_r\alpha_i(y_j\cdot z_j)+\frac{1}{2}(\alpha_r^2-\alpha_i^2)(|y_i|^2-|z_j|^2)\right]+$$
$$+\sin 2n\varphi\left[(\alpha_i^2-\alpha_r^2)(y_j\cdot z_j)+\alpha_r\alpha_i(|y_j|^2-|z_j|^2)\right]$$

For $|q^{n+}|$ when $x_r \cdot x_i = 0$ and $|x_r|^2+|x_i|^2=1$ we can take $|x_r| = \cos v$, $|x_i| =$ $\sin v$, and then we have

(B.18)
$$|q^{n+}|^2 = \beta^{2n}\left\{\frac{1}{2}(\alpha_r^2+\alpha_i^2)+\left[\frac{1}{2}(\alpha_r^2-\alpha_i^2)\cos 2n\varphi+\alpha_r\alpha_i\sin 2n\varphi\right]\cos 2v\right\}$$

In the M-plane $\left(q^{n+}\in M \text{ for } n=0,1\ldots\right)$ the polar coordinates of q are r and θ, here we have:

(B.19)
$$\left.\begin{aligned} r^{n+} &= |q^{n+}| \\ \theta^{n+}-\theta &= \arccos\left(\frac{q^{n+}\cdot q}{|q^{n+}||q|}\right)\end{aligned}\right\}$$

where $|q^{n+}|$ is given by (B.18).

(B.20)
$$|q|^2 = \alpha_r^2\cos^2 v+\alpha_i^2\sin^2 v$$
$$q^{n+}\cdot q = \beta^n(|q|^2\cos n\varphi+\alpha_r\alpha_i\cos 2v\sin n\varphi)$$

For the simplest case $\alpha_i = 0$ we have:

$$\left.\begin{aligned} |q^{n+}|^2 &= (r^{n+})^2 = \beta^{2n}\frac{\alpha_r^2}{2}(1+\cos 2\nu\cos 2n\varphi) \\ \theta^{n+} &= \arccos\left(\frac{\sqrt{2}\cos\nu\cos n\varphi}{\sqrt{1+\cos 2\nu\cos 2n\varphi}}\right) \end{aligned}\right\} \tag{B.21}$$

References

1. Y. Agnon & M. Glozman, *Periodic Solutions of Complex Hamiltonian Systems: New Standing Water Waves,* accepted to Wave Motion (1995).
2. Bryant & Stiassnie, *Different forms for nonlinear standing waves in deep water*. J. Fluid Mech., **272**, (1994), 135-156.
3. W. Craig and C. Salem, *Numerical simulation of gravity waves*. J. Comp. Phys., **108**, (1993), 73-83.
4. D.G. Dommermuth and D.K.P. Yue, *A high order spectral method for the study of nonlinear gravity waves*. J. Fluid Mech., **184**, (1987), 267-288.
5. E. Fermi, J.R. Pasta and S. Ulam, *Studies of nonlinear problems*. In: Collected Works of Enrico Fermi, **2**, 978. University of Chicago Press, Chicago, 1965.
6. M. Glozman, Y. Agnon and M. Stiassnie, *High-order formulation of the water wave problem. Physica D., Nonlinear Phenomena*, **66**, (1993), 347-367.
7. M. Glozman , *Energy Transfer in Water Waves*,Ph. D. Thesis, Technion - Israel Institute of Technology, 1994, 217p.
8. H. Goldstein, *Classical Mechanics*, 2nd Edition, Addison Wesley, Reading, Massachusetts 1980.
9. A.J. Lichtenberg and M.A. Lieberman, *Regular and stochastic motion*. 2nd Edition. Springer-Verlag, New York, 1992.
10. G.N. Mercer and A.J. Roberts, *Standing waves in deep water: Their stability and extreme form*. Phys. Fluids A **4** (2), (1992), 259-269.
11. M. Okamura, *Instabilities of weakly nonlinear standing gravity waves*. J. Phys. Soc. Jpn, **53**, (1984), 3788-3796.
12. G. Walker and J. Ford, *Amplitude instability and ergodic behavior for conservative nonlinear oscillator systems*. Phys. Rev. **188**, (1969), 416-432.
13. V.E. Zakharov, *Stability of Periodic Waves of Finite Amplitude on the Surface of Deep Fluid*, J. Appl. Mech. Tech. Phys. **2** (1968), 190-194.

Coastal and Marine Engineering Research Institute (CAMERI), Technion, Haifa, 32000

E-mail address: **michael@cameri2.technion.ac.il**

Department of Civil Engineering, Technion, Haifa, 32000, Israel.

E-mail address: **agnon@tx.technion.ac.il.**

Contemporary Mathematics
Volume **200**, 1996

Solitary Wave Interactions for the Fifth-Order KdV Equation

B. Malomed and J.-M. Vanden-Broeck

ABSTRACT. The interaction of solitary waves for the fifth order KdV equation is studied numerically. It is shown that after interaction some of the solitary waves regain essentially their shape while others split into several solitary waves of different types.

1. Introduction.

Solitary waves with decaying oscillatory tails have received a lot of attention in recent years (see for example [1]-[9] and also [10] for a review and more references). They occur for example at the free surface of a fluid when both gravity and surface tension are taken into account. Such waves were calculated numerically by Vanden-Broeck and Dias [5] for the fully nonlinear water wave equations in water on infinite depth. Their existence was shown analytically in water of finite depth by Iooss and Kirchgassner [6]. In particular it was found that when the Bond number

$$\tau = \frac{\sigma}{\rho g H^2} \tag{1}$$

is close to $\frac{1}{3}$, the shape $Y = \eta(X,T)$ of the free surface satisfies approximately the fifth order KdV equation

$$\eta_T + \frac{3}{2}\eta\eta_X + \frac{1}{2}(\frac{1}{3} - \tau)\eta_{XXX} + \frac{1}{90}\eta_{XXXXX} = 0 \tag{2}$$

Here σ is the surface tension, g the acceleration of gravity, ρ the density of the fluid and H the depth of the fluid. In (2) the unit of length is H and the unit of time is $(H/g)^{1/2}$.

For waves travelling at a constant velocity C to the right, (2) reduces to the ordinary differential equation

$$-C\eta' + \frac{3}{2}\eta\eta' + \frac{1}{2}(\frac{1}{3} - \tau)\eta''' + \frac{1}{90}\eta''''' = 0 \tag{3}$$

1991 *Mathematics Subject Classification.* Primary 76B25, 76B45.

The work of the second author was supported in part by the National Science Foundation Grant #DMS-9500956, as well as the Lady Davis Foundation. He gratefully acknowledges the hospitality of the Hebrew University of Jerusalem, Tel Aviv Univerisity and the Technion.

where $\zeta = X - CT$ and the primes denote derivatives with respect to ζ.

Equation (3) was derived by Hunter and Vanden-Broeck [2] by a formal perturbation expansion for waves of small amplitude. The corresponding derivation for the time dependent equation (2) was done by Hunter and Scheurle [4], who proved the existence of travelling-wave solution for this equation.

In this paper, we study the interaction of solitary waves with oscillatory tails for the equation (3). Following Hunter and Scheurle [4], we first introduce the scaling

$$u(x,t) = \frac{3}{2}\eta(X,T) \tag{4}$$

$$x = (\frac{6}{1-3\tau})^{1/2} X \tag{5}$$

$$t = (\frac{6}{1-3\tau})^{1/2} T \tag{6}$$

Substituting (4)-(6) into (2), we obtain

$$u_t + uu_x + u_{xxx} + \epsilon^2 u_{xxxxx} = 0 \tag{7}$$

where

$$\epsilon^2 = \frac{2}{5(1-3\tau)^2} \tag{8}$$

We stress that ϵ will not be treated as a small parameter. On the contrary, it must be large enough to provide for the existence of solitary waves with oscillatory tails. All the numerical calculations presented in this paper were done with $\epsilon^2 = 40$.

Similarly (3) becomes after scaling

$$-cu' + uu' + u''' + \epsilon^2 u''''' = 0 \tag{9}$$

where

$$c = C(\frac{6}{1-3\tau})^{1/2} \tag{10}$$

and the primes denote derivatives with respect to $\nu = x - ct$.

The numerical procedure to solve (7) and (9) are described in Section 2. These schemes are used in Section 3 to study the interaction of solitary waves. One difficulty in the problem is that there are an infinite number of solitary waves with decaying oscillatory tails which satisfy (9) (see Zufiria [7] and Champneys and Toland [8]). Therefore we do not attempt an exhaustive study of the interaction of solitary waves with decaying oscillatory tails. Instead we present a few numerical examples of interactions. Our results exhibit two different kinds of interaction. In the first, each of the solitary waves regain essentially its identity after interaction. In the second, one of the solitary wave regains essentially its shape while the other splits into several solitary waves of different types.

2. Numerical procedures

We solve the equation (7) numerically. We use a fourth order Runge-Kutta time integrating procedure to march in time.

To achieve high accuracy, we approximate the derivatives with respect to x by high order finite difference formulas. First we introduce the mesh points

$$x_I = -\frac{(N-1)e}{12} + (I-1)e, \qquad I = 1,\ldots,N \tag{11}$$

and the corresponding unknowns

$$u_I = u(x_I, t). \tag{12}$$

Here e is the interval of discretization.

We approximate the first, third and fifth derivatives in (7) by the seven point formulas

$$u'(x_I,t) = \frac{-2/5!u_{I-3} + 18/5!u_{I-2} - 3/4u_{I-1} + 3/4u_{I+1} - 18/5!u_{I+2} + 2/5!u_{I+3}}{e} \tag{13}$$

$$u'''(x_I,t) = \frac{3/4!u_{I-3} - u_{I-2} + 13/8u_{I-1} - 13/8u_{I+1} + u_{I+2} - 3/4!u_{I+3}}{e^3} \tag{14}$$

$$u'''''(x_I,t) = \frac{-1/2u_{I-3} + 2u_{I-2} - 5/2u_{i-1} + 5/2u_{I+1} - 2u_{I+2} + 1/2u_{I+3}}{e^5} \tag{15}$$

For the end points, we assume that $u_I = 0$ for $I < 1$ and $I > N$. Since our scheme is explicit, we need to choose the time step Δ sufficiently small to insure numerical stability (the stability condition is $\Delta < ce^5$, where c is a constant). Most of the calculations were performed with $e = 3.1$ and $\Delta = 2.17$.

Next we describe the procedure to compute the solitary waves with decaying oscillatory tails. This is done by solving the ordinary differential equation (9) numerically. We first introduce the mesh points

$$\nu_I = -\frac{(M-1)E}{2} + (I-1)E, \quad I = 1, \ldots, M \tag{16}$$

and the corresponding unknown $\eta^I = \eta(\nu_I)$, where E is the interval of discretization. As before we approximate the first, third and fifth derivatives in (9) by seven point finite difference formulas. Substituting these expressions in (9) evaluated at the mesh points (16), yield a system of M nonlinear equations for the $M+1$ unknowns η^I, $I = 1, \ldots, M$ and c. The last equation obtained by fixing the value of η at $\nu = 0$. Thus we write

$$\eta^{\frac{N+1}{2}} = A \tag{17}$$

where A is a prescribed amplitude parameter. This system of $M+1$ equations with $M+1$ unknowns is solved by Newton's method.

3. Discussion of the results

We first computed four different types of solitary waves by using the Newton's iterations scheme described at the end of Section 2. In the following discussion it is convenient to describe these four types as waves of type 1, 2 and 3 and 4.

Two waves of type 1 and of type 2 are shown in Figures 1a and 2a respectively. A wave of type 3 is shown on the left of Figure 4a (the wave on the right of Figure 4a is a wave of type 1). A wave of type 4 is shown on the right of Figure 6a. For these calculations $E = 3.1$ and $N = 131$ were used. We checked by increasing N and decreasing E that the profiles shown are correct to at least graphical accuracy. For example profiles calculated with $E = 1.55$ and $N = 262$ agree with those calculated with $E = 3.1$ and $N = 131$ within 0.6% (the formula (15) implies that our scheme is second order accurate).

All these profiles have oscillatory decaying tails. In most of them the amplitude of the decaying tail is too small to be seen on the graph but they appear on an expanded scale.

Next we use the time marching procedure described at the beginning of Section 2. We first check the scheme by following in time each of the four types of waves. As expected we found that the waves move in space without change of shape. This is illustrated in Figure 7 for a wave of type 4.

We now study intreactions between solitary waves of types 1, 2, 3 and 4. As mentioned in the introduction, we observed two different kinds of interactions.

The first kind of interaction is illustrated in Figures 1, 2 and 3. The initial conditions are shown in Figures 1a, 2a and 3a. They consists of two waves of type 1 (Figure 1a), two waves of type 2 (Figure 2a) and one wave of type 1 and one wave of type 2 (Figure 3a). Resulting profiles obtained after marching in time are shown in Figures 1b, 2b and 3b. In these calculations, we used $e = 3.1$, $\Delta = 2.17$ and $N = 4200$. We repeated the calculations with $\Delta = 1.085$ and twice the number of time steps and obtained essentially the same results. The range of x extends from $x = 1000$ to $x = 14000$. In all the Figures only a small interval of x is shown, so that the profiles appear clearly. Outside these intervals the values of u are essentially zero. Comparing Figures 1b, 2b and 3b with Figures 1a, 2a and 3a, we see that the waves have moved to the left. Since the wave on the right of Figures 1a, 2a and 3a moves faster than the one on the left, the waves interact. The profiles shown in Figures 1b, 2b and 3b correspond to a long time after the interactions take place (typically 60000 time steps were used). Figures 1 and 3 show that the waves of type 1 have essentially recovered their shape after the interaction. In Figures 2 and 3, the waves of type 2 have also regained their identity but they are distorted. Furthermore there is an oscillatory tail on the right.

The second kind of interactions is illustrated in Figures 4-6. The initial conditions are shown in Figures 4a-6a. They consist of one wave of type 3 and one wave of type 2 (Figures 4a), one wave of type 3 and one wave of type 1 (Figure 5a) and one wave of type 3 and one wave of type 4 (Figure 6a). The profiles obtained a long time after the interaction are shown in Figures 4b-6b. In Figure 4b, the wave of type 2 has regained essentially its shape. However the wave of type 3 has split into 3 waves of type 1. In Figure 5b, the wave of type 1 has regained its shape while the wave of type 3 has split into a wave of type 2 and a wave of type 1. Similarly in Figure 6b, the wave of type 4 has regained its shape while the wave of type 3 has split into three waves of type 1.

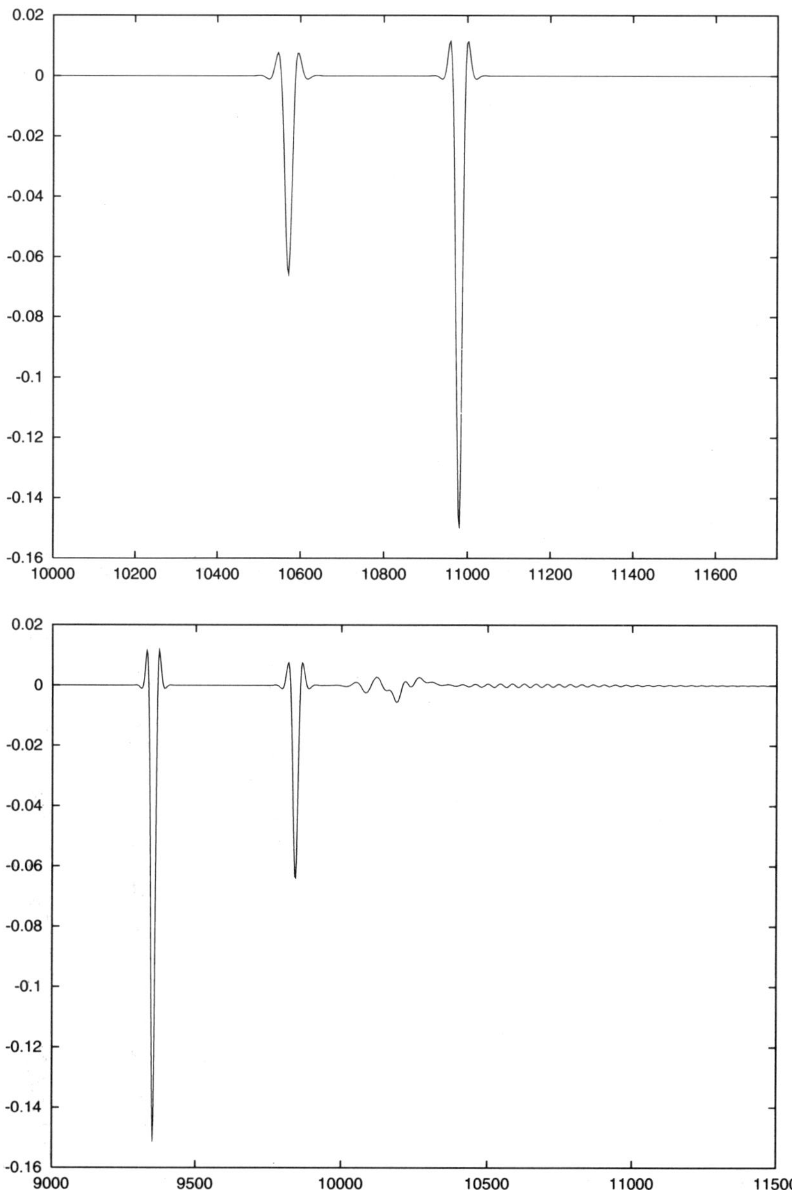

FIGURE 1. Interaction of two solitary waves of type 1. Figure 1a (top) shows the initial configuration. The resulting solution a long time after the interaction is shown in Figure 1b (bottom).

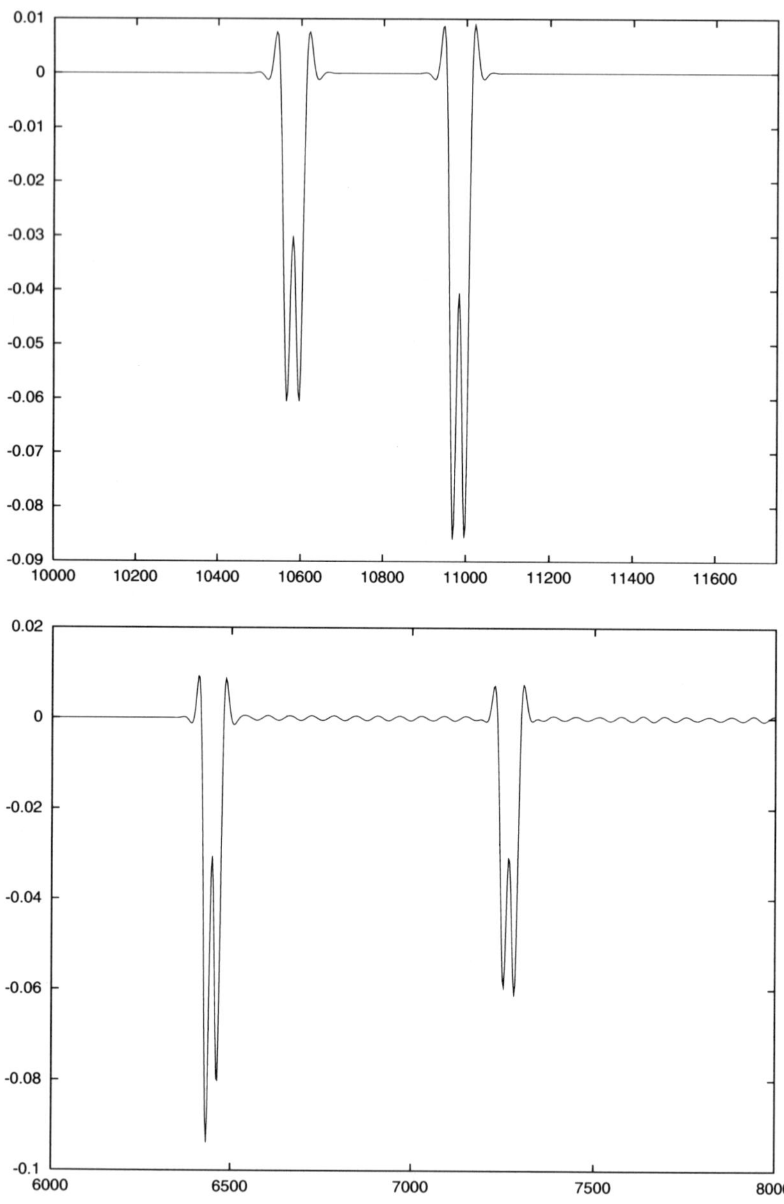

FIGURE 2. Same as Figure 1 with two solitary waves of type 2.

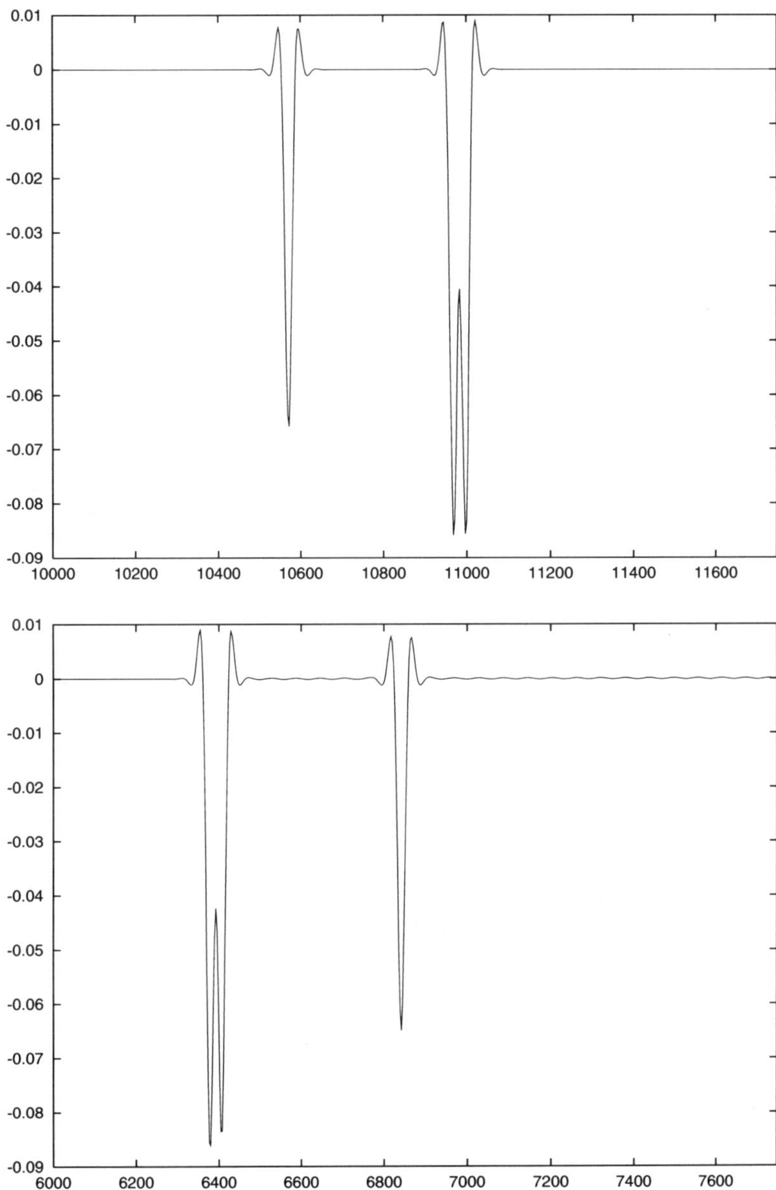

FIGURE 3. Same as Figure 1 with one solitary wave of type 1 and one solitary wave of type 2.

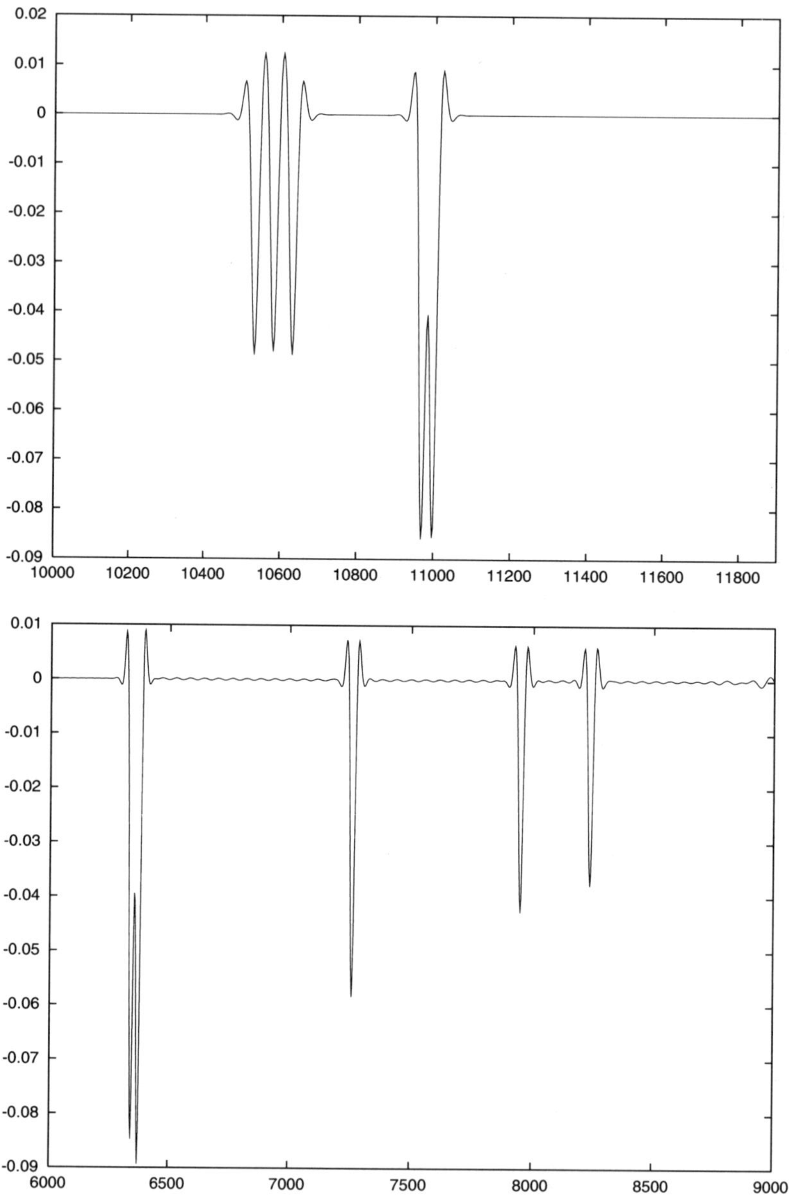

FIGURE 4. Same as Figure 1 with one solitary wave of type 3 and one solitary wave of type 2.

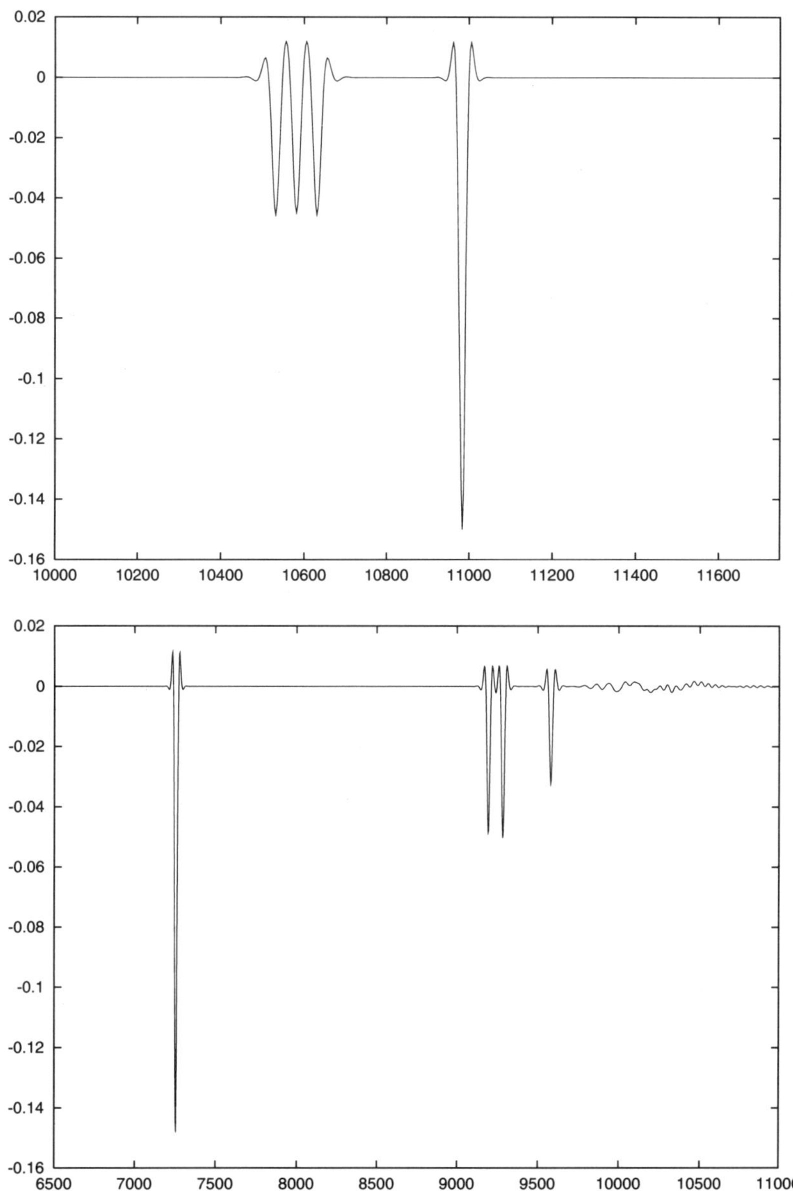

FIGURE 5. Same as Figure 1 with one solitary wave of type 3 and one solitary wave of type 1.

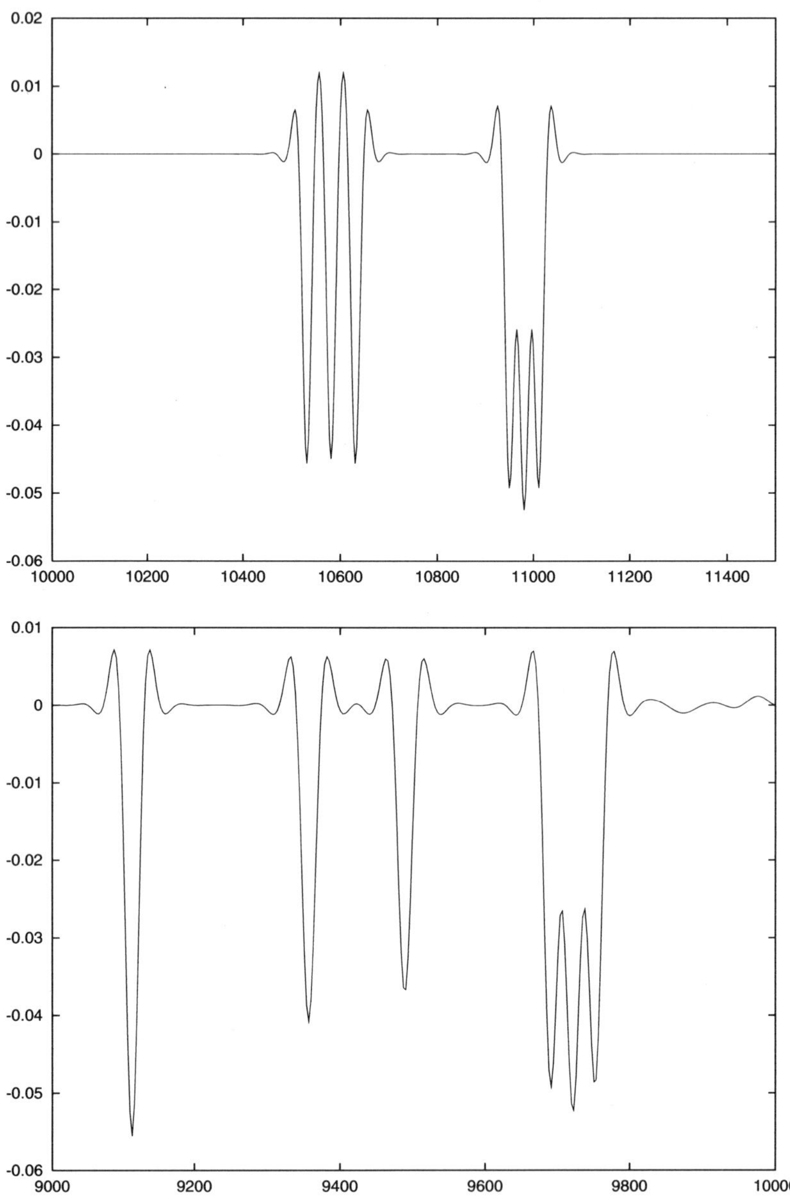

FIGURE 6. Same as Figure 1 with one solitary wave of type 3 and one solitary wave of type 4.

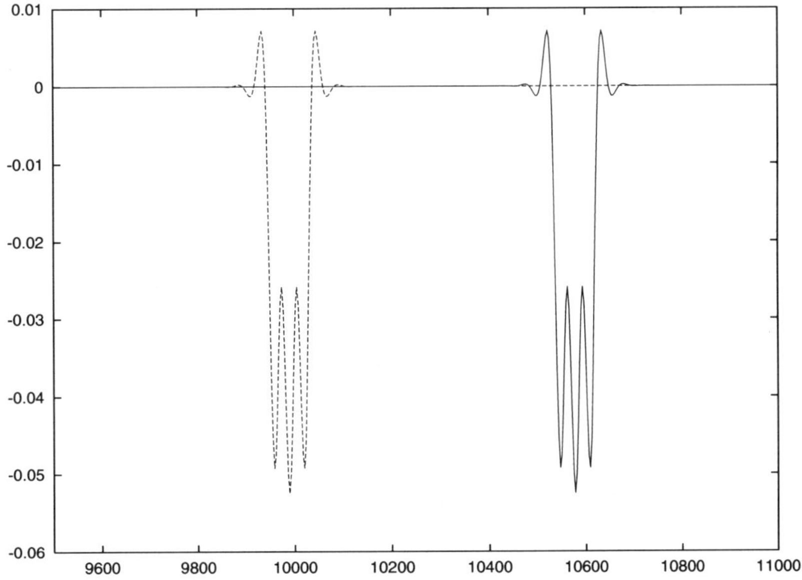

FIGURE 7. The time evolution of a solitary wave of type 4. The solid curve is the initial profile and the broken curve the computed solution after 13000 time steps.

References

[1] T. Kawahara, *Oscillatory solitary waves in dispersive media*, J. Phys. Soc. Jpn **33** (1972), 260–264.

[2] J. K. Hunter and J.-M. Vanden-Broeck, *Solitary and periodic gravity-capillary waves of finite amplitude*, J. Fluid Mech. **134** (1983), 205–219.

[3] Y. Pomeau, A. Ramani and B. Grammaticos, *Structural stability of the Korteweg-de Vries solitons under perturbation*, Physica D **31** (1988), 127–134.

[4] J. K. Hunter and J. Scheurle, *Existence of perturbed solitary wave solutions to a model equation for water waves*, Physica D **32** (1988), 253–268.

[5] J.-M. Vanden-Broeck and F. Dias, *Gravity-capillary solitary waves in water of infinte depth and related free surface flows*, J. Fluid Mech. **240** (1992), 549–557.

[6] G. Iooss and K. Kirchgassner, *furcation d'ondes solitaires en presence d'une faible tension superficielle*, C. R. Acad. Sci. Paris **311 I** (1990), 265–268.

[7] J. A. Zufiria, *Symmetry breaking in periodic ans solitary gravity-capillary waves on water of finite depth*, J. Fluid Mech. **184** (1987), 183–206.

[8] A. R. Champneys and J. F. Toland, *Bifurcation and coalescence of a plethora of large amplitude homoclinic orbits for Hamiltonian systems*, Nonlinearity **6** (1993), 665–721.

[9] F. Dias, D. Menasce and J.-M. Vanden-Broeck, *Numerical study of capillary-gravity solitary waves*, Europ. J. Mech. B/Fluids **15** (1996), 17–36.

[10] G. Iooss, *Capillary-gravity water-waves as a dynamical system*, Proceedings, IUTAM/SIMM symposium on structure and dynamics of nonlinear waves in fluids, World Scientific, 1995, pp. 42–57.

DEPARTMENT OF APPLIED MATHEMATICS, SCHOOL OF MATHEMATICAL SCIENCES, TEL AVIV UNIVERSITY, RAMAT AVIV 69978, ISRAEL

E-mail address: malomed@math.tau.ac.il

DEPARTMENT OF MATHEMATICS AND CENTER FOR THE MATHEMATICAL SCIENCES, UNIVERSITY OF WISCONSIN-MADISON, WI 53706, U.S.A.

E-mail address: vandenbr@math.wisc.edu

Contemporary Mathematics
Volume **200**, 1996

Forced Benjamin-Ono Equations and Related Topics

Yoshimasa Matsuno

Abstract. A review is made on the forced Benjamin-Ono (fBO) equations and related topics while centering on my recent works. We first introduce the fBO equation which models the propagation of long interfacial waves in a simple two-layer fluid system with uneven bottom topography. We then elucidate the effect of small topographic forcing on the dynamics of an algebraic soliton which is particular solution to the BO equation. For the purpose, a singular perturbation method using the multiple time scale expansion is employed to derive the evolution equations that describe the change of the amplitude and position of the soliton. It is shown that the system of equations for these soliton parameters becomes to be completely integrable for a simple bottom profile. The dynamics of the soliton are found to exhibit a variety of phenomena, such as the capture and repulsion of the soliton by topography and the occurrence of solitonlike phase shift. Lastly, we discuss some open problems concerning the fBO and related equations.

1. Introduction

The forced soliton equations have been paid much attention in recent years owing to their practical importance. In the context of the water wave problems, the forced Korteweg-de Vries (fKdV) equation has been derived as a typical model equation where the forcing terms are induced by moving pressure distribution [1-4] and/or uniform flow over a localized bottom topography [5-7]. While various types of forcings appear in accordance with problems under consideration, the forcing terms usually break down the complete integrability of the soliton equations which is characterized by the properties such as multisoliton solutions, an infinite number of conservation laws and Bäcklund transformation. Therefore, one must recourse to approximate methods and/or numerical analysis to solve their initial value problems. Along this line, a number of works have been made for the study of the forced soliton equations [8-11]. If the forcing terms can be regarded as small perturbations to the soliton equations, we can apply asymptotic methods for which several different approaches are known at present. It should be emphasized, however that the

---

1991 *Mathematics Subject Classification.* Primary 35Q51 45K05; Secondary 76C10.

mathematical analysis of the forced soliton equations has not been fully established as far as the properties of solutions such as the existence, uniqueness and stability are concerned.

The purpose of this paper is to review the forced soliton equations and related topics while centering on my recent works [12-19]. In particular, we shall focus our attention on the forced Benjamin-Ono (fBO) equation. In the following discussion, we shall consider the following fBO equation [19] describing the propagation of long interfacial waves in a two-layer fluid system with uneven bottom topography where the upstream condition is imposed such that the constant flow is in the same direction in both layers. It can be written in an appropriate dimensionless form as

$$\eta_t + (F-1)\eta_x - \frac{3\alpha}{2}\eta\eta_x - \frac{\Delta\delta}{2}H\eta_{xx} = \frac{\gamma\alpha F^2}{2}B_x, \tag{1.1a}$$

with the operator H being the Hilbert transform defined by

$$H\eta(x,t) = \frac{1}{\pi}P\int_{-\infty}^{\infty}\frac{\eta(y,t)}{y-x}dy. \tag{1.1b}$$

Here, $\eta = \eta(x,t)$ is the interfacial elevation, F is the Froude number, γB characterizes the unevenness of bottom topography and $\Delta(<1)$ is the density ratio of both layers. The small parameters α and δ measure the magnitude of nonlinearity and dispersion, respectively and they are balanced according to the relation $\delta = O(\alpha)$. The fBO equation (1.1) has been derived by employing a novel asymptotic perturbation theory [12, 14] developed recently. The right-hand side of Eq. (1.1) represents the effect of bottom topography and it acts as an external force. This is a reason why it is called the fBO equation. Particular concern here is the situation when the flow is nearly critical, namely the Froude number F is close to unity. Furthermore, we assume that the topographic effect is small compared with nonlinear and dispersive effects. This enables us to perform a perturbation analysis.

The fBO equation has been derived in various physical contexts as asymptotic model equation. For instance, it describes wave disturbances due to a ship in a shallow thermocline [20], coastally trapped waves in a continuously stratified ocean of great depth [21] and resonantly generated internal gravity waves [22]. While Eq. (1.1) incorporates the lowest-order nonlinearity and dispersion and hence its applicability is restricted to the description of small amplitude waves, the higher-order effects are easily introduced. In this respect, we note that the higher-order BO equations have been obtained in a stratified fluid system [23] and in a two-layer fluid system [16] where the bottom of fluid is assumed to be flat in both cases. Within the framework of the perturbation analysis, higher-order nonlinear and dispersive terms are regarded as small corrections to the BO equation. A higher-order BO equation derived in [16] has been used to elucidate the effect of higher-order terms on the interaction process of two algebraic solitary waves of the BO equation [17, 18].

In Sec. 2, we shall briefly summarize the multisoliton perturbation theory for the BO equation [17, 18]. Then, it is applied to Eq. (1.1) to study the dynamics of an algebraic soliton under small topographic forcing. In Sec. 3, some open problems are discussed concerning the fBO and related equations.

2. Soliton dynamics

2.1. Multisoliton perturbation theory. Even if the forcing terms are regarded as small perturbations to the BO equation, the complete integrability of the equation would be lost due to forcing. Hence, we must employ approximate methods of solution to analyze the fBO equations. Since we are particularly interested in the effect of forcing on the dynamics of algebraic solitons, we find it appropriate to use the soliton perturbation theory. Several different approaches are known at present to the analytical description of soliton dynamics. These are a method based on the inverse scattering transform (IST) method [24], a direct method using multiple time scale expansion [25], a mixture of the above two methods [26], a technique using variational principle [27], and a generalized perturbation method [28]. Within the framework of these methods, the dynamics of solitons have been treated satisfactorily for a wide class of soliton equations. Nevertheless, there exists a class of perturbed soliton equations which prevent us the applications of the methods mentioned above. Typical examples are the nonlocal nonlinear evolution equations such as the fBO and forced intermediate long wave equations which are characterized by the dispersive terms expressed in terms of definite integrals. The mathematical structure of these equations is found to be quite different from that of the local equations such as the fKdV and forced nonlinear Schrödinger equations [29, 30]. Indeed, due particularly to the presence of the nonlocal operators, the above equations share many properties common to the two-dimensional nonlinear evolution equations such as the forced Kadomtsev-Petviashvili equation.

Here, we shall briefly summarize a direct multisoliton perturbation theory [17, 18] which is applicable to a wide class of nonlinear integrodifferential evolution equations. In doing so, we shall exemplify it in the case of the fBO equation only for the sake of simplicity.

First, we write the fBO equation in a standard form

$$u_t + 4uu_x + Hu_{xx} = \epsilon R[u], \ u = u(x,t). \tag{2.1}$$

Here $\epsilon R[u]$ represents the perturbation and ϵ is a small positive parameter that measures the magnitude of the perturbation.

Among various techniques for analyzing singular perturbation problems, we shall use the method of multiple scales [31, 32]. It has been applied to a variety of problems in physics, engineering, and applied mathematics. The essence of the method is to obtain a uniformly valid expansion by developing the derivatives as well as the dependent variables in powers of the small parameter.

Following its spirit, we first introduce the different time scales t_j by $t_j = \epsilon^j t$, $(j = 0, 1, 2, ...)$ and expand u into an asymptotic series of the form $u = \sum_{j=0}^{\infty} \epsilon^j u_j$, $u_j = u_j(x, t_0, t_1, ...)$ where the dependent variables u_j have been assumed to be functions of the multiple times t_j as well as the spatial coordinate x. The time derivative is then transformed according to $\partial/\partial t = \sum_{j=0}^{\infty} \epsilon^j \partial/\partial t_j$. Substituting these expressions into (2.1) and equating coefficients of like powers of ϵ, we obtain the system of equations for u_j, the first two members of which read as follows:

$$u_{0,t_0} + 4u_0 u_{0,x} + Hu_{0,xx} = 0, \tag{2.2}$$

$$u_{1,t_0} + 4(u_0 u_1)_x + Hu_{1,xx} = R[u_0] - u_{0,t_1}. \tag{2.3}$$

Equation (2.2) is just the BO equation while u_j with $j \geq 1$ satisfy the linear inhomogeneous equations.

In order to solve the above system of equations, we must first specify the solution of (2.2). Since we are interested in the dynamics of solitons, we take for it the N-soliton solution [29, 30, 33] which includes the N amplitude parameters a_j and the N position parameters ξ_j $(j = 1, 2, ..., N)$. In the absence of the perturbation, a_j is constant independent of t_0 and $\xi_j = a_j t + \xi_{j0}$ where ξ_{j0} is the initial position of the jth soliton. It is important to note that for large time, u_0 can be represented by a superposition of N algebraic solitons as [33]

$$u_0 \sim \sum_{j=1}^{N} \frac{a_j}{a_j^2(x-\xi_j)^2+1}, \quad (t_0 \to \pm\infty). \tag{2.4}$$

The above expression shows that the BO solitons exhibit no phase shift after collision between them. This is a remarkable characteristic that has never been observed in collision process of solitons expressed in terms of exponential functions. The detail description of the interaction process of the BO solitons has been given in [29, 30, 33]. Now, due to the action of the small perturbation, the soliton parameters would be modulated slowly on the time scale of order ϵ^{-1}. Hence, it is reasonable to assume that $a_j = a_j(t_1, t_2, ...)$, $\xi_{j0} = \xi_{j0}(t_1, t_2, ...)$, $(j = 1, 2, ..., N)$. Thus, the problem under consideration reduces to determining the time evolution of these parameters.

With the N-soliton solution of Eq. (2.2), we are ready to solve Eq. (2.3). Though it is a linear equation, the nonlocal term expressed by the Hilbert transform makes the analytical treatment more difficult. The method based on IST has not been applied as yet and a direct method using Green's function has worked well only for u_0 with one-soliton solution. Nevertheless, as far as the leading-order analysis is concerned, we can obtain the equations that determine the time evolution of the soliton parameters by requiring that the correction term u_1 is no more singular than the leading-order solution u_0. This condition turns out to be equivalent to the elimination of secular terms which would appear provided that a straightforward expansion is carried out. Thus, we can obtain the evolution equations for the soliton parameters. The detail of their derivation is referred to the original papers [17, 18]. We quote only the final result as follows:

$$\frac{da_j}{dt} = -\frac{4\epsilon}{\pi}(g_{j+N}, R[u_0]), \quad \frac{d\xi_j}{dt} = a_j + \frac{4\epsilon}{\pi}(g_j, R[u_0]), \quad (j = 1, 2, ..., N). \tag{2.5a, b}$$

Here, g_j and g_{j+N} are given by

$$g_j = \int_{-\infty}^{x} \frac{\partial u_0}{\partial a_j} dx, \quad g_{j+N} = \int_{-\infty}^{x} \frac{\partial u_0}{\partial \xi_j} dx, \quad (j = 1, 2, ..., N), \tag{2.6a, b}$$

and the round brackets on the right-hand side of Eqs. (2.5) denote the inner product.

The solutions of the above equations describe the slow changes of the amplitude and the position of each soliton which are induced by the perturbation. It should be emphasized that for the purpose of calculating the inner products in (2.5), we only need the information of the N-soliton solution which can be usually obtained without recourse to IST. The leading-order solution u_0 is valid uniformly over the long-time interval of order ϵ^{-1}. Beyond this interval, however, one must take account of the higher-order modulation effects, namely it is necessary to introduce other time scales $t_2, t_3, ...$ to keep the expansion uniformly valid. The present analysis disregards the emission of radiation as well as the distortion of the shape of solitons due to perturbation. These effects can be elucidated by extending the analysis to a higher-order approximation, i.e. by solving Eq. (2.3) with u_0 being the N-soliton solution of Eq. (2.2). However, at present, we have no analytical means to resolve the problem. This is an interesting subject to be left for a future work.

Lastly, we remark that for the one-soliton case, the similar perturbation analysis has been performed by several authors [29, 34, 35].

2.2. Effect of localized topography. While the fBO equation (1.1) has wide applications, we shall investigate the effect of forcing on the dynamics of algebraic solitons which are particular solutions to the BO equation. In particular, we perform an analytical study of the interaction process of an algebraic soliton with localized bottom topography. By employing the soliton perturbation theory, we derive a system of ordinary differential equations describing the change of the amplitude and position of a soliton for a simple bottom profile. It is found that the solutions of these equations exhibit a variety of phenomena such as the capture and repulsion of the soliton by topography and the occurrence of soliton-like phase shifts due to the interaction of the soliton with topography [19]. These predictions are quite interesting because of the possibility of their appearance in lower atmospheric systems, for instance where a mountain may be modeled by a localized topography.

2.2.1. *Basic equation.* First, we suppose that the flow is nearly critical, namely the upstream uniform velocity is close to the phase velocity of linear shallow water wave. Mathematically, it is represented by the relation

$$F = 1 + \alpha\Gamma, \tag{2.7}$$

where Γ is a parameter which measures the departure of the Froude number from unity. In addition, we assume that the effect of topography is weak in comparison with other terms so that the inequality $\alpha << \gamma << 1$ holds. Under these conditions, the second to fourth terms on the left-hand side of Eq. (1.1) enter at the same order whereas the forcing term becomes higher order and hence it can be treated as a small perturbation to the BO equation. Furthermore, we rescale the variables according to $t \to (\Delta\delta/\alpha^2)t, x \to (\delta/\alpha)x, \eta \to (8/3)u$ and $B \to (16/3F^2)B$ in which t is transformed into a slow time in view of the ordering $\delta = O(\alpha)$. Then, Eq. (1.1) takes the following form

$$u_t + \Gamma u_x - 4uu_x - Hu_{xx} = \gamma B_x, \ u = u(x,t). \tag{2.8}$$

Note that this equation has a Hamiltonian structure

$$u_t = -\frac{\partial}{\partial x}\frac{\delta H}{\delta u}, \tag{2.9a}$$

with the Hamiltonian H given by

$$H = \int_{-\infty}^{\infty} \left[\frac{1}{2}\Gamma u^2 - \frac{2}{3}u^3 - \frac{1}{2}uHu_x - \gamma Bu\right] dx, \tag{2.9b}$$

where $\delta H/\delta u$ denotes the Frechet derivative of H. Since the Hamiltonian H does not depend on time explicitly, it is a constant of motion. We also remark that the mass $M \equiv \int_{-\infty}^{\infty} u dx$ is conserved if the function $B(x)$ is localized, i.e. $B(\pm\infty) = 0$.

To proceed the analysis further, we must specify the profile of bottom topography. While it can be chosen to be arbitrary provided that it is sufficiently smooth and localized, we consider the following simple form

$$B(x) = \frac{\lambda b}{(bx)^2 + 1}, \quad (\lambda = \pm 1), \tag{2.10}$$

where b^{-1} is a typical length scale of topography and λ characterizes the nature of forcing. In the limit of $b \to \infty$, $B(x)$ approaches a delta function and this case may be called a narrow forcing. The other limit $b \to 0$ yields a broad forcing where a soliton behaves like a delta function. An advantage of introducing rather specific profile of topography is that the resulting motion of a soliton is described by a completely integrable Hamiltonian system as will be shown below.

2.2.2. *Perturbation equations for soliton parameters.* In the problem under consideration, we take the following one-soliton solution as a lowest-order solution of Eq. (2.2):

$$u_0 = \frac{a}{a^2(x-\xi)^2 + 1}. \tag{2.11}$$

Here, a and ξ are the amplitude and position of the soliton, respectively. The topographic forcing induces slow changes of these parameters. If we use (2.5) and (2.6), we find that a and ξ evolve according to the equations

$$\frac{da}{dt} = -\frac{4\gamma}{\pi}(g_2, B_x), \quad \frac{d\xi}{dt} = \Gamma - a + \frac{4\gamma}{\pi}(g_1, B_x), \tag{2.12a, b}$$

where

(2.13a, b)

$$g_1 = \int_{-\infty}^{x} \frac{\partial u_0}{\partial a} dx = \frac{x-\xi}{a^2(x-\xi)^2+1}, \quad g_2 = \int_{-\infty}^{x} \frac{\partial u_0}{\partial \xi} dx = -\frac{a}{a^2(x-\xi)^2+1}.$$

It is pointed out that the amplitude equation (2.12a) coincides with the equation derived directly from the energy balance equation $\partial(u,u)/\partial t = 2\gamma(u, B_x)$ with u given by (2.11).

Performing the inner products in (2.12), we finally obtain the following system of ordinary differential equations for the parameters a and ξ:

$$\frac{da}{dt} = -\frac{8\lambda\gamma(\frac{1}{a}+\frac{1}{b})\xi}{[\xi^2 + (\frac{1}{a}+\frac{1}{b})^2]^2}, \quad \frac{d\xi}{dt} = \Gamma - a + \frac{4\lambda\gamma}{a^2}\frac{\xi^2 - (\frac{1}{a}+\frac{1}{b})^2}{[\xi^2 + (\frac{1}{a}+\frac{1}{b})^2]^2}. \tag{2.14a, b}$$

In the absence of the perturbation, i.e. $\gamma = 0$, a remains its initial value whereas $\xi = (\Gamma - a)t + \xi_0$ with ξ_0 being the initial position. Remarkably, we find that the above system of equations is completely integrable. To show this, we first introduce the new dependent variable θ by the relation $\theta = (\frac{1}{a}+\frac{1}{b})^{-1}\xi$ to rewrite Eqs. (2.14) in the form

$$\frac{da}{dt} = -\frac{8\lambda\gamma}{(\frac{1}{a}+\frac{1}{b})^2}\frac{\theta}{(\theta^2+1)^2}, \quad \frac{d\theta}{dt} = \frac{\Gamma - a}{\frac{1}{a}+\frac{1}{b}} - \frac{4\lambda\gamma}{a^2(\frac{1}{a}+\frac{1}{b})^3}\frac{1}{\theta^2+1}. \tag{2.15a, b}$$

Furthermore, if we define a new variable A by $A = a/b + \ln a$, we can transform Eqs. (2.15) into the following Hamiltonian form

$$\frac{dA}{dt} = -\frac{\partial H}{\partial \theta}, \quad \frac{d\theta}{dt} = \frac{\partial H}{\partial A}, \tag{2.16a, b}$$

where the Hamiltonian H is given by

$$H = H(A,\theta) = \Gamma a - \frac{a^2}{2} - \frac{1}{\frac{1}{a}+\frac{1}{b}}\frac{4\lambda\gamma}{\theta^2+1}, \quad (a = a(A)). \tag{2.17}$$

While the explicit expression of a in terms of A is not obtainable, it is important to stress that a is a single-valued function of A. It readily follows from (2.16) and (2.17) that the Hamiltonian is a constant of motion, which we write in the form

$$[(a-\Gamma)^2 + p](\frac{1}{a} + \frac{1}{b}) = -\frac{8\lambda\gamma}{\theta^2+1}, \tag{2.18}$$

for later use, where p is an integration constant which can be determined by the initial conditions for a and θ.

2.2.3. *Properties of solutions.* Before discussing the properties of the solution , we shall briefly consider stationary solutions of Eqs. (2.14) and their linear stability. In the following analysis, we put $b = 1$ without loss of generality. The stationary solutions a_s and ξ_s are readily obtained by solving the equations $da_s/dt = d\xi_s/dt = 0$ as

$$\Gamma - a_s = \frac{4\lambda\gamma}{(a_s+1)^2}, \quad \xi_s = 0. \tag{2.19a, b}$$

Figure 1 is a plot of the stationary amplitude a_s as a function of the parameter Γ for $\gamma = 0.05$ and $\lambda = \pm 1$.

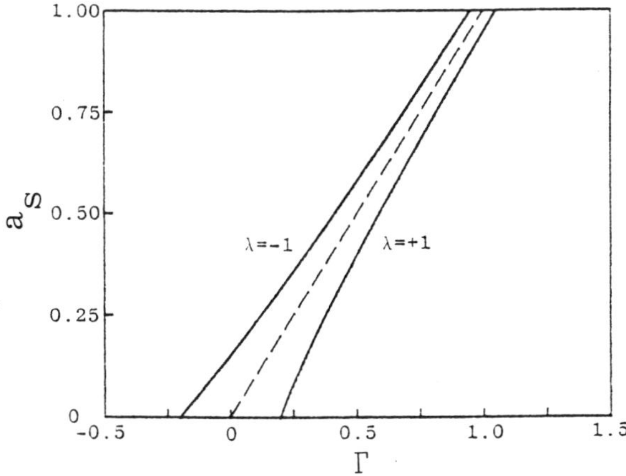

FIG. 1. Plot of the stationary amplitude a_s as a function of Γ.

In the case of $\lambda = -1$, we can see that for any value of Γ within the range $\Gamma > -4\gamma$, there exists one stationary amplitude. On the other hand, in the case of

$\lambda = +1$, there exists one stationary amplitude only if $\Gamma > 4\gamma$. The linear stability analysis of the solution (2.19) is easily performed and gives the following result: The solution corresponding to $\lambda = -1$ ($\lambda = +1$) is stable (unstable) against infinitesimal perturbations. In the case of $\lambda = -1$, the stable points represent centers since the corresponding eigenvalues are found to be pure imaginary whereas for $\lambda = +1$, the unstable points are unstable nodes with one of the two eigenvalues being real and positive.

Lastly, we point out that Eq. (2.8) exhibits an *exact* stationary solution for a special value of Γ. Indeed, careful investigation of the stationary form of Eq. (2.8) shows that an exact solution which satisfies the boundary condition $u \to 0$ as $|x| \to \infty$ exists when the relation $\Gamma = 1 + \lambda\gamma$ holds. Explicitly, it reads in the form

$$u_s = \frac{1}{x^2 + 1}. \tag{2.20}$$

It turns out that this solution corresponds to the stationary solutions $a_s = 1$ and $\xi_s = 0$ of Eq. (2.14) as seen from (2.19).

Next, we shall investigate the properties of the general solution. Since the solution (2.18) is a first integral of the system of equations (2.15), we must integrate the equations once more to obtain the explicit time dependence of the soliton parameters. Its integration is, however, too difficult to perform analytically. Instead of practicing the integration numerically, we shall here analyze the solution in phase plane (a, θ). The following four different combinations of the parameters λ and Γ are considered separately:

a. $\lambda = -1$, $\Gamma > -4\gamma$. This case represents a bottom topography which is concave downward. Figure 2 shows a typical phase portrait of the solution (2.18) for various values of the constant p where the parameters Γ and γ have been chosen as 1.0 and 0.05, respectively. The arrow in each trajectory indicates the direction of the motion of a soliton. Note in this case that there exists only one linearly stable stationary amplitude for a given Γ as seen from Fig. 1.

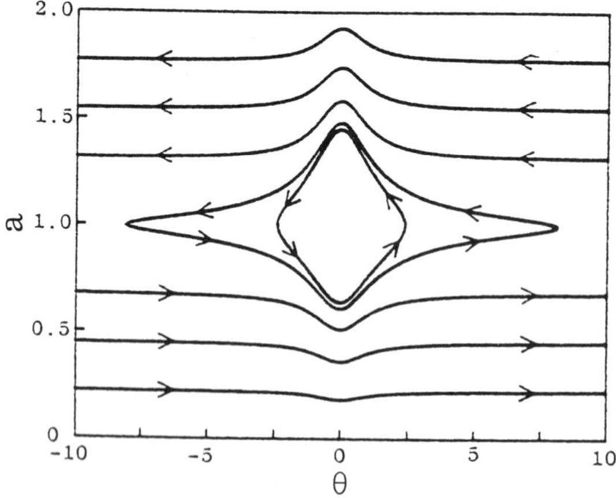

FIG. 2. Phase portrait of the solution (2.18) with $\lambda = -1, \Gamma = 1.0$ and $\gamma = 0.05$.

If the initial amplitude of the soliton is either much larger or much less than unity, the soliton interacts with the bottom topography whereby it suffers a small change of the amplitude near the origin $\theta = 0$. After passing through topography, the soliton gradually recovers its initial amplitude. While there is no net change in the amplitude after the interaction, the soliton exhibits a phase shift which may be given by the relation

$$\Delta\xi = \int_{-\infty}^{\infty} \left[\frac{d\xi}{dt} - (\Gamma - a_0)\right] dt, \tag{2.21}$$

where a_0 is the initial amplitude of the soliton. The second term $\Gamma - a_0$ in the brackets represents the soliton velocity in the absence of the interaction and hence (2.21) yields a net phase shift due to the interaction. The integration of (2.21) can be performed numerically by using (2.14) and (2.18). Instead of doing this, we evaluate it by a successive approximation and give a result correct up to order γ as follows:

$$\Delta\xi = -\frac{4\pi\lambda\gamma}{(\Gamma - a_0)^2}\operatorname{sgn}(\Gamma - a_0). \tag{2.22}$$

It is important to observe that that this quantity is proportional to the cross-sectional area of the bottom topography because $\int_{-\infty}^{\infty} B(x)dx = \pi\lambda$ from (2.10). The phase shift of the soliton predicted by (2.22) is quite interesting since algebraic solitons usually exhibit no phase shift after their collisions [36].

Figure 2 also indicates another fascinating aspect of the solutions. To be more specific, if the soliton enters from the right with an amplitude a bit larger than unity, it interacts with topography and then propagates to the left direction while decreasing its amplitude. After a lapse of time, the soliton changes its direction and goes back to the right. The motion is repeated in the same way. One observes that the orbit of the soliton becomes periodic around a stationary state $a_s = 1.048$ and $\xi_s = 0$ which turns out to be a center in the present example. This phenomenon implies that the soliton has been captured by topography while oscillating around the stationary state indicated above.

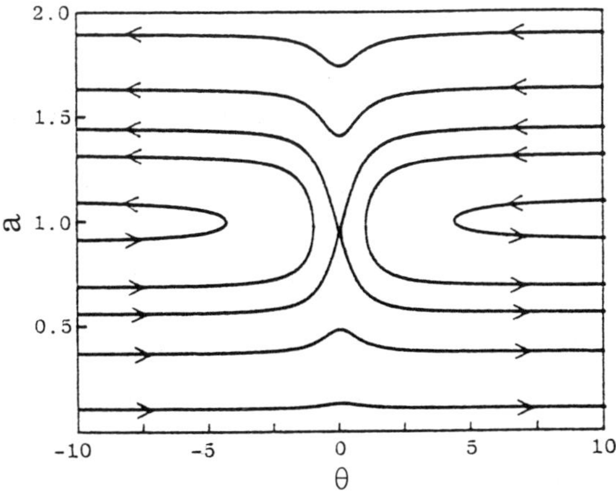

FIG. 3. As in FIG. 2 except $\lambda = +1.0$.

b. $\lambda = -1$, $\Gamma < -4\gamma$. Different from case a, there exist no stationary amplitudes in the parameter range $\Gamma < -4\gamma$. This case is not so interesting because solitons propagate to the left direction irrespective of their initial amplitudes and positions. In fact, one can see that the velocity of the soliton given by (2.14b) always becomes to be negative provided that $\gamma << 1$. The only effect due to the interaction with topography is a phase shift given by (2.22).

c. $\lambda = +1$, $\Gamma > 4\gamma$. This corresponds to a bottom topography which is convex upward. In the parameter range $\Gamma > 4\gamma$, there is only one linearly unstable stationary amplitude. A typical phase portrait is depicted in Fig. 3 for various values of p where we have put $\Gamma = 1.0$ and $\gamma = 0.05$.

As seen from the figure, solitons with larger or smaller amplitudes compared to unity pass through topography while suffering small phase shifts. On the other hand, solitons with amplitudes a bit larger than unity decrease gradually their amplitudes in the course of the propagation to the left direction and eventually they are repelled by topography so that they propagate back to the right direction with amplitudes smaller than unity.

d. $\lambda = +1$, $\Gamma < 4\gamma$. In the parameter range $\Gamma < 4\gamma$, there exist no stationary amplitudes. The properties of the orbits of solitons are similar to those of case b. When solitons enter from the right, they always pass through topography. The net effect is a phase shift given by (2.22).

2.2.4. *Remark.* We showed that a simple choice of topographic function yields an integrable system of equations for the soliton parameters. Nevertheless, the result obtained here is highly suggestive of the dynamics of an algebraic soliton under the action of a wide class of forcing functions with similar profiles. As for analogous works based on the fKdV equations, see, for example [8-11].

3. Outlook

A number of works are left for the study of the fBO equations. We shall conclude this review by listing some of them. The most important theme is to develop general methods for solving initial value problems. Although IST method works well for many of the soliton equations including the BO equation, its applicability to the forced soliton equations is severely restricted. In fact, for the purpose of analyzing them, only the perturbative approach is now available [24]. However, it should be emphasized that the perturbation theory of the fBO equation by means of IST has not been established even at the present time. An effective technique is an asymptotic method using the multiple time scale expansion developed in Sec. 2, even though it has been limited to the leading-order analysis. From purely mathematical point of view, various problems associated with initial value problems such as existence and uniqueness of solutions are of great interest.

The term *soliton* is widely used when a solitary wave solution to certain nonlinear evolution equation is stable against perturbations. It is seldom that the forced soliton equations exhibit exact solutions except some stationary solutions. In the context of the fBO equation (1.1), the solution (2.20) is indeed an exact stationary solution. This indicates that the effect of forcing yields a pinned solitary wave profile in the vicinity of the forcing region. Its stability characteristic is worth studying in direct relevance to the wave phenomenon such as the periodic generation of solitary waves caused by forcing [3].

Throughout this paper, the discussion has been confined to one-dimensional forced soliton equations. In view of their tractability, these simplified model equations are often employed to simulate underlying physical phenomena. We can pursue several possibilities to make these equations more realistic. One of them has been accomplished by taking account of higher-order nonlinearity and dispersion into the BO equation [16, 23]. Another important direction is the introduction of higher-dimensional effects [15, 37, 38]. The mathematical structure of higher-dimensional forced soliton equations is an important subject to be searched in a future study.

Acknowledgment. I would like to express my cordial thanks to the organizing committee, in particular Professor Jean-Claude Saut for inviting me this workshop.

4. References

[1] T.R. Akylas, J. Fluid Mech. **141**, 455 (1984).
[2] C.C. Mei, J. Fluid Mech. **162**, 53 (1986).
[3] T.W. Wu, J. Fluid Mech. **184**, 75 (1987).
[4] S.J. Lee, G.T. Yates and T.Y. Wu, J. Fluid Mech. **199**, 569 (1989).
[5] S.L. Cole, Wave Motion **7**, 579 (1985).
[6] R. Grimshaw and N. Smyth, J. Fluid Mech. **169**, 429 (1986).
[7] W.K. Melville and K.R. Helfrich, J. Fluid Mech. **178**, 31 (1987).
[8] A. Patoine and T. Warn, J. Atmos. Sci. **39**, 1018 (1982).
[9] T. Warn and B. Brasnett, J. Atmos. Sci. **40**, 28 (1983).
[10] B.A. Malomed, Physica D **32**, 393 (1988).
[11] R. Grimshaw, E. Pelinovsky and X. Tian, Physica D **77**, 405 (1994).
[12] Y. Matsuno, Phys. Rev. Lett. **69**, 609 (1992).
[13] Y. Matsuno, J. Fluid Mech. **249**, 121 (1993).
[14] Y. Matsuno, J. Phys. Soc. Jpn. **62**, 1902 (1993).
[15] Y. Matsuno, Phys. Rev. E **47**, 4593 (1993).
[16] Y. Matsuno, Phys. Rev. E **49**, 2091 (1994).
[17] Y. Matsuno, Phys. Rev. Lett. **73**, 1316 (1994).
[18] Y. Matsuno, Phys. Rev. E **51**, 1471 (1995).
[19] Y. Matsuno, Phys. Rev. E **52**, 6333 (1995).
[20] T. Miloh and M.P. Tulin, Tech. Rep. 88-31, Ocean Eng. Lab. Univ. of California at Santa Barbara (1988).
[21] H. Mitsudera and R. Grimshaw, Pure Appl. Geophys. **133**, 635 (1990).
[22] H. Mitsudera and R. Grimshaw, Stud. Appl. Math. **84**, 183 (1991).
[23] R. Grimshaw, Phys. Fluids **24**, 1611 (1981).
[24] D.J. Kaup, SIAM J. Appl. Math. **31**, 121 (1976); D.J. Kaup and A.C. Newell, Proc. R. Soc. London A **361**, 413 (1978); V.I. Karpman and E.M. Maslov, Phys. Lett. A**60**, 307 (1977); Soviet Phys. JETP **46**, 281 (1977); **48**, 252 (1978); V.I. Karpman, Physica Scripta **20**, 462 (1979); E.M. Maslov, Theor. Math. Phys. **42**, 237 (1980).
[25] M. Tanaka, J. Phys. Soc. Jpn. **49**, 807 (1980); K.A. Gorshkov and L.A. Ostrovsky, Physica D **3**, 428 (1981); Y. Kodama and M.J. Ablowitz, Stud. Appl. Math. **64**, 225 (1981); R. Grimshaw and H. Mitsudera, Stud. Appl. Math. **90**, 75 (1993).
[26] J.P. Keener and D.W. McLaughlin, Phys. Rev. A **16**, 777 (1977); R.L. Herman, J. Phys. A: Math. Gen. **23**, 2327 (1990); L.A. Kalyakin, Theor. Math. Phys. **92**, 736 (1992); V.V. Konotop and V.E. Vekslerchik, Phys. Rev. **E49**, 2397 (1994).

[27] A. Bonderson, M. Lisak and D. Anderson, Physica Scripta **20**, 479 (1979); J.-C. Fernandez, C. Froeschle and G. Reinisch, Physica Scripta **20**, 545 (1979).
[28] M. Oikawa and N. Yajima, J. Phys. Soc. Jpn. **34**, 1093 (1973); Prog. Theor. Phys. Suppl. **55**, 36 (1974).
[29] Y. Matsuno, *Bilinear Transformation Method* (Academic, New York, 1984).
[30] Y. Matsuno, Int. J. Mod. Phys. B **9**, 1985 (1995).
[31] A.H. Nayfeh, *Perturbation Methods* (Wiley, New York, 1973).
[32] A. Jefferey and T. Kawahara, *Asymptotic Methods in Nonlinear Wave Theory* (Pitman, London, 1982).
[33] Y. Matsuno, J. Phys. A: Math. Gen. **12**, 619 (1979); **13**, 1519 (1980).
[34] R. Grimshaw, Proc. Roy. Soc. London Ser. A **376**, 319 (1981).
[35] P.M. Edwin and B. Roberts, Wave Motion **8**, 151 (1986).
[36] H. Airault, H.P. Mckean and J. Moser, Comm. Pure Appl. Math. **30**, 95 (1977); S.V. Manakov, V.E. Zakharov, L.A. Bordag, A.R. Its and V.B. Matveev, Phys. Lett. A **63**, 205 (1977); M.J. Ablowitz and J. Satsuma. J. Math. Phys. **19**, 2180 (1978); J. Satsuma and M.J. Ablowitz, J. Math. Phys. **20**, 1496 (1979); Y. Matsuno, J. Phys. Soc. Jpn. **57**, 1577 (1988); **58**, 1948(1989); J. Math. Phys. **31**, 2904 (1990); C.P. Gilson and J.J.C. Nimmo, Phys. Lett. A**147**, 472 (1990).
[37] M.J. Ablowitz and H. Segur, Stud. Appl. Math. **62**, 249 (1980).
[38] W. Choi, J. Fluid Mech. **295**, 381 (1995).

Department of Applied Science, Faculty of Engineering, Yamaguchi University, Ube 755, Japan

E-mail address: matsuno@po.cc.yamaguchi-u.ac.jp

Contemporary Mathematics
Volume **200**, 1996

Nonlinear Evolution of Water Surface Waves: the Frequency Down-Shift Phenomenon

C. Skandrani† C. Kharif† and J. Poitevin‡

ABSTRACT. In this paper a generalized mode coupling approach is used to analyze numerically nonlinear evolutions of uniformly-traveling trains of Stokes' waves in deep water. It is well known that these trains are subject to sideband instability producing amplitude and frequency modulations. Recently Poitevin & Kharif [1] using exact boundary conditions showed, for short waves, how joint effects of viscosity and surface tension can promote growth of the lower sideband of the modulational instability. Following Ruvinsky et *al.* [2], their model is improved by taking account of the vorticity generated by the viscosity in the vicinity of the free surface. A shift to the lower carrier frequency is observed after the first modulation-demodulation cycle and an explanation of this phenomenon is given.

Introduction

Evolution of a uniformly-traveling train of Stokes' waves has been extensively studied for the three last decades. One of the first detailed investigations of the time evolution of nonlinear deep water waves was performed by Lighthill [3] who used the Witham's theory. Zakharov [4,5] developed a Hamiltonian formulation of the water wave problem and demonstrated that an uniform continuous wavetrain is unstable to modulational perturbations. Independently, Benjamin & Feir [6] showed that a Stokes wave train is unstable to sideband perturbations. Later, numerical computations have been used to analyze the stability of water waves. The main advantage is that there exists in principle no restriction concerning the length of disturbances and the wave steepness value. Early numerical work was limited to two-dimensional instability (Longuet-Higgins [7,8]) but an extension to three-dimensional perturbations was made by McLean [9] (see also McLean et *al.* [10]).

Experimentally, the frequency down-shifting phenomenon has been first reported by Lake et *al.* [11]. Their experiments were performed using initi ally uniform, or nearly uniform, wavetrains. At some distance from the source the

---

1991 *Mathematics Subject Classification.* Primary 54C40, 14E20; Secondary 46E25, 20C20.
This research was financially supported by the ONR grant N 00014-93-1-05 00. Most of the computations were performed on the Meteo-France supercomputer, Cray C90-8, France.

modulational instability appears as the growth of two dominant sidebands. Approximate recurrence cycles were observed in which the wavetrains became strongly modulated until they were again nearly uniform. A more careful inspection, however, reveals that the dominant component is the component at the frequency of the lower sideband of the original carrier: this is the frequency down-shift phenomenon.

Trulsen and Dysthe [12] showed using a modified Schrödinger equation correct to fourth order in the wave steepness that breaking characterized by semi- empirical hypothesis is essential in the process of the frequency down-shift. Uchiyama & Kawahara [13] considering a modified nonlinear Schrödinger equation proved that a dissipative correction term representing the damping of mean flow can change drastically the characteristic of the sideband instability, suggesting a possible mechanism for frequency down-shift in nonlinear wave modulation. Hara & Mei [14] coupling the air flow above the waves with Dysthe's extension of the cubic Shrödinger equation, found the frequency down-shift in narrowbanded surface gravity waves. Very recently, they examined in [15] a similar problem for gravity-capillary waves and showed that persistent downshift of frequency can occur for relatively long waves, but upshift happens for very short waves.

Herein, we shall ignore the influence of wind on the surface to focus our attention on the frequency down-shift due to wave-wave interactions. Laboratory results of Lake et *al.* [11] clearly showed that downshifting can occur even in the absence of wind. In addition we shall assume that waves do not break.

In section 1 we formulate our mathematical model. Following West [16], Poitevin & Kharif [1] considered water as a weakly viscous fluid and neglected the vorticity generated by viscosity. So it was assumed that the dominant characteristics of the motion can be described by irrotational flow alone. However this model is not quite correct since the damping coefficient is half the exact value for linear waves. In the case of low viscosity, Ruvinsky et *al.* [2] proposed a quasi -potential approximation for the description of nonlinear gravity-capillary waves on the surface of a viscous fluid. In their equations they expressed the vortex component of the fluid velocity through the potential component. However their approximate model develops unphysical instabilities. Herein we correct the vorticity equation to develop a set of equations with dynamics which are basically stable in time. The numerical algorithm based on the pseudo spectral method is presented in section 2. The validity and accuracy of the model are tested using exact solutions of the inviscid problem. In the last section some results are discussed. The classical Fermi-Pasta-Ulam recurrence phenomenon is reproduced in the inviscid case, characterized by symmetric behavior of the two sidebands. In this case the frequency down-shift occurs only temporarily at maximum of modulation. The modulational instability is suspected to give rise to the frequency down-shift but other ingredients participate to this mechanism. The aim of the present work is : (i) to determine what are the ingredients which break the symmetric behavior of the sidebands and (ii) to give an explanation of the frequency down-shift of short waves.

1. Mathematical formulation

For free surface problem in water, viscous effects are generally weak producing a thin rotational layer adjacent to the potential flow. In the case of low viscosity Ruvinsky et *al.* [2] derived from the linearized vorticity equation a simple evolution for the vortical component of the flow and arrived at the set of equations (1.1)-(1.5).

Equation (1.5) is not exactly the equation they obtained. In the right hand side the diffusive term they neglected avoids unphysical instabilities to develop with time. Their model is numerically correct only for short time evolutions.

$$\nabla^2\phi(x,z,t) = 0, \quad z \in]-\infty, \eta] \tag{1.1}$$

$$\lim_{z\to-\infty} \nabla\phi = 0 \tag{1.2}$$

$$\phi_t + \frac{1}{2}\nabla\phi.\nabla\phi + g\eta - \frac{\alpha}{\rho}\frac{\eta_{xx}}{[1+\eta_x^2]^{\frac{3}{2}}} - 2\nu\phi_{xx} = 0 \quad \text{on} \quad z = \eta, \tag{1.3}$$

$$\eta_t + \eta_x\phi_x = \phi_z + v \quad \text{on} \quad z = \eta, \tag{1.4}$$

$$v_t = 2\nu(\phi_{xxz} + v_{xx}) \quad \text{on} \quad z = \eta. \tag{1.5}$$

Let x, z, t be abscissa, ordinate and time respectively. The origin is located at the mean water level. The potential velocity $\phi(x,z,t)$, satisfying the classical Laplace's equation (1.1), vanishes at infinity. Equations (1.3) and (1.4) are the dynamic and kinematic boundary conditions at the free surface $z = \eta(x,t)$. The vertical velocity, v, is induced by the vortical motion. Parameters α, ν, ρ and g are the surface tension, the kinematic viscosity, the water density and the gravitational acceleration respectively. The subscripts mean partial derivatives and $\nabla = (\partial/\partial x, \partial/\partial z)$.

Following Zakharov [5] we introduce the potential velocity at the free surface $\phi^s(x,t) = \phi(x,\eta(x,t),t)$ into equations (1.3), (1.4) and (1.5).

$$\phi_t^s = -\eta + vW - \frac{1}{2}\phi_x^{s2} + \frac{1}{2}W^2[1+\eta_x^2] + \kappa\frac{\eta_{xx}}{[1+\eta_x^2]^{\frac{3}{2}}} + 2R_e^{-1}\frac{\phi_{xx}^s - W\eta_{xx}}{1+\eta_x^2} \tag{1.6}$$

$$\eta_t = -\eta_x\phi_x^s + W[1+\eta_x^2] + v \tag{1.7}$$

$$v_t = \frac{2}{R_e}(W_{xx} + v_{xx}) \tag{1.8}$$

where $W = \phi_z(x, z = \eta(x,t), t)$.

Equations (1.6)-(1.8) are given in dimensionless form. Reference length and reference velocity are, $1/k_0$ and $(g/k_0)^{1/2}$, respectively. The reference wavenumber is defined as $k_0 = 2\pi/\lambda_0$ where λ_0 is the length of the initial uniform wavetrain.

Note that the reference velocity corresponds to the linear phase velocity of a gravity wave of wavenumber k_0. Dimensionless parameters $\kappa = k_0^2\alpha/\rho g$ and $Re = g^{1/2}/\nu k_0^{3/2}$ characterize the effect of surface tension and viscosity.

The main difficulty is the computation of the vertical velocity, W, due to the potential part of the flow. Following Dommermuth & Yue [17] the potential $\phi(x,z,t)$ is written in a finite perturbation series up to a given order M.

$$\phi(x,z,t) = \sum_{m=1}^{M} \phi^{(m)}(x,z,t) \tag{1.9}$$

The term $\phi^{(m)}$ is of $O(\epsilon^m)$ where ϵ, a small parameter, is a measure of the wave steepness. Then expanding each $\phi^{(m)}$ evaluated on $z = \eta$ in a Taylor series about $z = 0$, we obtain :

$$\phi^s(x,t) = \phi(x, z = \eta, t) = \sum_{m=1}^{M} \sum_{l=0}^{M-m} \frac{\eta^l}{l!} \frac{\partial^l}{\partial z^l} \phi^{(m)}(x, 0, t) \tag{1.10}$$

The validity of the above expansion is restricted by the radius of convergence (distance from $z = 0$) of the potential velocity ϕ which is limited by the occurrence of the first singularity. For gravity waves, this singularity corresponds to the 'Stokes corner-flow'. The appearance of this extreme wave is the early stage of wave breaking. The numerical model used herein cannot simulate breaking waves and fails to be valid in this case. At a given instant of time ϕ^s and η are known, so that from (1.10), at each order, we can calculate $\phi^{(m)}$:

$$\begin{cases} \phi^{(1)}(x, 0, t) &= \phi^s(x, t) \\ \phi^{(m)}(x, 0, t) &= -\sum_{l=1}^{m-1} \frac{\eta^l}{l!} \frac{\partial^l}{\partial z^l} \phi^{(m-l)}(x, 0, t) \quad m \geq 2 \end{cases} \tag{1.11}$$

As in Zakharov approach each component $\phi^{(m)}$ is represented as an eigenfunction expansion of free modes. For 2π-periodic boundary conditions in infinite depth, $\phi^{(m)}$ is given by the following truncated series:

$$\phi^{(m)}(x, z, t) = \sum_{n=-N}^{N} \phi_n^{(m)}(t) exp[|n|z + inx], \qquad z \leq 0 \tag{1.12}$$

Substitution of(1.12) into the set of equations (1.11) gives modes $\phi_n^{(m)}(t)$. The vertical velocity, W, is then:

(1.13)

$$W = \phi_z(x, \eta, t) = \sum_{m=1}^{M} \sum_{l=0}^{M-m} \frac{\eta^l}{l!} \sum_{n=-N}^{N} \phi_n^{(m)}(t) exp(inx) \frac{\partial^{l+1}}{\partial z^{l+1}} exp(|n|z) \qquad (z = 0)$$

Substitution of (1.13) into the boundary conditions (1.3)-(1.5) yields the evolution equation for ϕ^s, η and v in term of the modal amplitudes $\phi_n^{(m)}$ which are given by the values of ϕ^s and η. We have generalized the Dommermuth & Yue's approach to the problem of weakly viscous fluids.

2. Numerical method

The numerical method used to solve the evolution equations (1.6)-(1.8) is similar to that developed by Dommermuth and Yue [17]. Equations (1.6)-(1.8) are integrated using N wave modes and retaining nonlinear terms up to order M.

2.1. Pseudo spectral treatment. Once known the surface elevation $\eta(x,t)$ and the potential $\phi^s(x,t)$ at time t, the modal amplitudes may be computed. The spatial derivatives of $\phi^{(m)}$, ϕ^s, η, W and v are calculated in the spectral space, while nonlinear terms are evaluated in the physical space at a discrete set of collocation points x_j. Fast Fourier Transforms are used to link the spectral and physical spaces. Equations (1.11) are solved in the spectral space. The number of operations in pseudo spectral method is, at each order, $O(N\ln N)$. For equations with nonlinear terms retained up to a fixed order M, the number of operations is $O(MN\ln N)$ per time step.

2.2. Time integration. Evolution equations for ϕ^s, η and v are integrated in time using a fourth order Runge Kutta integrator with constant time step.

Starting from given initial conditions we apply the two step procedures 2.1 and 2.2.

2.3. Numerical convergence tests. The convergence and accuracy of the model is tested using exact solutions of equations (1.1)-(1.4) with $\alpha = 0$, $\nu = 0$ and $v = 0$. Surface tension and viscosity have been ignored. In this case it is well known, since the work of Stokes, that there exists exact solutions which are nonlinear periodic waves traveling with permanent form (Stokes waves). Computations of these waves are performed with a very efficient iterative algorithm due to Longuet-Higgins [18]. For a wavelength of 2π, the eigenfunctions are given by (1.12) with $n = 0, \pm1, \pm2, ..., \pm N$.. Figures 1(a, b, c) display the maximum absolute errors in $W = \phi_z|_{z=\eta}$ of a Stokes wave of steepness ϵ and fundamental wavenumber $k = 1$, for different values of mode number N, as a function of the order of approximation M. Numerically computed values are represented by W_{num} while W_{th} is the exact value given by Longuet-Higgins' algorithm. For $M < 10$, the error decreases exponentially fast with N and M as they increase. We select M so that a given accuracy is achieved and then minimize N to restrict round-off errors. In order to avoid aliasing errors we use $2N$ complex conjugate pairs of Fourier coefficients for N alias-free modes.

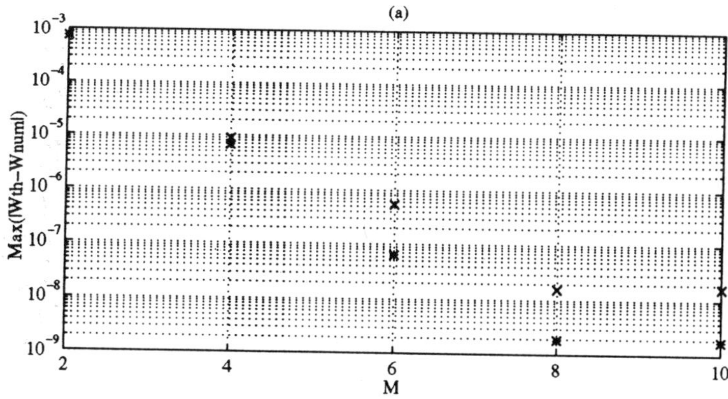

FIGURE 1. (a). For caption see next page.

We also used Stokes waves to estimate the integration truncation error of the numerical scheme. Table 1 gives maximum absolute errors in the free surface elevation of a progressive Stokes wave of steepness ϵ ($k = 1$) and fundamental period T after integration time $t = 10T$ for different values of time step. For $\epsilon = 0.10$ and $\epsilon = 0.20$ we fixed $(N, M) = (16, 6)$ and$(N, M) = (16, 8)$ respectively. We note that the expected O $(\triangle t^4)$ error is attained provided $\triangle t$ is decreased. Parameters M, N and $\triangle t$ are chosen following the procedure : (i) the desired accuracy δ is fixed; (ii) M is selected so that the error $\delta \approx \epsilon^M$; (iii) then the minimum value of N is chosen to satisfy the desired accuracy and (iv) $\triangle t$ is selected in table 1 subject to the Courant stability condition.

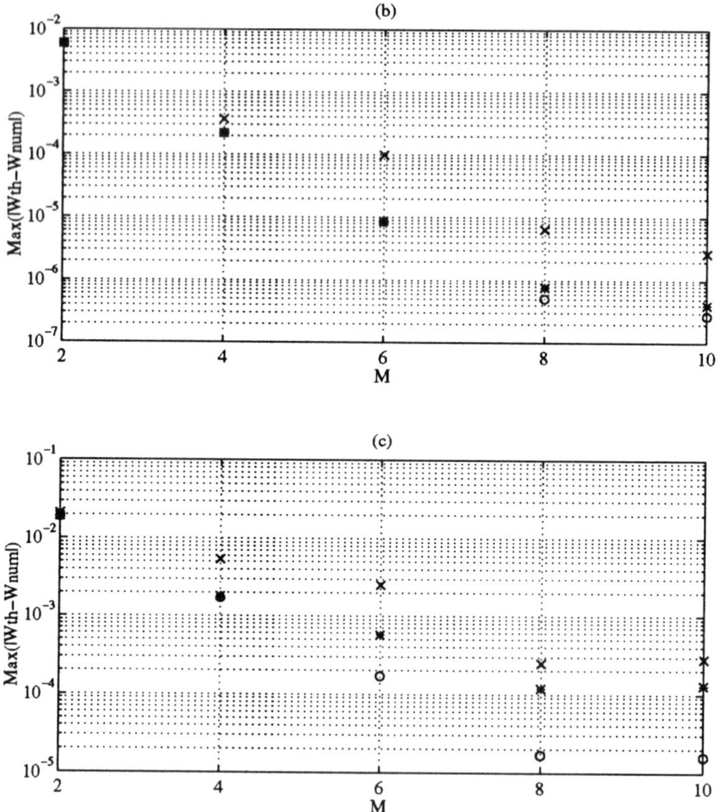

FIGURE 1. Maximum absolute error in the free-surface vertical velocity of a Stokes wave of steepness ϵ for different value of mode number N as a function of order of approximation M. $N = 8$(x), $N = 16$(*) and $N = 32$(o). (a) for $\epsilon = 0.1$; (b) for $\epsilon = 0.2$; and (c) for $\epsilon = 0.3$.

TABLE 1. Maximum absolute error in the free-surface elevation, η, of a progressive Stokes wave of steepness ϵ and period T after integration time, t, for different values of time step $\triangle t$

		$T/\triangle t$				
ϵ	t/T	20	30	40	50	60
0.1	10	0.34×10^{-3}	0.46×10^{-4}	0.11×10^{-4}	0.40×10^{-4}	0.10×10^{-5}
0.2	10	0.11×10^{-2}	0.31×10^{-3}	0.17×10^{-3}	0.70×10^{-4}	

Further numerical experiments on the evolution of Stokes wavetrain of wave steepness $\epsilon = 0.20$ and wavenumber $k = 9$ prove the validity of the model. The wave steepness ϵ is defined as $\frac{1}{2}k(\eta_{max} - \eta_{min})$. We set $\kappa = 0$, $R_e = \infty$ and $v = 0$

into the evolution equations, $(N, M) = (144, 8)$ and $\Delta t = T/50$ where T is the fundamental period of the Stokes wave. As it is expected amplitudes of the fundamental ($k = 9$) and harmonics ($k = 18, 27, ..$) of the Stokes wavetrain remain constant (see figure 2). Figure 3 shows how well the numerical method conserves energy and volume as the wave evolves with time.

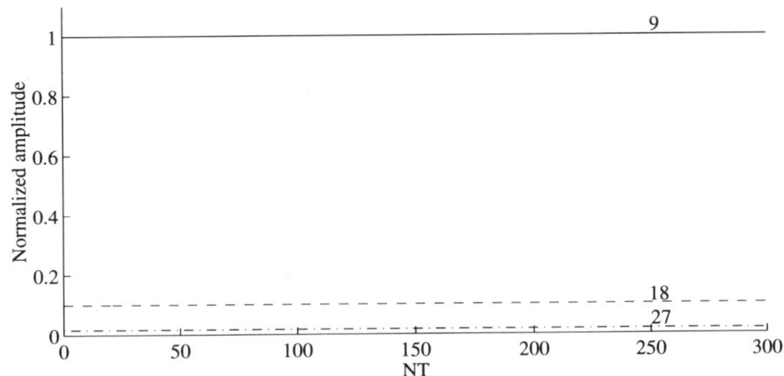

FIGURE 2. Time histories of the amplitude of the fundamental ($k = 9$), first harmonic ($k = 18$) and second harmonic ($k = 27$) relative to the initial amplitude of the fundamental for an evolving Stokes wave ($\epsilon = 0.20, k = 9$). NT is the number of fundamental periods.

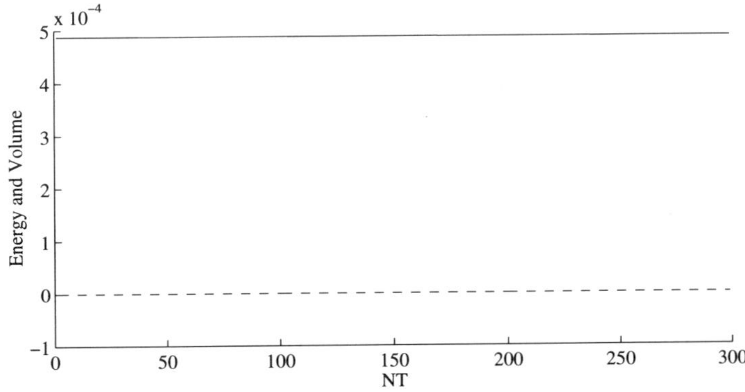

FIGURE 3. Total energy (—) and volume (- - -) of an evolving Stokes wave ($\epsilon = 0.20, k = 9$) as a function of time. NT is the number of fundamental periods.

3. Results and discussion

In this section we shall study the long time evolution of Stokes wavetrains. First we consider the inviscid case ($R_e = \infty$) without taking into account surface tension effects ($\kappa = 0$). For initially weakly nonlinear Stokes waves, numerical experiments exhibit the classical Fermi-Pasta-Ulam recurrence phenomenon while for steeper trains numerical solutions break down due to the occurrence of the singularity. This singularity corresponds to the first stage of breaking of a wave in the train. Wave breaking often occur when the wavetrains are strongly modulated. Lake et *al.* [11] suggested that these events which dissipate energy could favor the frequency downshifting. They also suggested that formation of capillary waves on the steepest crests could participate in dissipation of wave energy. Our model is unable to simulate breaking waves. On the other hand it describes rather well nonlinear evolutions of short gravity waves influenced by both effects of viscosity and surface tension. This case will constitute the third part of this section. The wavetrains considered in the experiments of Lake et *al.* [11] are short waves with wavelength near 13 cm which are subject to viscous and surface tension effects. The initial condition used for the following numerical simulations is a Stokes wave disturbed by its most unstable infinitesimal perturbation.

3.1. Modulational instability of Stokes waves. The procedure used to calculate the linear stability of Stokes waves is similar to the method described by Kharif & Ramamonjiarisoa [19].

Let $\eta = \overline{\eta} + \eta'$ and $\phi = \overline{\phi} + \phi'$ be the perturbed elevation and perturbed velocity potential where $(\overline{\eta}, \overline{\phi})$ and (η', ϕ') correspond respectively, to the unperturbed Stokes wave and infinitesimal perturbative motions ($\eta' \ll \overline{\eta}$, $\phi' \ll \overline{\phi}$). Substituting these decompositions in the boundary conditions linearized about the unperturbed motion and using the following forms:

$$\eta' = exp(\lambda t + ipx) \sum_{-\infty}^{\infty} a_j exp(ijx) \tag{3.1}$$

$$\phi' = exp(\lambda t + ipx) \sum_{-\infty}^{\infty} b_j exp(ijx + \gamma_j z) \tag{3.2}$$

$$\gamma_j = |p + j|$$

an eigenvalue problem for λ with eigenvector $\mathbf{u} = (a_j, b_j)^t$ is obtained:

$$(\mathbf{A} - \lambda \mathbf{B})\mathbf{u} = 0 \tag{3.3}$$

where $\mathbf{A}$ and $\mathbf{B}$ are complex matrices depending on the unperturbed wave steepness ϵ and the arbitrary real number p. The eigenvalue, λ, satisfies

$$det(\lambda \mathbf{B} - \mathbf{A}) = 0 \tag{3.4}$$

The physical disturbances are obtained from the real parts of the complex expression η' and ϕ' at $t = 0$.

McLean et *al.* [10] and McLean [9] showed numerically that the dominant instability of a uniformly-traveling train of Stokes' waves in deep water is the two-dimensional modulational instability (class I) provided that its steepness is less than $\epsilon = 0.30$. For higher values of the wave steepness three-dimensional instabilities (class II) become dominant, phase locked to the unperturbed wave. Herein we shall focus our study on the nonlinear evolution of the first kind of instability using the previous method briefly described.

3.2. Fermi-Pasta-Ulam recurrence. Experiments in wave tank by Lake et *al.* [11] showed that the unstable modulations grew to a maximum and then subsided. They performed numerical computations using the nonlinear Schrödinger equation to confirm this phenomenon. They also observed numerically that the evolution of an unstable wavetrain is a series of modulation-demodulation cycles, at least in the absence of dissipation. Other authors, using the nonlinear Schrödinger equation or modified nonlinear Schrödinger equations, confirmed theoretically or numerically this phenomenon. Stiassnie & Schemer [20] extended to higher order the Zakharov equation to study nonlinear evolutions of Stokes waves subject to different kinds of instability. They rediscovered the Fermi-Pasta-Ulam recurrence when modulational instability is considered alone. In contrast to single evolution of modulational instability, the coupled behavior of the two kinds of instability is non-periodic.

In order to reproduce the series of modulation-demodulation cycles we set $R_e = \infty$, $\kappa = 0$ and $v = 0$ into equations (1.6)-(1.8). Initial condition is a Stokes wave of steepness $\epsilon = 0.13$ disturbed by its most unstable perturbation which corresponds to $p \approx 0.22$. The fundamental wavenumber of the Stokes wave is chosen so that integral numbers of the sidebands can be fitted into the computational domain. For $p \approx 0.22$ the fundamental harmonic of the Stokes wave is $k = 9$ and the dominant sidebands are $k = 7$ and $k = 11$ for subharmonic and superharmonic part of the perturbation respectively. There exists higher harmonics present in the interactions which are not presented here. The normalized amplitude of the perturbation relative to Stokes wave amplitude is initially taken equal to 10^{-2}. The time histories of the fundamental mode $k = 9$ and dominant sidebands $k = 7, 11$ are shown in figure 4(a). Following Dommermuth & Yue [17], we set $M = 4$, $N = 64$ and $\triangle t = T/50$ to compute the long time evolution. They used these values to compare with Stiassnie & Shemer's fourth-order simulation. Amplitudes of sideband modes increase exponentially as time increases and then saturate owing to nonlinearity to reach maxima while amplitude of the fundamental is minimum. At this stage the modulation is maximum, then the wave train becomes almost fully demodulated and the initial condition is almost reconstructed. Like Dommermuth & Yue [17] our simulation breaks down after few modulation- demodulation cycles. In Stiassnie & Shemer simulations, only a limited approximation with five free waves was used. Better agreement with their results is obtained when we fix $N = 45$ instead of $N = 64$ (see figure 4 b). The return of amplitudes to their initial states is better satisfied when the number of modes N is decreased. Breakdown doesn't occur in this case. At the maximum of modulation high wavenumber free modes play a crucial role in the behavior of the upper sideband and the fundamental. Unlike the lower sideband the upper sideband stops its growth to decrease and then increase with the fundamental. When the group of waves becomes again weakly modulated the sidebands develop almost similar evolutions. Nonlinear interactions between higher harmonics nk and $n(k \pm \delta k)$ of the fundamental and the sidebands contribute to the generation of higher components , $n\delta k$, of the modulation which becomes strongly nonlinear producing waves near breaking. So on the steepest part of the train the amplitude of some crests are near their limiting value. This very localized effect, which leads to breaking, can be reduced if the number of free modes is diminished. The wavelet transform, which gives the spatial distribution of the local phase, amplitude and energy of various components is applied to the profile $\eta(x, t)$ at the time t corresponding to the maximum of modulation. The transform shows a spatial localization of the superharmonic components, $k > 9$, of the train on the steepest waves while the subharmonic components $k < 9$ are uniformly distributed all along the train. The asymmetrical behavior between the lower and upper sidebands when strong modulation occur is due to the spatial localization of the superharmonic components. On the steepest part the fundamental and the upper sideband contribute to the generation of high harmonics of the modulation. In the next subsection it will be shown numerically that the decrease of the superharmonic and the fundamental is enhanced by surface tension.

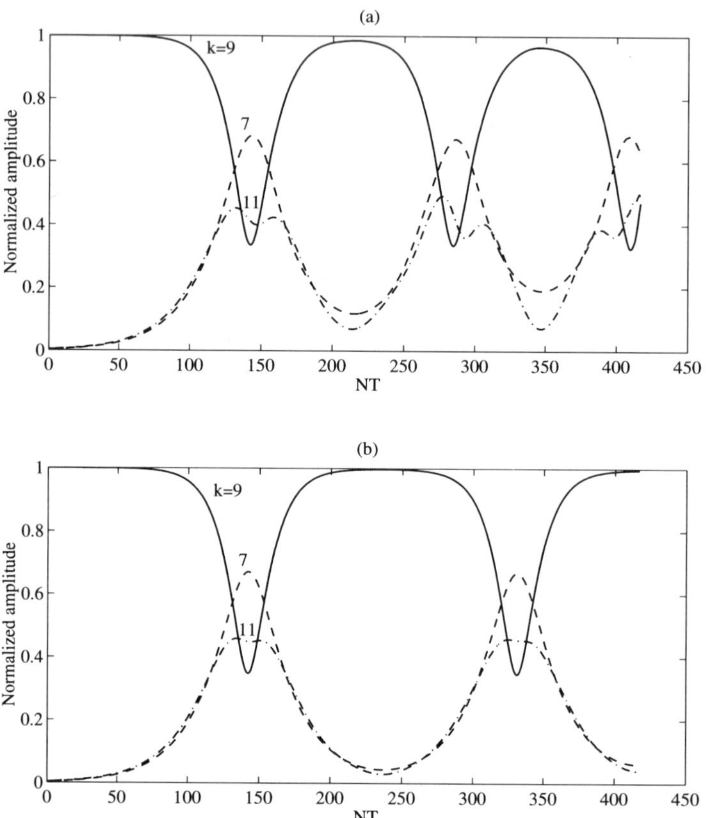

FIGURE 4. Time histories of the normalized amplitude of the fundamental ($k = 9$) , subharmonic ($k = 7$) and superharmonic ($k = 11$) modes for an evolving perturbed Stokes wave ($\epsilon = 0.13, k = 9$). $\kappa = 0$ and $R_e = \infty$ (a) Number of free modes $N = 64$. (b) $N = 45$.

For steepness $\epsilon = 0.20$ our numerical solution breaks down after few modulation-demodulation cycles. Breakdown corresponds to occurrence of the first singularity which indicates the appearance of breaking waves. The waves slope becomes too high so that $\phi^{(m)}$ is no more of $O(\epsilon^m)$. The present numerical method can not simulate this process which concerns multivalued free surfaces and at later stages two-phase flows. Within the author's knowledge, numerical simulations of spilling breakers currently observed in open seas is still up to now an open problem. The main difficulty is the treatment of this two-phase flow event.

In this subsection we have shown numerically that nonlinear gravity waves propagating on the free surface of an inviscid fluid are either subject to the Fermi-Pasta-Ulam recurrence or breaking. The result is not new but it confirms previous theoretical and numerical works.

Lake et al [11] suggested that the essential ingredient participating in the frequency-downshift mechanism could be dissipation in wave energy. They mentioned two candidates : wave breaking (spilling breaker) or formation of capillary waves. In the next subsection we shall introduce coupled effect of viscosity and surface tension.

3.3. Frequency down-shift phenomenon. To compare our numerical simulations with the experiments of Lake et *al.* [11] we choose a train of waves of length $\lambda = 10cms$. The long time evolution of these short waves is strongly influenced by both effect of surface tension and viscosity. For water and air/water interface, viscosity, density and surface tension are $\nu = 0.01$, $\rho = 1$ and $\alpha = 74$ in c.g.s units. The gravity acceleration is $g = 980cms^{-2}$.

Firstly we test the validity of our code when viscous effect is introduced into the governing equations. Figure 5 displays theoretical and numerical evolution of the normalized total energy of weakly damped nonlinear gravity waves of wavelength $\lambda_0 = 10cms$ ($R_e \approx 6.3 \times 10^3$) as a function of time. The initial condition is an unperturbed Stokes waves of steepness $\epsilon = 0.20$.

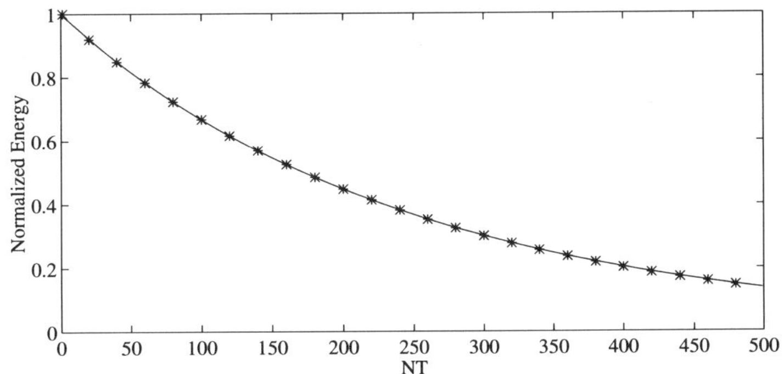

FIGURE 5. Evolution of the normalized total energy of weakly damped nonlinear gravity waves. Theoretical values $\frac{E}{E_0} = exp(-4\nu k^2 t)$ (solid line). Numerical values (*).

We consider now a wavetrain which has nine waves in the computational domain $[0, 2\pi]$, i.e $(\epsilon, k) = (0.13, 9)$.

Previous values of the parameters computed now with different reference length and reference velocity become $\kappa \approx 3.7 \times 10^{-4}$ and $R_e \approx 1.7 \times 10^5$. They correspond to a train of length $\lambda_0 = 90cms$.

Before introducing both capillarity and viscosity we consider only the effect of surface tension. Figure 6 shows clearly that the decrease of the superharmonic and fundamental is strongly enhanced by surface tension. Capillarity increases the transfer of energy from the fundamental and the dominant superharmonic component of the instability to high wavenumber modes. Breakdown occurs during the second modulation.

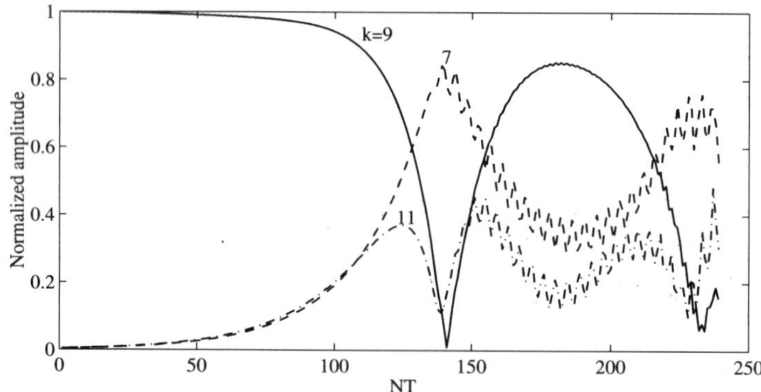

FIGURE 6. Same as figure 4(a).$\kappa \approx 3.7 \times 10^{-4}$ and $R_e = \infty$, $N = 64$, $M = 6$.

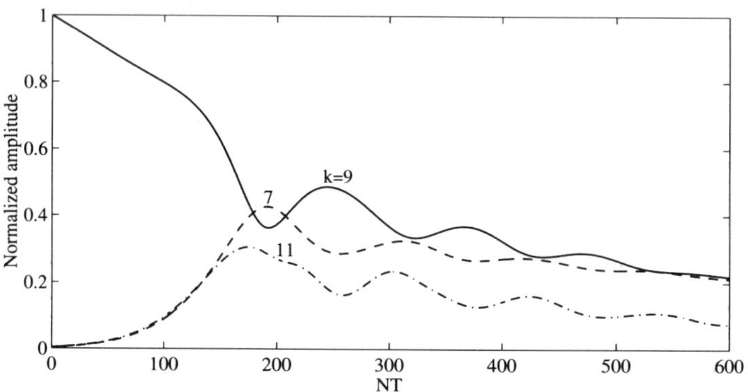

FIGURE 7. Same as figure 4 with $N = 64$ and $M = 6$. $\kappa \approx 3.7 \times 10^{-4}$ and $R_e \approx 1.7 \times 10^5$

Figure 7 describes the time development of spectral components corresponding to the fundamental and dominant modes of the modulational instability of a weakly nonlinear train of perturbed Stokes waves ($\epsilon = 0.13$) when viscosity and surface tension are taken into account. A series of damped modulation-demodulation cycles is observed. For weakly nonlinear wavetrains, the frequency down-shift phenomenon is not observed even if viscosity and surface tension are considered.

We consider now a train of Stokes waves of steepness $\epsilon = 0.20$ perturbed by its most unstable disturbance ($p = 3/9$). For $\kappa \approx 3.7 \times 10^{-4}$ and $R_e \approx 1.7 \times 10^5$, the long time evolution of the normalized amplitudes of the fundamental ($k = 9$) and sidebands ($k = 6, k = 12$) are shown in figure 8. The shift to the lower sideband component is obtained after the first modulation-demodulation cycle and persists during the evolution of the train. This phenomenon is the frequency down-shift. Figures 9(a, b) illustrate the transformation of the initial group of nine waves into a final group of six waves. For nonlinear short wavetrains, the frequency down-shift phenomenon occurs if viscosity and surface tension are considered. Evolutions of the nondimensionalized total energy of modulated nonlinear waves subject to capillarity and viscosity are shown in figure 10. The solid line corresponds to an initially

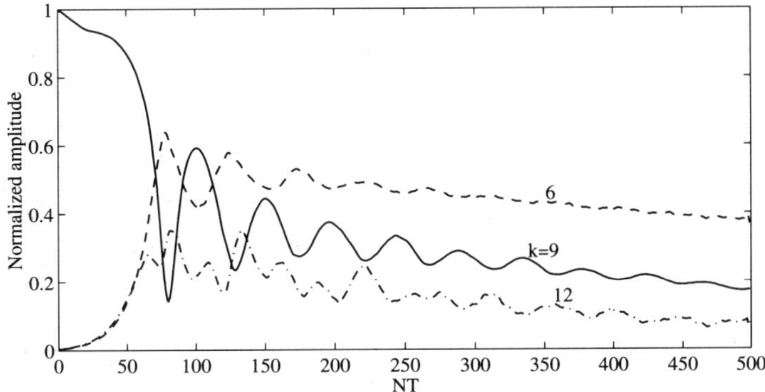

FIGURE 8. Same as figure 7 for $\epsilon = 0.2$ with $N = 54$, $M = 8$.

unperturbed wave train while the dashed lines corresponds to perturbed waves. When surface tension is considered we can observe a strong nonlinear damping occurring at the first and second modulations due to the generation of high harmonics. This nonlinear damping of the energy is lower when capillarity is neglected. The

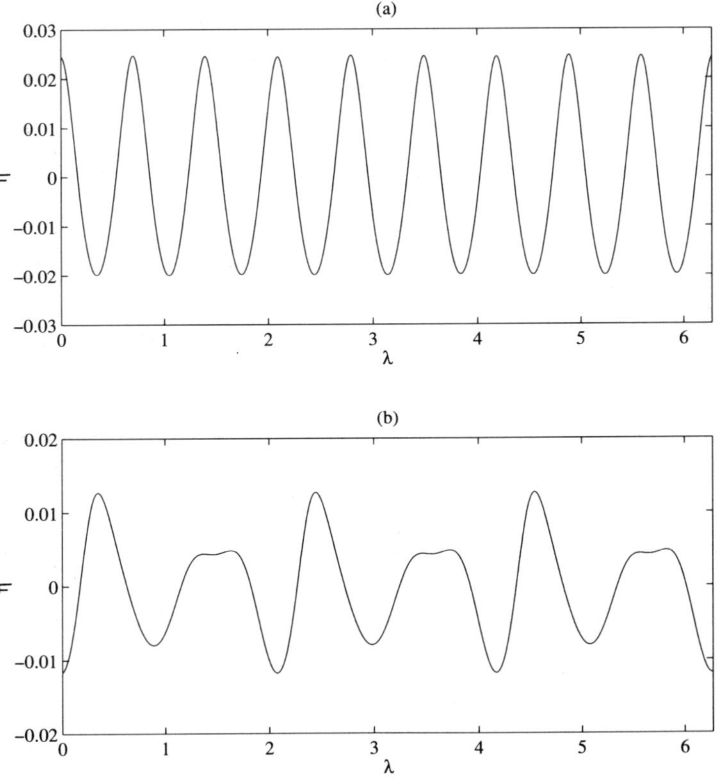

FIGURE 9. Instantaneous free-surface elevation of an evolving Stokes wavetrain ($\epsilon = 0.2$). (a) $NT = 0$. (b) $NT = 500$

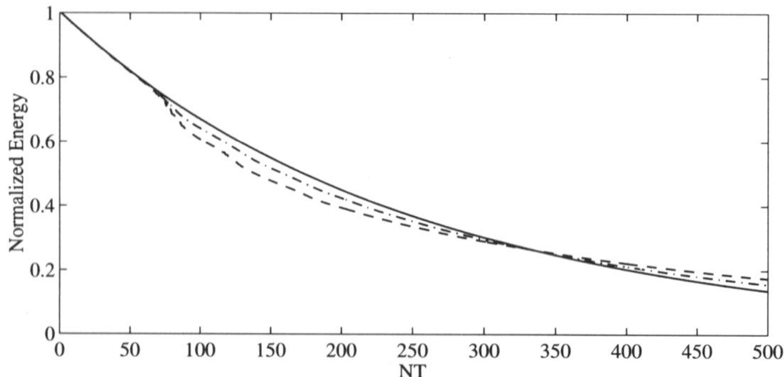

FIGURE 10. Evolutions of the normalized total energy. Solid line: unperturbed Stokes wave train, $\kappa = 0$ and $R_e \approx 1.7 \times 10^5$; dashed doted line: perturbed Stokes wave train, $\kappa = 0$ and $R_e \approx 1.7 \times 10^5$; dashed line: perturbed Stokes wave train, $\kappa \approx 3.7 \times 10^{-4}$ and $R_e \approx 1.7 \times 10^5$.

nonlinear damping observed at the maximum of modulation is mainly due to the generation of short scales on the steepest parts of the train by interaction between the fundamental and the upper sideband. Additional numerical simulations have been performed to achieve this study. For $\epsilon = 0.20$, $\kappa \approx 3.7 \times 10^{-4}$ and $R_e = \infty$ the solution breaks down after several modulation-demodulation cycles. For $\kappa = 0$ and $R_e = 1.7 \times 10^5$ a series of damped modulation-demodulation cycles is observed.

In conclusion we have shown numerically that the shift to lower frequency is promoted by a damping mechanism. High wavenumber modes, generated by nonlinear interactions between the fundamental and the upper sideband when the modulation reaches its maximum, play a crucial role in this transition. Generation of these small scales which are strongly damped by viscosity is enhanced by capillarity and prevent breaking of the waves.

References

[1] POITEVIN, J. & KHARIF, C., *Subharmonic transition of a nonlinear wave train on deep water*, IN MATHEMATICAL AND NUMERICAL ASPECTS OF WAVE PROPAGATION PHENOMENA, (G. COHEN, L. HALPERN & P. JOLY (EDS)), SIAM, PHILADELPHIA, 1991, 567-576.

[2] RUVINSKY, K. D., FELDSTEIN, F. I. & FREIDMAN, G. I. , *Numerical simulations of the quasi-stationary stage of ripple excitation by steep gravity-capillary waves*, J. FLUID MECH. **230** (1991), 339-353.

[3] LIGHTHILL, M. J., *Contributions to the theory of waves in nonlinear dispersive system*, J. INST. MATH. APPL., **1** (1965), 269-306.

[4] ZAKHAROV, V. E., *The instability of waves in nonlinear dispersive media*, ZH. EKSP. TEOR. FIZ. **51** (1966), 1107-1114. SOV. PHYS. JETP. 24, 4 (1967).

[5] ZAKHAROV, V. E., *Stability of periodic waves of finite amplitude on the surface of a deep fluid*, J. APPL. MECH. TECH. PHYS. **27** (1968), 86-94.

[6] BENJAMIN, T. B. & FEIR, J. E., *The disintegration of wave trains in deep water. Part 1. Theory*, J. FLUID MECH. **27** (1967), 417-430.

[7] LONGUET-HIGGINS, M. S., *The instabilities of gravity waves of finite amplitude in deep water. I. Superharmonics*, PROC. ROY. SOC. **A360** (1978), 417-488.

[8] LONGUET-HIGGINS, M. S., *The instabilities of gravity waves of finite amplitude in deep water. II. Subharmonics*, PROC. ROY. SOC. **A360** (1978), 489-505.

[9] McLEAN J. W. *Instabilities of finite-amplitude water waves* , J. FLUID MECH. **114** (1982), 315-330.

[10] McLEAN, J. W., MA, Y. C., MARTIN, D. U., SAFFMAN, P. G. & YUEN, H. C., *Three-dimensional instability of finite amplitude gravity waves*, PHYS. REV. LET. **46** (1981), 817-820.

[11] LAKE, B. M., YUEN, H. C., RUNGALDIER, H. & FERGUSON, W. E., *Nonlinear deep-water waves; theory and experiments. Part 2. Evolution of a continuous wave train*, J. FLUID MECH., **83** (1977), 49-74.

[12] TRULSEN, K. & DYSTHE, K. B., *Frequency down-shift through self modulation and breaking*, WATER WAVES KINEMATICS (A. TφRUM & O. T. GUDMESTAD (EDS.)) KLUWER ACADEMIC PUBLISHERS. (1990), 561-572.

[13] UCHIYAMA, Y., KAWAHARA T., *A possible mechanism for frequency down-shift in nonlinear wave modulation*, WAVE MOTION., **20** (1994), 99-110.

[14] HARA, T. & MEI, C. C., *Frequency downshift in narrowbanded surface waves under the influence of wind*, J. FLUID MECH., **230** (1991), 429-477.

[15] HARA, T. & MEI, C. C., *Wind effects on the nonlinear evolution of slowly varying gravity-capillary waves*, J. FLUID MECH ., **267** (1994), 221-250.

[16] WEST, B. J., *Statistical properties of water waves. Part 1. Steady-state distribution of wind-driven gravity-capillary waves*, J. FLUID MECH., **117** (1982), 187-210.

[17] DOMMERMUTH, D. G., YUE, D. K. P., *A high-order spectral method for the study of nonlinear gravity waves*, J.FLUID MECH., **184** (1987), 267-288.

[18] LONGUET-HIGGINS, M. S., *Bifurcation in gravity waves*, J. FLUID MECH., **151** (1985), 457-475.

[19] KHARIF, C. & RAMAMONJIARISOA, A., *Deep-water gravity wave instabilities at large steepness*, PHYS. FLUIDS., **31** (1988), 1286-1288.

[20] STIASSNIE, M. & SHEMER, L., *Energy computations for evolution of class I and II instabilities of Stokes waves*, J. FLUID MECH., **174** (1987), 299-312.

†INSTITUT DE RECHERCHE SUR LES PHENOMENES HORS EQUILIBRE, 163 AVENUE DE LUMINY, 13288 MARSEILLE CEDEX 9, FRANCE.

E-mail address: chafih@castor.univ-mrs.fr
E-mail address: kharif@castor.univ-mrs.fr

‡METEO-FRANCE, 42 AVENUE GASPARD CORIOLIS 31057 TOULOUSE CEDEX, FRANCE
E-mail address: Joel.poitevin@meteo.fr

Contemporary Mathematics
Volume **200**, 1996

Solitary Waves on the Free Surface of a Rotating Cylindrical Flow

S. M. Sun

Abstract. This paper concerns the existence of solitary waves moving with constant speed on the free surface of a rotating cylindrical flow. By assuming that the fluid is inviscid and incompressible with constant density and the flow is axisymmetric and rotating with constant angular speed around its longitudinal axis, it is found that there are infinitely many critical wave speeds and it is rigorously proved that there exist solitary wave solutions of the exact equations, which decay to uniform rotating flows at infinity, when the wave speed is near the largest critical speed. The main difficulties from the unknown free surface and a singularity at the longitudinal axis have to be overcome.

1. Introduction

It has been generally recognized that the flow of a fluid with density stratification and fluid flow in a rotating frame of reference are analogous. In a rotating system the centrifugal force and Coriolis force are the body forces to replace the usual gravitational force in a two-dimensional stratified fluid. Such analogy was studied intensively in many hydrodynamical problems. The general theory for rotating fluids was developed and summerized in the books by Greenspan [**7**] and Yih [**21**] and the references therein. This paper deals with the existence of solitary waves on the free surface of a long cylindrical column of inviscid and incompressible fluid of constant density. The problem arises in many physical situations. The most well-known physical example related to this problem is the vortex breakdown phenomenon observed frequently in the flow fields of slender delta wings. Many experiments have been carried out and the existence of solitary waves in the rotating flow has been observed [**9, 14**]. This problem was also studied formally by using asymptotic method and the well-known Korteweg-de Vries equation (KdV equation) was derived for the change of surface from the equilibrium state under the long-wave approximation of solutions of the governing equations [**2, 3, 11**]. Therefore the existence of solitary wave solutions, which are the solutions of the KdV equations, was obtained. Further discussion about the vortex breakdown phenomenon can be found in [**4, 12**]. Grimshaw [**8**] studied a rotating fluid flow in a cylinder past an obstacle placed on the axis of the cylinder and formally derived

1991 *Mathematics Subject Classification.* Primary 76C10; Secondary 35J60.

the forced KdV equation and solitary wave solutions for the flow. Solitary waves are also derived for a rotating jet or rotating two-fluid flow in [**15**]. Recently, the existence of solitary waves in a rotating stratified fluid bounded by a rigid cylinder is rigorously proved in [**16**]. However, there is no rigorous result to show the existence of solitary waves on the free surface of a rotating jet since the problem with a free surface is more difficult to study mathematically than the problem with a fixed boundary. We note that it is important to know whether the exact equations governing the flow have solitary wave solutions, and the solitary wave solutions derived from the KdV equation are the approximations of some solutions of the exact equations, since the solitary wave solutions derived in [**3, 11**] are the solutions of the KdV equation and the KdV equation is only a model equation for the exact equations.

Motivated by the physical problems, we consider a simple rotating jet with a constant angular velocity at infinity [**15**] and shall rigorously prove the existence of solitary waves moving with constant speed on the free surface for the exact equations governing the flow. The result can be stated as follows. Let the fluid be inviscid and incompressible and the flow be axisymmetric and rotating with an angular speed ω around the longitudinal axis. Assume that there is a wave moving with a constant speed c on the free surface. Then a nondimensional wave speed $Ro = (c/2\omega R)$, which is called Rossby number and determines the motion of the flow, can be defined, where R is the equilibrium radius of the free surface at infinity. It is found that there are infinitely many critical values $(1/\mu_1) > (1/\mu_2) > (1/\mu_3) > \cdots > 0$ for Ro. By a formal method, solitary waves can be derived when Ro is near each of the critical values [**15**]. Then the existence of the solitary wave solutions for the exact equations, when Ro is near the largest critical value $1/\mu_1$, is stated in the following Theorem.

MAIN THEOREM. *Let the wave speed c satisfy* $(1/Ro)^2 = (2\omega R/c)^2 = \mu_1^2 + \lambda_1\epsilon$ *with* $\lambda_1 < 0$ *and* $\epsilon > 0$ *small enough. Then in a moving cylindrical coordinate system* (r^*, θ^*, x^*) *there exists a solution for the exact equations such that the free surface is determined by*

$$r^* = R + (R/2)\epsilon J_1(\mu_1)S(\epsilon^{1/2}x^*/R) + O(\epsilon^2),$$

where μ_1 *is the smallest positive root of* $\lambda J_1(\lambda) + 4J_0(\lambda) = 0$, $J_0(z), J_1(z)$ *are the Bessel functions of the first kind of order zero and one* [**22**] *respectively, and*

$$S(z) = (-3\lambda_1\alpha_2/2\alpha_3)sech^2((-\lambda_1\alpha_1/4\alpha_2)^{1/2}z)\,,$$

with

$$\alpha_1 = \int_0^1 J_1^2(\mu_1 r)rdr > 0, \quad \alpha_2 = \int_0^1 J_0^2(\mu_1 r)rdr > 0,$$
$$\alpha_3 = (3J_1^3(\mu_1)/4)\mu_1^2(1 + (\mu_1^2/16)) \neq 0.$$

Therefore the solitary wave solution $R + (R/2)\epsilon J_1(\mu_1)S(\epsilon^{1/2}x^*/R)$ is a first order approximation of the free surface for a solution of the exact equations when ϵ is small.

The idea to prove the theorem is the following. Since the free surface is part of the solution, the domain of the fluid is unknown. By using the stream function to replace r^* together with the longitudinal distance x^* as independent variables and introducing the so-called streamline function as the only dependent variable, we transform the fluid domain into a fixed strip domain. Then we invert the differential equations governing the motion of the flow to a system of two integro-differential equations with two unknowns. To prove the existence of solutions of the differential equations is now reduced to finding fixed points of the corresponding integro-differential operators. In order to use the contraction mapping theorem to find the fixed points, we need estimates for the integro-differential operators. However when the equations are transformed into the corresponding ones in cylindrical coordinates a singularity in the governing equations arises, which makes the estimates complicated. Here we use a method introduced in [**13**] to overcome such difficulties. The idea is that when the operators are extended to these in a higher dimensional space the singularity disappears. One of the operators becomes an elliptic operator and general theory for elliptic operators can be applied. After obtaining the estimates for these operators, we are able to show that the operators are contractions in certain Banach space. Thus the existence is proved by contraction mapping theorem. Our approach is influenced by the work of Friedrichs and Hyers [**5**], Yih [**21**], and Ni [**13**], and by our early work in [**16, 18**]. When *Ro* is near the second largest critical value $1/\mu_2$, the generalized solitary wave solutions will appear [**1, 10, 17**] and the existence proof of such solutions will be different and more complicated. This is subject to further study.

The paper is organized as follows. In Section 2 the problem is formulated and the governing equations are transformed into simplified and mathematically tractable equations. Then the solitary wave solutions are formally derived. The rigorous proof of the Main Theorem is presented in Sections 3 to 6. In Section 3 the governing equations and boundary conditions are further reduced to more manageable forms and several Banach spaces are defined. In Section 4 the estimates for solutions of a nonlocal ordinary differential equation is given and the solutions of the partial differential equations are obtained using eigenfunction expansions from a corresponding eigenvalue problem. In Section 5 the difficulties from the singularity at the longitudinal axis are overcome and the required estimates are obtained for the solutions of the partial differential equations. In Section 6 the differential equations are inverted into integro-differential equations and the existence of solitary wave solutions is proved by the contraction mapping theorem. Then the solutions are transformed into the ones in terms of dimensional variables and the regularity of the solutions is obtained.

2. Governing equations and formal derivation of solutions

We consider a long rotating axisymmetric flow of an inviscid and incompressible fluid of constant density ρ^* with free surface. Assume that there exists a permanent axisymmetric surface wave moving with a constant speed c in the direction of the longitudinal axis of the flow. In reference to a cylindrical coordinate system rotating with a constant angular velocity ω with x^*-axis coincident with the axis of rotation and moving in x^*-direction with a constant speed c, the motion of the flow is steady. Therefore in coordinate system (r^*, θ^*, x^*), the governing equations are the following [**7**]: in $0 \leq r^* < R + \eta^*, \quad -\infty < x^* < \infty$,

$$(r^*u^*)_{r^*} + (r^*w^*)_{x^*} = 0\,, \tag{1}$$

$$\rho^*(u^*u^*_{r^*} + w^*u^*_{x^*} - (r^*)^{-1}(v^*)^2 - 2\omega v^* - \omega^2 r^*) = -p^*_{r^*}\,, \tag{2}$$

$$\rho^*(u^*v^*_{r^*} + w^*v^*_{x^*} + (u^*v^*/r^*) + 2\omega u^*) = 0\,, \tag{3}$$

$$\rho^*(u^* w^*_{r^*} + w^* w^*_{x^*}) = -p^*_{x^*}\,; \tag{4}$$

on the free surface $r^* = R + \eta^*(x^*)$,

$$w^* \eta^*_{x^*} - u^* = 0\,, \quad p^* = P^*_0\,, \tag{5}$$

where $u^*(x^*, r^*)$, $v^*(x^*, r^*)$ and $w^*(x^*, r^*)$ are the components of the velocity in the directions of r^*, θ^* and x^* respectively, $p^*(x^*, r^*)$ is the pressure, $R + \eta^*(x^*)$ is the free surface, and P^*_0 is the constant pressure outside the flow.

We first make the variables nondimensional by using appropriate units. Let

$$\begin{aligned} &(\tilde{r}, x) = (r^*, x^*)/R, \quad Ro = c/(2\omega R), \quad P_0 = P^*_0/(4\rho^*(\omega R)^2), \\ &(u, v, w) = (u^*, v^*, w^*)/(2\omega R), \quad p = p^*/(4\rho^*(\omega R)^2), \quad \eta = \eta^*/R, \end{aligned} \tag{6}$$

where Ro is called Rossby number and is a nondimensional wave speed. In terms of these nondimensional variables, (1) to (5) become

$$(\tilde{r}u)_{\tilde{r}} + (\tilde{r}w)_x = 0\,, \tag{7}$$

$$uu_{\tilde{r}} + wu_x - \tilde{r}^{-1}v^2 - v - (\tilde{r}/4) = -p_{\tilde{r}}\,, \tag{8}$$

$$uv_{\tilde{r}} + wv_x + (uv/\tilde{r}) + u = 0\,, \tag{9}$$

$$uw_{\tilde{r}} + ww_x = -p_x\,; \tag{10}$$

on the free surface $\tilde{r} = 1 + \eta(x)$,

$$w\eta_x - u = 0, \qquad p = P_0. \tag{11}$$

One can easily find an equilibrium $u = v = 0, w = -Ro, \eta = 0$ and $p = P_0 + (\tilde{r}^2 - 1)/8$ for (7) to (11). Since the domain $0 \le \tilde{r} < 1 + \eta$ is unknown and $\eta(x)$ is part of the solution, we have to transform the domain into a fixed domain. By (7) there exists a stream function $\psi^*(x, \tilde{r})$ such that

$$\tilde{r}u = \psi^*_x, \qquad \tilde{r}w = -\psi^*_{\tilde{r}}. \tag{12}$$

By (9) we find that

$$\psi^*_x(\tilde{r}v + (\tilde{r}^2/2))_{\tilde{r}} - \psi^*_{\tilde{r}}(\tilde{r}v + (\tilde{r}^2/2))_x = 0,$$

which implies that $\tilde{r}v = F(\psi^*) - (\tilde{r}^2/2)$ where $F(\psi^*)$ is a function only dependent upon ψ^*. Also (8) and (10) become

$$w(u_x - w_{\tilde{r}}) - (F^2(\psi^*)/\tilde{r}^3) = -(p + (u^2 + w^2)/2)_{\tilde{r}}\,, \tag{13}$$

$$u(w_{\tilde{r}} - u_x) = -(p + (u^2 + w^2)/2)_x\,. \tag{14}$$

Multiply (13) by u and (14) by w and add them together to have

$$-\psi^*_x(p+((u^2+w^2)/2)+(F^2(\psi^*)/2\tilde{r}^2))_{\tilde{r}}+\psi^*_{\tilde{r}}(p+((u^2+w^2)/2)+(F^2(\psi^*)/2\tilde{r}^2))_x = 0$$

by using (12). Therefore there exists a function $G(\psi^*)$ dependent upon ψ^* only so that

$$p + ((u^2 + w^2)/2) + (F^2(\psi^*)/2\tilde{r}^2) = G(\psi^*), \tag{15}$$

which is called Bernoulli's equation. Cross differentiation of (13) and (14) and the use of (12) produces

$$-\psi^*_{\tilde{r}}\Big(((u_x - w_{\tilde{r}})/\tilde{r}) + F(\psi^*)F'(\psi^*)/\tilde{r}^2\Big)_x + \psi^*_x\Big(((u_x - w_{\tilde{r}})/\tilde{r}) + F(\psi^*)F'(\psi^*)/\tilde{r}^2\Big)_t r = 0.$$

Thus there exists a function $H(\psi^*)$ with

$$((u_x - w_{\tilde{r}})/\tilde{r}) + (F(\psi^*)F'(\psi^*)/\tilde{r}^2) = H(\psi^*). \tag{16}$$

Since $F(\psi^*), G(\psi^*)$ and $H(\psi^*)$ are functions of ψ^* only, they can be determined by conditions of u, v, w, η and p at infinity. There are two ways to find $F(\psi^*), G(\psi^*)$ and $H(\psi^*)$, either by giving u, v, w, η and p at infinity or by assuming forms of $F(\psi^*), G(\psi^*)$ and $H(\psi^*)$ and determining u, v, w, η and p accordingly. In this paper, the former will be used. Let $u = v = 0, w = -Ro, \eta = 0$ and $p = P_0 + (\tilde{r}^2 - 1)/8$ at infinity, which implies that the flow at infinity is the equilibrium. Therefore at infinity

$$\begin{aligned} &F(\psi^*) = (\tilde{r}^2/2), \quad -\tilde{r}Ro = -\psi^*_{\tilde{r}}, \quad \psi^* = (\tilde{r}^2 Ro/2), \quad v = (\psi^*/\tilde{r}Ro) - (\tilde{r}/2), \\ &G(\psi^*) = P_0 + (\psi^*/2Ro) + (Ro^2/2) - (1/8), \quad F(\psi^*) = (\psi^*/Ro), \end{aligned} \tag{17}$$

where $\psi^* = 0$ at $\tilde{r} = 0$. The free surface is $\tilde{r} = 1$ at infinity and $\psi^* = (Ro/2) = \psi_s$. On the free surface $\psi^* = \psi_s$, by (15) we have

$$p + ((u^2 + w^2 + (1/2\tilde{r})^2)/2) = P_0 + (Ro^2/2) + (1/8). \tag{18}$$

Thus from (16), $H(\psi^*) = (1/2Ro)$ at infinity and

$$(u_x - w_{\tilde{r}})\tilde{r}^{-1} + (\psi^*/\tilde{r}^2 Ro^2) = (1/2Ro). \tag{19}$$

By using (12), (18) and (19) we obtain the following equations, which are equivalent to (7) to (11),

$$\begin{aligned} &\tilde{r}((\psi^*_x/\tilde{r})_x + (\psi^*_{\tilde{r}}/\tilde{r})_{\tilde{r}}) + (\psi^*/Ro^2) = \tilde{r}^2/(2Ro) \\ &\qquad \text{at } x \in (-\infty\infty), 0 \le \psi^* < \psi_s, \end{aligned} \tag{20}$$

$$(\psi^*_x)^2 + (\psi^*_{\tilde{r}})^2 + (1/4) = ((1/4) + Ro^2)\tilde{r}^2 \quad \text{at } \psi^* = \psi_s, \tag{21}$$

$$\tilde{r} = 0 \quad \text{at} \quad \psi^* = 0. \tag{22}$$

Let $\psi = \psi^*/\psi_s$ and use x, ψ as independent variables and $f(x, \psi)$ as dependent variable so that

$$\psi(x, f) = \text{constant} \tag{23}$$

is a stream line, where f is called streamline function. Thus

$$\psi_{\tilde{r}} = 1/f_\psi\,, \qquad \psi_x = -f_x/f_\psi\,. \tag{24}$$

From (24), (20) to (22) become

$$\begin{aligned} &(f_x/f_\psi)_\psi(f_x/f_\psi) - (f_x/f_\psi)_x - (f_{\psi\psi}/f^3_\psi) - (1/ff_\psi) \\ &\qquad + ((\psi - f^2)/Ro^2) = 0, \end{aligned} \tag{25}$$

$$(Ro^2/4)(1 + f^2_x)(ff_\psi)^{-2} + (1/4f^2) = Ro^2 + (1/4) \quad \text{at } \psi = 1, \tag{26}$$

$$f = 0 \quad \text{at} \quad \psi = 0. \tag{27}$$

To let (25) to (27) be mathematical treatable, we use change of variable

$$f(x,\psi) = (g(x,\psi))^{1/2}. \tag{28}$$

(25) to (27) are transformed into

$$\begin{aligned} &g_{\psi\psi}(g + (g_x^2/4)) + (g_\psi^2 g_{xx}/4) - (g_x g_\psi g_{x\psi}/2) \\ &\qquad\qquad + (g_\psi^3/4Ro^2)(g - \psi) = 0, \end{aligned} \tag{29}$$

$$Ro^2(4g + g_x^2) + g_\psi^2 = g g_\psi^2(4Ro^2 + 1) \quad \text{at } \psi = 1, \tag{30}$$

$$g = 0 \qquad \text{at} \qquad \psi = 0. \tag{31}$$

It is easy to see that (29) to (31) have an equilibrium state $g = \psi$. In the following, we shall show that (29) to (31) have a solitary wave solution near the equilibrium. Let

$$g(x,\psi) = \psi + w(x,\psi), \tag{32}$$

and write (29) to (31) as

$$\begin{aligned} \psi w_{\psi\psi} + (1/4)w_{xx} &= -[w w_{\psi\psi} + (w_x^2 w_{\psi\psi}/4) + ((2w_\psi + w_\psi^2)w_{xx}/4) \\ &\quad - (w_x(1 + w_\psi)w_{\psi x}/2)] - (\mu^2/4)w(1 + w_\psi)^3, \end{aligned} \tag{33}$$

$$\begin{aligned} w_\psi + (\mu^2/8)w &= (w_x^2/8) + (\mu^2 w_\psi^2/8) \\ &\quad - ((1/2) + (\mu^2/8))(2w w_\psi + w_\psi^2 + w w_\psi^2) \quad \text{at } \psi = 1, \end{aligned} \tag{34}$$

$$w = 0 \qquad \text{at} \qquad \psi = 0, \tag{35}$$

where $\mu = 1/Ro$. Therefore we shall show equivalently that (33) to (35) have a solitary wave solution.

First we use a formal method to derive solitary wave solutions. Use stretching variable $\xi = \epsilon^{1/2}x$ and following asymptotic expansions of w and μ to obtain solutions of (33) to (35);

$$w(x,\psi) = \epsilon w_1(\xi,\psi) + \epsilon^2 w_2(\xi,\psi) + \cdots, \tag{36}$$

$$\mu^2 = \lambda_0^2 + \epsilon\lambda_1, \tag{37}$$

where $\epsilon > 0$ is a small parameter. Substitute (36) and (37) into (33) to (35) to have a sequence of equations and boundary conditions for the successive approximations. The first order approximation is

$$\begin{aligned} &\psi w_{1\psi\psi} + (\lambda_0^2/4)w_1 = 0, \\ &w_{1\psi} + (\lambda_0^2/8)w_1 = 0 \quad \text{at} \quad \psi = 1, \\ &w_1 = 0 \qquad \text{at} \qquad \psi = 0. \end{aligned}$$

Thus $w_1 = \psi^{1/2}J_1(\lambda_0\psi^{1/2})S(\xi)$ and λ_0 satisfies

$$\lambda_0 J_1(\lambda_0) + 4J_0(\lambda_0) = 0, \tag{38}$$

where $J_0(z)$ and $J_1(z)$ are the Bessel functions of the first kind of order zero and one [**22**]. (38) has infinitely many positive solutions $0 < \mu_1 < \mu_2 < \mu_3 < \cdots$. Let

$\lambda_0 = \mu_n$. In order to find $S(\xi)$, we have to use second order approximation. The second order approximate equation and boundary conditions are

$$\psi w_{2\psi\psi} + (\mu_n^2/4)w_2 = (-1/4)w_{1xx} - (\lambda_1/4)w_1 - (3\mu_n^2/4)w_1 w_{1\psi} - w_1 w_{1\psi\psi}, \tag{39}$$

$$w_{2\psi} + (\mu_n^2/8)w_2 = -(\lambda_1/8)w_1 + (\mu_n^2 w_{1\psi}^2/8) - ((1/2) + (\mu_n^2/8))(2w_1 w_{1\psi} + w_{1\psi}^2) \quad \text{at} \quad \psi = 1, \tag{40}$$

$$w_2 = 0 \qquad \text{at } \psi = 0. \tag{41}$$

Multiply (39) by $J_1(\mu_n \psi^{1/2})/\psi^{1/2}$, integrate it from zero to one and use (40) and (41) to obtain

$$J_1(\mu_n)(-(\lambda_1/8)w_1 + (\mu_n^2 w_{1\psi}^2/8) - ((1/2) + (\mu_n^2/8))(2w_1 w_{1\psi} + w_{1\psi}^2)) + \int_0^1 ((1/4)w_{1xx} + (\lambda_1/4)w_1 + (3\mu_n^2/4)w_1 w_{1\psi} + w_1 w_{1\psi\psi})\psi^{-1/2} J_1(\mu_n \psi^{1/2}) d\psi = 0.$$

Substitution of w_1 and (38) yields

$$\alpha_1 S_{\xi\xi} + \lambda_1 \alpha_2 S + \alpha_3 S^2 = 0, \tag{42}$$

where

$$\alpha_1 = \int_0^1 J_1^2(\mu_n r) r dr, \quad \alpha_2 = \int_0^1 J_1^2(\mu_n r) r dr - (J_1^2(\mu_n)/4),$$
$$\alpha_3 = 3J_1^3(\mu_n)\mu_n^2(1 + (\mu_n^2/16))/4.$$

By using the formulas for the Bessel functions [**22**] and (38), $\alpha_2 = \int_0^1 J_0^2(\mu_n r) r dr > 0$. Thus (42) has a solitary wave solution

$$S(\xi) = (-3\lambda_1 \alpha_2 / 2\alpha_3) \operatorname{sech}^2((-\lambda_1 \alpha_1 / 4\alpha_2)^{1/2} \xi), \tag{43}$$

and formally

$$w_1(\xi, \psi) = \psi^{1/2} J_1(\mu_n \psi^{1/2}) S(\xi) \tag{44}$$

is the first order approximation of the solution of (33) to (35) where $\lambda_1 < 0$ and μ_n satisfies (38). In the following, we shall rigorously show that w_1 in (44) is a first order approximation of a solution of (33) to (35) for ϵ small and $n = 1$. For $n \geq 2$, the generalized solitary wave solutions [**1, 10, 17**] must be introduced and the proof is different, which is subject to further study.

3. Preliminaries

From the form of the approximate solution $w_1(\epsilon^{1/2}x, \psi)$ in (44), we first transform (33) to (35) to more convenient forms before we prove the existence of a solitary wave solution for the equations. Let $\psi = r^2$ and $w(x, \psi) = r^2 u(x, r)$ in (33) to (35) to have

$$
\begin{aligned}
&u_{rr} + (3/r)u_r + u_{xx} + \mu^2 u = \Big((u + r^2u_x^2/4)(u_{xx} + \mu^2 u) \\
&\quad - ((u + ru_r/2)^2 + 2u + ru_r)u_{xx} + u_x(ru_{rx} + 2u_x)(1 + u + ru_r/2) \\
&\quad - \mu^2 u(3(u + ru_r/2) + 3(u + ru_r/2)^2 + (u + ru_r/2)^3)\Big)(1 + u + r^2u_x^2/4)^{-1} \\
&\quad = F_1(u)\,, \qquad -\infty < x < \infty,\ 0 < r < 1\,,
\end{aligned} \tag{45}
$$

$$
\begin{aligned}
&u_r + (2 + \mu^2/4)u = (u_x^2/4) + (\mu^2/4)(u + u_r/2)^2 - (1 + \mu^2/4)(uu_r \\
&\quad + 2u^2 + (u + u_r/2)^2 + u(u + u_r/2)^2) = \Psi_1(u) \qquad \text{at} \quad r = 1,
\end{aligned} \tag{46}
$$

$$
u \text{ is finite} \quad \text{at } r = 0. \tag{47}
$$

Motivated by the formal solutions in Section 3, we let

$$
\mu^2 = \mu_1^2 + \lambda_1 \epsilon \tag{48}
$$

with $\lambda_1 < 0$ and

$$
u(x, r) = \epsilon(r^{-1}J_1(\mu_1 r)(S(\epsilon^{1/2}x) + \omega(x)) + \theta(x, r))\,, \tag{49}
$$

where $\omega(x)$ and $\theta(x, \psi)$ are small perturbation terms, and

$$
\int_0^1 r^2 J_1(\mu_1 r)\theta(x, r)dr = 0 \qquad \text{for all } x \in \mathbf{R}. \tag{50}
$$

In terms of $\omega(x)$ and $\theta(x, \psi)$, (45) to (47) can be rewritten as

$$
\begin{aligned}
&\theta_{rr} + (3/r)\theta_r + \theta_{xx} + \mu_1^2\theta = -r^{-1}J_1(\mu_1 r)(S(\epsilon^{1/2}x) + \omega(x))_{xx} \\
&\quad - \lambda_1\epsilon(r^{-1}J_1(\mu_1 r)(S + \omega) + \theta) + \epsilon^{-1}F_1(\epsilon(r^{-1}J_1(\mu_1 r)(S + \omega) + \theta)),
\end{aligned} \tag{51}
$$

$$
\begin{aligned}
&\theta_r + (2 + \mu_1^2/4)\theta = -(\lambda_1\epsilon/4)(J_1(\mu_1)(S + \omega) + \theta) \\
&\quad + \epsilon^{-1}\Psi_1(\epsilon(r^{-1}J_1(\mu_1 r)(S + \omega) + \theta)) \quad \text{at } r = 1,
\end{aligned} \tag{52}
$$

$$
\theta \text{ is finite at } r = 0. \tag{53}
$$

If we multiply both sides of (51) by $r^2 J_1(\mu_1 r)$, integrate it from zero to one and use (50), we have a solvability condition of (51) to (53)

$$
\begin{aligned}
&\int_0^1 \Big(- r^{-1}J_1(\mu_1 r)(S + \omega)_{xx} - \lambda_1\epsilon(r^{-1}J_1(\mu_1 r)(S + \omega) + \theta) \\
&\qquad + \epsilon^{-1}F_1(\epsilon(r^{-1}J_1(\mu_1 r)(S + \omega) + \theta))\Big) r^2 J_1(\mu_1 r)dr \\
&\quad = J_1(\mu_1)\Big(- (\lambda_1\epsilon/4)(J_1(\mu_1)(S + \omega) \\
&\qquad + \theta(x, 1)) + \epsilon^{-1}\Psi_1(\epsilon(r^{-1}J_1(\mu_1 r)(S + \omega) + \theta))\Big)\Big|_{r=1},
\end{aligned}
$$

which implies

$$
\alpha_1\omega_{xx} + \lambda_1\alpha_2\epsilon\omega + 2\alpha_3\epsilon S(\epsilon^{1/2}x)\omega = g_1^*(S, \omega, \theta), \tag{54}
$$

where $\alpha_1, \alpha_2, \alpha_3$ are defined in (42) and

$$\begin{aligned} g_1^*(S,\omega,\theta)(x) = &-\alpha_3\epsilon\omega^2 + J_1(\mu_1)\Big((\lambda_1\epsilon/4)\theta(x,1) - \epsilon^{-1}\Psi_1(\epsilon(r^{-1}J_1(\mu_1 r)(S+\omega) \\ &+\theta)) - \epsilon(1+\mu_1^2/4)J_1(\mu_1)\mu_1 J_0(\mu_1)(S+\omega)^2 - (S+\omega)^2(\mu_1^2 J_0^2(\mu_1)/4)\Big) \\ &+\int_0^1 \Big(-\lambda_1\epsilon\theta + \epsilon^{-1}F_1(\epsilon(r^{-1}J_1(\mu_1 r)(S+\omega)+\theta)) \\ &+(S+\omega)^2\mu_1^2\epsilon J_1(\mu_1 r)((3\mu_1/2r)J_0(\mu_1 r) - (J_1(\mu_1 r)/r^2))\Big)r^2 J_1(\mu_1 r)dr. \end{aligned}$$

By dividing both sides of (54) by α_1, we have

$$\omega_{xx} + \lambda_1 A_1\epsilon\omega + A_2\epsilon S(\epsilon^{1/2}x)\omega = g_1(S,\omega,\theta), \tag{55}$$

where $A_1 = \alpha_2/\alpha_1, A_2 = 2\alpha_3/\alpha_1$, and $g_1 = g_1^*/\alpha_1$. In the following we shall prove that (51) to (53) and (55) have a solution θ, ω in some Banach spaces and θ and ω will go to zero when ϵ goes to zero. Thus by (49), $\epsilon r^{-1}J_1(\mu_1 r)S(\epsilon^{1/2}x)$ is a first order approximation of $u(x,r)$.

Before we prove the existence, we first define several Banach spaces for later use. Let $0<\lambda<1$, m, n be nonnegative integers,

$$\begin{aligned} D &= \{(x,r) \mid 0 \le r \le 1, -\infty < x < \infty\}, \\ C(f(x,r)) &= \sup_{(x,r)\in D} (|f(x,r)|\exp(d\epsilon^{1/2}|x|)), \end{aligned}$$

and

$$H_\lambda(f(x,r)) = \sup_{\substack{0\le r_1,r_2\le 1\\ x\in\mathbf{R},|\delta|<1}} (|f(x+\delta,r_1) - f(x,r_2)|(\delta^2+(r_1-r_2)^2)^{-\lambda/2}\exp(d\epsilon^{1/2}|x|)),$$

where $d>0$ is a small fixed constant and less than $(-\lambda_1\alpha_1/4\alpha_2)^{1/2}$. Define

$$\begin{aligned} B_{n,\lambda} = \{&f(x,r) \in C^n((-\infty,\infty)\times[0,1]) \big| f(x,r) = f(-x,r), \\ &\|f\|_{B_{n,\lambda}} = \sum_{m=0}^{n}\sum_{k=0}^{m} C(\partial^m f/\partial x^k \partial r^{m-k}) + \sum_{m=0}^{n} H_\lambda(\partial^n f/\partial x^m \partial r^{n-m}) < \infty\}. \end{aligned}$$

Here the definition of $B_{n,\lambda}$ can also be used for functions of x only.

4. Estimates of solutions for differential equations

First we consider estimates for a solution of (55). Let

$$\mathcal{L}\omega = \omega_{xx} + \lambda_1 A_1\epsilon\omega + A_2\epsilon S(\epsilon^{1/2}x)\omega = h(x). \tag{56}$$

It is easy to see that a solution of $\mathcal{L}\omega = 0$ is $dS(\epsilon^{1/2}x)/dx = \psi_1(x)$, which is an odd function and decays to zero at $|x|=\infty$. By the method of variation of parameters, another linearly independent even solution $\psi_2(x)$ can be found, which tends to infinity as $|x| \to +\infty$. Note that we need $\psi_1'(x)\psi_2(x) - \psi_2'(x)\psi_1(x) = -1$. If $h(x)$ and $\omega(x)$ are even, (56) can be transformed into

$$\begin{aligned} \omega(x) &= \int_0^\infty k(x,s)h(s)ds \quad \text{for } x \geq 0\,, \\ &= \int_0^\infty k(-x,s)h(s)ds \quad \text{for } x < 0\,, \\ &= \mathcal{L}^{-1}(h(s))(x)\,, \end{aligned} \tag{57}$$

where

$$k(x,s) = \begin{cases} \psi_1(x)\psi_2(s)\,, & \text{for } x > s\,, \\ \psi_1(s)\psi_2(x) & \text{for } x < s\,. \end{cases}$$

Then we have the following Lemma.

LEMMA 1. *For $h(x) \in B_{m,\lambda}$ with $m \geq 0$,*

$$\|\mathcal{L}^{-1}(h(s))(x)\|_{B_{m+2,\lambda}} \leq K\epsilon^{-1}\|h(x)\|_{B_{m,\lambda}}\,,$$

where $K > 0$ is a generic constant independent of ϵ.

The proof of Lemma 1 is a standard one and can be found in [**17, 19**], which is omitted here. From (55), we can check term by term in $g_1(S,\omega,\theta)$ and obtain that if $\|\omega\|_{B_{m,\lambda}} \leq K\epsilon^{1/2}$, $\|\theta\|_{B_{m,\lambda}} \leq K\epsilon$ (or $K\epsilon^{1/2}$) for $m \geq 2$, then $g_1 \in B_{m-2,\lambda}$ and $\|g_1(S,\omega,\theta)\|_{B_{m-2,\lambda}} \leq K\epsilon^2$ (or $K\epsilon^{3/2}$). Thus by Lemma 1, we have

THEOREM 1. *If $\omega, \theta \in B_{m,\lambda}$ for $m \geq 2$ and $\|\omega\|_{B_{m,\lambda}} \leq K\epsilon^{1/2}$, $\|\theta\|_{B_{m,\lambda}} \leq K\epsilon$ (or $K\epsilon^{1/2}$), then $\mathcal{L}^{-1}g_1(S,\omega,\theta) \in B_{m,\lambda}$ and $\|\mathcal{L}^{-1}g_1(S,\omega,\theta)\|_{B_{m,\lambda}} \leq K\epsilon$ (or $K\epsilon^{1/2}$).*

Now we study the solutions of the following partial differential equation

$$v_{rr} + v_{xx} + (3/r)v_r + \mu_1^2 v = \varphi_1(x,r) \quad \text{at} \quad -\infty < x < \infty,\ \ 0 < r < 1\,, \tag{58}$$

$$v_r + (2 + \mu_1^2/4)v = \varphi_2(x) \quad \text{at } r = 1\,, \tag{59}$$

$$r^{1/2}v \quad \text{is finite at } r = 0\,, \tag{60}$$

where φ_1, φ_2 satisfy

$$\int_0^1 \varphi_1(x,r)r^2 J_1(\mu_1 r)dr - J_1(\mu_1)\varphi_2(x) = 0\,, \tag{61}$$

and tend to zero exponentially as $|x| \to \infty$. Before studying (58) to (60), we first consider the corresponding homogeneous problem

$$v_{rr} + v_{xx} + (3/r)v_r + \mu_1^2 v = 0 \quad \text{at} \quad -\infty < x < \infty,\ \ 0 < r < 1\,, \tag{62}$$

$$v_r + (2 + \mu_1^2/4)v = 0 \quad \text{at } r = 1\,, \tag{63}$$

$$r^{1/2}v \text{ is finite at } r = 0\,. \tag{64}$$

If we use separation of variables and let $v = U(r)X(x)$, then $U(r)$ and $X(x)$ satisfy

$$X'' + \tilde{\lambda}X = 0 \quad \text{at} \quad -\infty < x < \infty\,, \tag{65}$$

$$U_{rr} + (3/r)U_r + \mu_1^2 U - \tilde{\lambda}U = 0 \quad \text{at} \quad 0 < r < 1\,, \tag{66}$$

$$U_r + (2 + \mu_1^2/4)U = 0 \quad \text{at } r = 1 \tag{67}$$

$$r^{1/2}U \text{ is finite at } r = 0\,. \tag{68}$$

(66) to (68) are an eigenvalue problem with singularity at $r=0$. If we let $U = r^{-3/2}u$ in (66) and (67), the equations (66) and (67) are transformed into

$$u_{rr} + (\mu_1^2 - \tilde{\lambda} - (3/4r^2))u = 0 \quad \text{in} \quad 0<r<1\,, \tag{69}$$

$$u_r + ((2+\mu_1^2)/4)u = 0 \quad \text{at} \quad r=1\,. \tag{70}$$

The eigenvalue problem (69) and (70) is a special case studied in [**20**]. It is shown in [**20**] that the eigenfunctions of (69) and (70) form an orthonormal basis of $L^2(0,1)$ and there is a decreasing sequence of simple eigenvalues $\tilde{\lambda}_0, \tilde{\lambda}_1, \cdots$ such that the eigenfunction associated with $\tilde{\lambda}_m$ has just m zeros in $0<r<1$. Therefore the eigenfunctions of (66) and (67) form an orthonormal basis for a Hilbert space $H = \{f(r) \mid \|f\|_w = \langle f,f\rangle_w^{1/2} < \infty\}$ with weighted inner product $\langle f,g\rangle_w = \int_0^1 f(r)g(r)r^3dr$. By calculating the eigenfunctions of (66) and (67) in H explicitly, we can see that the eigenfunctions also satisfy (68). However $\tilde{\lambda}=0$ is an eigenvalue of (66) to (68) with eigenfunction $r^{-1}J_1(\mu_1 r)$, which is always positive in $(0,1)$ since the smallest positive root of (38) is less than the first positive zero of $J_1(z)$. Thus the other eigenvalues of (66) to (68) are negative. Therefore we have

THEOREM 2. *(66) to (68) possess countably infinitely many nonpositive simple eigenvalues and the corresponding eigenfunctions are complete in H.*

We denote the eigenvalues by $0=\sigma_0 > -\sigma_1^2 > -\sigma_2^2 > \cdots > -\sigma_n^2 > \cdots$ and the corresponding eigenfunctions by $e_0(r) = r^{-1}J_1(\mu_1 r), e_1(r), e_2(r), \cdots$ respectively, where $\sigma_n, n=1,2,3,\cdots$ are the positive roots of

$$(\mu_1^2+\sigma_n^2)^{1/2}J_0((\mu_1^2+\sigma_n^2)^{1/2}) + (\mu_1^2/4)J_1((\mu_1^2+\sigma_n^2)^{1/2}) = 0, \tag{71}$$

and

$$\begin{aligned} e_n(r) &= (C_n^*)^{1/2}r^{-1}J_1((\mu_1^2+\sigma_n^2)^{1/2}r), \\ C_n^* &= \left(\int_0^1 J_1^2((\mu_1^2+\sigma_n^2)^{1/2}r)rdr\right)^{-1} \\ &\sim \left((\pi(\mu_1^2+\sigma_n^2)^{1/2})^{-1} + O(|\sigma_n|^{-2}\ln|\sigma_n|)\right)^{-1}, \end{aligned} \tag{72}$$

$$(\mu_1^2+\sigma_n^2)^{1/2} \sim (3\pi/4) + n\pi \qquad \text{for } n \text{ large}\,. \tag{73}$$

Now we can use these eigenfunctions to find a solution of (58) to (60). Multiplying both sides of (58) by $r^3e_n(r)$ for $n \geq 0$ and integrating the resulting equation by parts twice, we obtain

$$C_{nxx} - \sigma_n^2 C_n = d_n \quad \text{for } n=0,1,2,\cdots \tag{74}$$

where

$$C_n(x) = \int_0^1 v(x,r)e_n(r)r^3dr\,, \tag{75}$$

$$d_n(x) = \int_0^1 \varphi_1(x,r)e_n(r)r^3dr - e_n(1)\varphi_2(x) = d_n^{(1)}(x) - d_n^{(2)}(x)\,. \tag{76}$$

From (61), $d_0 = 0$. Thus we choose $C_0(x) \equiv 0$. For $\int_{-\infty}^{\infty}\int_0^1 \varphi_1^2(x,r)r^3 dr dx + \int_{-\infty}^{\infty}\varphi_2^2(x)dx < \infty$, by (74)

$$C_n(x) = (1/2\sigma_n)\int_{-\infty}^{\infty}\exp(-\sigma_n|x-\xi|)d_n(\xi)d\xi\,, \tag{77}$$

and

$$\begin{aligned}\int_{-\infty}^{\infty} C_n^2(x)dx &= (1/4\sigma_n^2)\int_{-\infty}^{\infty}\left(\int_{-\infty}^{\infty}\exp(-\sigma_n^{1/2}|x-\xi|)d_n^{(1)}(\xi)d\xi\right)^2 dx \\ &\quad + (1/4\sigma_n^2)\int_{-\infty}^{\infty}\left(\int_{-\infty}^{\infty}\exp(-\sigma_n^{1/2}|x-\xi|)d_n^{(2)}(\xi)d\xi\right)^2 dx \\ &\le K\sigma_n^{-4}(\int_{-\infty}^{\infty}(d_n^{(1)}(\xi))^2 d\xi + \int_{-\infty}^{\infty}\varphi_2^2(\xi)d\xi)\,,\end{aligned}$$

where K is a fixed positive constant. By a similar calculation,

$$\begin{aligned}\int_{-\infty}^{\infty}|D_x C_n(x)|^2 dx &\le K\sigma_n^{-2}\left(\int_{-\infty}^{\infty}(d_n^{(1)}(x))^2 dx + \int_{-\infty}^{\infty}\varphi_2^2(x)dx\right)\,, \\ \int_{-\infty}^{\infty}|D_x^2 C_n(x)|^2 dx &\le K\left(\int_{-\infty}^{\infty}(d_n^{(1)}(x))^2 dx + \sigma_n^{-2}\int_{-\infty}^{\infty}\varphi_{2x}^2(x)dx\right)\,,\end{aligned}$$

if $\int_{-\infty}^{+\infty}\varphi_{2x}^2(x)dx < +\infty$. From Theorem 2,

$$\sum_{n=1}^{\infty}\int_{-\infty}^{+\infty}(d_n^{(1)}(\xi))^2 d\xi \le \int_{-\infty}^{+\infty}\int_0^1 \varphi_1^2(x,r)r^3 dr dx < +\infty.$$

Let

$$v(x,r) = \sum_{n=1}^{\infty} C_n(x)e_n(r)\,. \tag{78}$$

For $\int_{-\infty}^{+\infty}\int_0^1 \varphi_1^2(x,r)r^3 dr dx + \sum_{j=0}^{1}\int_{-\infty}^{+\infty}(D_x^j\varphi_2(x))^2 dx < +\infty$, it is obtained that $D_x^j v(x,r) \in H$ for $j = 0,1,2$ and almost every $x \in \mathbf{R}$, and

$$\begin{aligned}\int_{-\infty}^{+\infty}\int_0^1 (D_x^j v(x,r))^2 r^3 dr &\le \int_{-\infty}^{+\infty}\int_0^1 \varphi_1^2(x,r)r^3 dr dx + \sum_{j=0}^{1}\int_{-\infty}^{+\infty}(D_x^j\varphi_2(x))^2 dx \\ &\equiv B(\varphi_1,\varphi_2).\end{aligned} \tag{79}$$

If $d_n(x)$ in (76) is not zero for only finitely many n, then $v(x,r)$ defined in (78) satisfies (58) to (60) and

$$\int_{-\infty}^{\infty}\int_0^1 (D_r v)^2 r^3 dr dx + \int_{-\infty}^{+\infty} v^2(x,1)dx \le KB(\varphi_1,\varphi_2)\,. \tag{80}$$

When $d_n(x), n = 1,2,3,\cdots$ have infinitely many nonzero terms, we can use finite terms of nonzero $d_n(x)$ to construct an approximation of $v(x,r)$ in (78) and then let the number of nonzero terms go to infinity to have (80). Thus (80) holds for

$v(x,r)$ defined in (78). In the following we call $w(x,r)$ a generalized solution of (58) to (60) if

$$\int_{-\infty}^{\infty}\int_{0}^{1}(\sum_{j=0}^{1}(D_x^j w)^2+(D_r w)^2)r^3 dr dx+\int_{-\infty}^{+\infty} w^2(x,1)dx<\infty, \tag{81}$$

and $w(x,r)$ satisfies

$$\begin{aligned}&\int_{-\infty}^{\infty}\left(\int_0^1 (D_r w(x,r)D_r z(x,r)+D_x w(x,r)D_x z(x,r)-\mu_1^2 w(x,r)z(x,r))r^3 dr\right)dx\\ &-\int_{-\infty}^{\infty}(\varphi_2(x)-(2+\mu_1^2/4)w(x,1))z(x,1)dx+\int_{-\infty}^{\infty}\int_0^1 \varphi_1(x,r)z(x,r)r^3 dr dx=0\,,\end{aligned} \tag{82}$$

for all $z(x,r)\in C^1(\mathbf{R}\times[0,1])$ with $z(\,\cdot\,,r)\in C_0^1(\mathbf{R})$. By the construction, $v(x,r)$ defined in (78) is a generalized solution of (58) to (60) if $B(\varphi_1,\varphi_2)<\infty$ and (61) holds.

5. Laplacian in $\mathbf{R}^5$

Equation (58) has a singularity at $r=0$. However, the first three terms of the left-hand side of (58) are the Laplacian in $\mathbf{R}^5$ for a function with cylindrical symmetry in cylindrical coordinates. This observation was first made by Ni [**13**] and was used in [**16, 18**]. Here we extend a weak solution of (58) to (60) to one of corresponding equations in $\mathbf{R}^5$. Let

$$\Omega=\left\{(x,y_1,y_2,y_3,y_4)=(x,\overline{y})\Big|-\infty<x<\infty,\rho=\big(\sum_{i=1}^{4}y_i^2\big)^{1/2}\le 1\right\}.$$

Then $w(x,\overline{y})$ is called a generalized solution of an elliptic equation in Ω

$$\begin{aligned}&\sum_{i=1}^{4}v_{y_iy_i}+v_{xx}+\mu_1^2 v=\varphi_1(x,\overline{y}),\\ &\frac{\partial v}{\partial n}+(2+(\mu_1^2/4))v=\varphi_2(x)\,,\end{aligned} \tag{83}$$

where $\partial/\partial n$ is the unit outward normal derivative of $\partial\Omega$, and φ_1,φ_2 satisfy

$$\int_\Omega \varphi_1^2 d\Omega+\sum_{j=0}^{1}\int_{-\infty}^{\infty}(D_x^j\varphi_2(x))^2dx<\infty\,, \tag{84}$$

if $w(x,\overline{y})$ in $H^1(\Omega)$ satisfies

$$\begin{aligned}&\int_\Omega\left(\sum_{i=1}^{4}w_{y_i}z_{y_i}+w_x z_x-\mu_1^2 wz\right)d\Omega-\int_{\partial\Omega}(\varphi_2(x)-(2+\mu_1^2/4)w(x,\overline{y}))z(x,\overline{y})d\Gamma\\ &+\int_\Omega \varphi_1 z d\Omega=0\,,\end{aligned} \tag{85}$$

for all $z(x,\overline{y}) \in C^1(\Omega)$ and $z(\cdot,\overline{y}) \in C_0^1(\mathbf{R})$. From (78), we let

$$v(x,\rho) = \sum_{n=1}^{\infty} C_n(x) e_n(\rho). \tag{86}$$

Since $v(x,r)$ satisfies (82), it is straightforward to check that $v(x,\rho)$ in (86) is a generalized solution of (83) in Ω if $\varphi_1(x,\overline{y}) = \varphi_1(x,\rho)$ is cylindrical symmetric in $\mathbf{R}^5$. In the following we shall make use of the general theory of regular elliptic equations in $\mathbf{R}^5$ [**6**] to derive a priori estimates for $v(x,\rho)$, which can be carried over to $v(x,r)$.

LEMMA 3. *Assume that $\varphi_1(x,\rho), \varphi_2(x)$ satisfy (61) with compact supports of x-variable in $(-10,10)$ and belong to $B_{0,\lambda}$ and $B_{1,\lambda}$ respectively for $0 < \lambda < 1$. Then $v(x,\rho) \in B_{2,\lambda}$ and*

$$\|v(x,\rho)\|_{B_{2,\lambda}} \le K(\|\varphi_1(x,\rho)\|_{B_{0,\lambda}} + \|\varphi_2(x)\|_{B_{1,\lambda}}).$$

PROOF. Since the supports of $\varphi_1(x,\rho)$ and $\varphi_2(x)$ are in $(-10,10)$, by (72), (73), (77) and (86) it is easy to see that for $|x| \ge 20$ we can take differentiations of $v(x,\rho)$ inside the integrals to obtain

$$\sup_{|x|\ge 20} |D^m v(x,\rho)| \le K(m) \left(\|\varphi_1\|_{B_{0,\lambda}} + \|\varphi_2\|_{B_{0,\lambda}}\right), \tag{87}$$

for $m \ge 0$ where $K(m)$ may depend on m. Since $\varphi_1(x,\rho) \in B_{0,\lambda}$, $\varphi_1(x,(y_1^2+y_2^2+y_3^2+y_4^2)^{1/2})$ is also Hölder continuous with exponent λ for x, y_i, $i = 1,\dots,4$. Then take a bounded domain $\tilde{\Omega}$ with $0 \in \tilde{\Omega}$ in $\mathbf{R}^5$ as follows; the boundary $\partial\tilde{\Omega}$ is smooth enough and contains Ω_1 with $\Omega_1 = \{|x| \le 30, \rho \le 1\}$. By letting $z(x,\overline{y}) \in C_0^1(\tilde{\Omega})$ in (83), we know that $v(x,\overline{y}) = v(x,\rho)$ in (86) is a solution of the equation (85) in $\tilde{\Omega}$. Since $\varphi_1(x,\overline{y})$ is Hölder continuous, by elliptic operator theory [**6**], $v(x,\overline{y})$ is Hölder continuous in the interior of $\tilde{\Omega}$ with exponent λ up to second derivatives and its interior Hölder norms are bounded by $\|\varphi_1(x,\rho)\|_{B_{0,\lambda}}$. Also inequality (87) holds on $\partial\tilde{\Omega} \cap \{|x| \ge 20\}$. On $\partial\tilde{\Omega} \cap \{|x| \le 20\}$, since $\varphi_2(x) \in B_{1,\lambda}$, the classical boundary estimates in elliptic operator theory imply that the Hölder norms of $v(x,\overline{y})$ up to second order derivatives are bounded by $\|\varphi_1(x,\rho)\|_{B_{0,\lambda}}$ and $\|\varphi_2(x)\|_{B_{1,\lambda}}$ in a small neighborhood of $\partial\tilde{\Omega} \cap \{|x| \le 20\}$. Thus by the interior and boundary estimates of $v(x,\rho)$ in $\tilde{\Omega}$ together with the estimates outside $\tilde{\Omega}$, and noticing that

$$v(x,\overline{y})\Big|_{y_2=y_3=y_4=0,\, y_1\ge 0} = v(x,y_1) = v(x,\rho),$$

we have that $v(x,\rho) \in B_{2,\lambda}$ and $\|v(x,\rho)\|_{B_{2,\lambda}} \le K(\|\varphi_1\|_{B_{0,\lambda}} + \|\varphi_2\|_{B_{1,\lambda}})$, since for $|x| \ge 20$, it is straightforward to show that $v(x,\rho)$ has the exponentially decay at infinity from (72), (73), (77) and (86). Thus we complete the proof of the Lemma.

In the following we shall prove Lemma 3 for general $\varphi_1(x,\rho), \varphi_2(x)$ satisfying (61).

LEMMA 4. *Let $\varphi_1(x,\rho) \in B_{0,\lambda}$, $\varphi_2(x) \in B_{1,\lambda}$ for $0 < \lambda < 1$ and $\varphi_1(x,\rho), \varphi_2(x)$ satisfy (61). Then $v(x,\rho) \in B_{2,\lambda}$ and*

$$\|v(x,\rho)\|_{B_{2,\lambda}} \le K(\|\varphi_1(x,\rho)\|_{B_{0,\lambda}} + \|\varphi_2(x)\|_{B_{1,\lambda}}).$$

PROOF. Let $x_0 \in \mathbf{R}$ be fixed first and $|x - x_0| < 1$. Define a cut-off even function $q(z) \in C^\infty(\mathbf{R})$ with $q(z) \equiv 1$ for $|z| \geq 10$ and $q(z) \equiv 0$ for $|z| \leq 5$. Write (77) as

$$\begin{aligned} C_n(x) = {} & (1/2\sigma_n) \int_{-\infty}^{+\infty} \exp(-\sigma_n |x - \xi|) q(\xi - x_0) d_n(\xi) d\xi \\ & + (1/2\sigma_n) \int_{-\infty}^{+\infty} \exp(-\sigma_n |x - \xi|)(1 - q(\xi - x_0)) d_n(\xi) d\xi \\ = {} & I_n(x) + II_n(x), \end{aligned} \tag{88}$$

and (86) as

$$\begin{aligned} v(x, \rho) &= \sum_{n=1}^{\infty} (I_n(x) + II_n(x)) e_n(\rho) = \sum_{n=1}^{\infty} I_n(x) e(\rho) + \sum_{n=1}^{\infty} II_n(x) e_n(\rho) \\ &= Y_1(x, \rho) + Y_2(x, \rho). \end{aligned} \tag{89}$$

By the definition of $q(z)$, we see that in the integral of $I_n(x)$ the integration variable ξ satisfies $|x - \xi| \geq 4$. Thus the series of $Y_1(x, \rho)$ is uniformly convergent and can be differentiated termwise up to any order. By a routine calculation [**19**] and the definition of $B_{n,\lambda}$, we have

$$\|Y_1(x, \rho)\|_{(B_{2,\lambda})^*} \leq K(\|\varphi_1(x, \rho)\|_{B_{0,\lambda}} + \|\varphi_2(x)\|_{B_{0,\lambda}}), \tag{90}$$

where K is independent of x_0 and $(B_{2,\lambda})^*$ is defined as $B_{2,\lambda}$ except the evenness of the function and Hölder norms taken only for $|x - x_0| < 1$. For estimates of $Y_2(x, \rho)$, we write

$$II_n(x) = (1/2\sigma_n) \int_{-\infty}^{+\infty} \exp(-\sigma_n |x - x_0 - z|)(1 - q(z)) d_n(z + x_0) dz.$$

Let $\tilde{x} = x - x_0$, $d_n^*(z) = (1 - q(z)) d_n(z + x_0)$ and

$$Y_2^*(\tilde{x}, \rho) = \sum_{n=1}^{\infty} (1/2\sigma_n) e_n(\rho) \int_{-\infty}^{+\infty} \exp(-\sigma_n |\tilde{x} - \xi|) d_n^*(\xi) d\xi,$$

which is a solution of (58) to (60) with inhomogeneous terms $\varphi_1^*(x, \rho) = (1 - q(x))\varphi_1(x + x_0, \rho)$ and $\varphi_2^*(x) = (1 - q(x))\varphi_2(x + x_0)$. $\varphi_1^*(x, \rho)$ and $\varphi_2^*(x)$ have compact supports in $(-10, 10)$ and satisfy (61) if $\varphi_1(x, \rho)$ and $\varphi_2(x)$ do. Thus by Lemma 3,

$$\begin{aligned} \|Y_2^*(\tilde{x}, \rho)\|_{B_{2,\lambda}} \leq K(& \|(1 - q(x))\varphi_1(x + x_0, \rho)\|_{B_{0,\lambda}} \\ & + \|(1 - q(x))\varphi_2(x + x_0)\|_{B_{1,\lambda}}), \end{aligned} \tag{91}$$

except the evenness of the functions, where K is independent of x_0. Since $1 - q(z) \equiv 0$ for $|z| \geq 10$, we have

$$\exp(-d\epsilon^{1/2}(|x_0| - 10)) \geq \exp(-d\epsilon^{1/2}|z + x_0|) \geq \exp(-d\epsilon^{1/2}(|x_0| + 10)),$$

and

$$\begin{aligned} (1 - q(z))\varphi_1(z + x_0, \rho) = {} & (1 - q(z)) \exp(-d\epsilon^{1/2}|z + x_0|)(\varphi_1(z + x_0, \rho) \\ & \times \exp(d\epsilon^{1/2}|z + x_0|)). \end{aligned}$$

Thus

$$\|(1-q(x))\varphi_1(x+x_0,\rho)\|_{B_{0,\lambda}} \le K\exp(-d\epsilon^{1/2}|x_0|)\|\varphi_1(x,\rho)\|_{B_{0,\lambda}},$$

and

$$\|(1-q(x))\varphi_2(x+x_0)\|_{B_{1,\lambda}} \le K\exp(-d\epsilon^{1/2}|x_0|)\|\varphi_2(x)\|_{B_{1,\lambda}}. \tag{92}$$

By the definition, $Y_2(x,\rho) = Y_2^*(x-x_0,\rho)$. We use (91) for $|\tilde{x}|<1$ and (92) to obtain

$$\|Y_2(x,\rho)\|_{(B_{2,\lambda})^*} \le K(\|\varphi_1(x,\rho)\|_{B_{0,\lambda}} + \|\varphi_2(x)\|_{B_{1,\lambda}}).$$

Therefore by (89),$\|v(x,\rho)\|_{(B_{2,\lambda})^*} \le K(\|\varphi_1(x,\rho)\|_{B_{0,\lambda}} + \|\varphi_2(x)\|_{B_{1,\lambda}})$, where K is independent of x_0. Since x_0 is arbitrary and $v(x,\rho)$ is even in x, we obtain

$$\|v(x,\rho)\|_{B_{2,\lambda}} \le K(\|\varphi_1(x,\rho)\|_{B_{0,\lambda}} + \|\varphi_2(x)\|_{B_{1,\lambda}}).$$

We have proved the Lemma.

Lemma 4 can be carried over to $v(x,r)$ and results are summarized as

THEOREM 3. *If $\varphi_1(x,r)\in B_{0,\lambda}$, $\varphi_2(x)\in B_{1,\lambda}$ with $0<\lambda<1$ and satisfy*

$$\int_0^1 \varphi_1(x,r)r^2 J_1(\mu_1 r)dr - J_1(\mu_1)\varphi_2(x) = 0,$$

then (58) to (60) have a solution defined by

$$v(x,r) = \sum_{n=1}^{\infty}(1/2\sigma_n)e_n(r)\int_{-\infty}^{\infty}\exp(-\sigma_n|x-\xi|)d_n(\xi)d\xi, \tag{93}$$

$$d_n(x) = \int_0^1 \varphi_1(x,\zeta)e_n(\zeta)\zeta^3 d\zeta - e_n(1)\varphi_2(x),$$

with $v(x,r)\in B_{2,\lambda}$ and

$$\|v(x,r)\|_{B_{2,\lambda}} \le K(\|\varphi_1(x,r)\|_{B_{0,\lambda}} + \|\varphi_2(x)\|_{B_{1,\lambda}}). \tag{94}$$

Now we are in a position to study (51) to (53). Rewrite (51) to (53) as

$$\theta_{rr}+\theta_{xx}+(3/r)\theta_r+\mu_1^2\theta = F_2(S,\omega,\theta) = F_3(\omega)+\tilde{F}_2(S,\omega,\theta), \tag{95}$$

$$\theta_r+(2+\mu_1^2/4)\theta = \Psi_2(S,\omega,\theta) = \Psi_3(\omega)+\tilde{\Psi}_2(S,\omega,\theta) \quad \text{at } r=1, \tag{96}$$

$$\theta \text{ is finite at } \quad r=0, \tag{97}$$

where $F_3(\omega)$ and $\Psi_3(\omega)$ include all the terms with ω_{xx}, $\epsilon\omega$ and $\epsilon S(\epsilon^{1/2})\omega$ in F_2 and Ψ_2 respectively and $\tilde{F}_2$ and $\tilde{\Psi}_2$ are the remainders. By the derivations of (54) and (55) from (51) to (53), we can see that the left hand side of (54) comes from $F_3(\omega)$ and $\Psi_3(\omega)$ only and the right hand side of (54) comes from $-\tilde{F}_2$ and $-\tilde{\Psi}_2$ only. Therefore if we use the notation in (56) and Theorem 1 and define

$$\Phi_1(S,\omega,\theta) = F_3(\mathcal{L}^{-1}g_1(S,\omega,\theta))+\tilde{F}_2(S,\omega,\theta),$$
$$\Phi_2(S,\omega,\theta) = \Psi_3(\mathcal{L}^{-1}g_1(S,\omega,\theta))+\tilde{\Psi}_2(S,\omega,\theta),$$

which are equivalent to F_2 and Ψ_2, then Φ_1 and Φ_2 satisfy the condition (61) since by the same derivation of (54) from (51) to (53) we have

$$\int_0^1 \Phi_1(S,\omega,\theta)r^2 J_1(\mu_1 r)dr - J_1(\mu_1)\Phi_2(S,\omega,\theta) \\ = \alpha_1(\mathcal{L}(\mathcal{L}^{-1}g_1(S,\omega,\theta))) - \alpha_1 g_1(S,\omega,\theta) = 0.$$

Then by checking $\Phi_1(S,\omega,\theta)$ and $\Phi_2(S,\omega,\theta)$ term by term carefully and noting that $(\mathcal{L}^{-1}g_1(S,\omega,\theta))_{xx}$ is of order ϵ since $(\mathcal{L}^{-1}g_1(S,\omega,\theta))$ satisfies the equation (54) by substituting it on the left hand side of (54), it is straightforward to show that if $\omega \in B_{2,\lambda}$, $\theta \in B_{2,\lambda}$ with $\|\omega\|_{B_{2,\lambda}} \le K\epsilon^{1/2}$ and $\|\theta\|_{B_{2,\lambda}} \le K\epsilon^{1/2}$, then $\Phi_1(S,\omega,\theta) \in B_{0,\lambda}$ and $\Phi_2(S,\omega,\theta) \in B_{1,\lambda}$ with

$$\|\Phi_1(S,\omega,\theta)\|_{B_{0,\lambda}} \le K\epsilon \quad \text{and} \quad \|\Phi_2(S,\omega,\theta)\|_{B_{1,\lambda}} \le K\epsilon\,. \tag{98}$$

By Theorem 3, there is a solution of (95) to (97) with F_2 and Ψ_2 replaced by Φ_1 and Φ_2 and the solution is

$$\theta(x,r) = \sum_{n=1}^{\infty}(1/2\sigma_n)e_n(r)\int_{-\infty}^{+\infty}\exp(-\sigma_n|x-\xi|)d_n(\xi)d\xi, \tag{99}$$

$$d_n(x) = \int_0^1 \Phi_1(S,\omega,\theta)e_n(\zeta)\zeta^3 d\zeta - e_n(1)\Phi_2(S,\omega,\theta)\,.$$

Denote (99) by

$$\theta(x,r) = \mathcal{G}(\Phi_1,\Phi_2)\,. \tag{100}$$

From (98) and Theorem 3, we have

THEOREM 4. *Assume that* $\omega \in B_{2,\lambda}$, $\theta \in B_{2,\lambda}$ *and* $\|\omega\|_{B_{2,\lambda}} + \|\theta\|_{B_{2,\lambda}} \le K\epsilon^{1/2}$ *with* $0 < \lambda < 1$. *Then* $\mathcal{G}(\Phi_1,\Phi_2) \in B_{2,\lambda}$ *and* $\|\mathcal{G}(\Phi_1,\Phi_2)\|_{B_{2,\lambda}} \le K\epsilon$.

6. Existence of solutions and conclusions

We write (51) to (53), which are equivalent to (100), and (54) as follows,

$$\omega(x) = \mathcal{L}^{-1}g_1(S,\omega,\theta) = \mathcal{T}_1^*(\omega,\theta), \tag{101}$$

$$\theta(x,r) = \mathcal{G}(\Phi_1,\Phi_2) = \mathcal{T}_2(\omega,\theta)\,. \tag{102}$$

To satisfy the conditions of Theorem 1 for (101), we substitute $\theta(x,r)$ given by (102) into (101) once to obtain

$$\omega(x) = \mathcal{T}_1^*(\omega,\mathcal{T}_2(\omega,\theta)) = \mathcal{T}_1(\omega,\theta)\,. \tag{103}$$

Define a closed convex set $\mathcal{S}_\ell$ in the Banach space $B_{2,\lambda}\times B_{2,\lambda}$ with the norm $\|X\| = \|\omega\|_{B_{2,\lambda}} + \|\theta\|_{B_{2,\lambda}}$ by

$$\mathcal{S}_\ell = \left\{X = (\omega(x),\theta(x,r)) \in B_{2,\lambda}\times B_{2,\lambda} \,\middle|\, \|X\| \le \ell\epsilon^{1/2}\right\}$$

for some fixed number $\ell > 0$ and $0 < \lambda < 1$. Define an operator $\mathcal{T}$ in $\mathcal{S}_\ell$ as $\mathcal{T}(\omega,\theta) = (\mathcal{T}_1(\omega,\theta),\mathcal{T}_2(\omega,\theta))$. By Theorems 1 and 4, it is easy to show that $\mathcal{T}$ maps $\mathcal{S}_\ell$ into $\mathcal{S}_\ell$ if ϵ is small enough. Also using the same proofs as Theorem 1 and 4, we can have

THEOREM 5. *For $X^{(1)}$ and $X^{(2)} \in \mathcal{S}_\ell$, $\|\mathcal{T}(X^{(1)}) - \mathcal{T}(X^{(2)})\| \le K\epsilon^{1/2}\|X^{(1)} - X^{(2)}\|$, where K is independent of ϵ.*

Therefore if $K\epsilon^{1/2} < 1/2$ in Theorem 5 and ϵ is so small that $\mathcal{T}$ maps $\mathcal{S}_\ell$ into $\mathcal{S}_\ell$, $\mathcal{T}$ is a contraction in $\mathcal{S}_\ell$ and by contraction mapping theorem $\mathcal{T}$ has a fixed point in $\mathcal{S}_\ell$. Finally we obtain

THEOREM 6. *The equations (51) to (53) and (54) possess a solution (ω, θ) with $\omega, \theta \in B_{2,\lambda}$ and*

$$\|\omega(x)\|_{B_{2,\lambda}} + \|\theta(x,r)\|_{B_{2,\lambda}} \le K\epsilon \tag{104}$$

for small $\epsilon > 0$ and $0 < \lambda < 1$, where K is independent of ϵ and may depend on the other parameters in the definition of $B_{2,\lambda}$.

Next we transform $\omega(x)$, $\theta(x,r)$ and r back to the original variables in (25) to (27). By (28), (32), (48), (49), $r = \psi^{1/2}$ and $w(x,\psi) = r^2 u(x,r)$, we have that if $Ro^{-2} = \mu_1^2 + \lambda_1\epsilon$ with $\lambda_1 < 0$ and μ_1 being the first positive root of $\mu J_1(\mu) + 4J_0(\mu) = 0$, then for small $\epsilon > 0$ equations (25) to (27) have a solution

$$f(x,\psi) = (\psi(1 + \epsilon(J_1(\psi^{1/2})\psi^{-1/2})(S(\epsilon^{1/2}x) + \omega(x)) + \epsilon\theta(x,\psi^{1/2})))^{1/2}, \tag{105}$$

where $\omega(x)$ and $\theta(x,r)$ satisfy (104) and

$$S(\epsilon^{1/2}x) = (-3\lambda_1\alpha_2/2\alpha_3)\mathrm{sech}^2((-\lambda_1\alpha_1\epsilon/4\alpha_2)^{1/2}x),$$

with

$$\alpha_1 = \int_0^1 J_1^2(\mu_1 r) r\,dr > 0, \quad \alpha_2 = \int_0^1 J_0^2(\mu_1 r) r\,dr > 0,$$
$$\alpha_3 = (3J_1^3(\mu_1)/4)\mu_1^2(1 + (\mu_1^2/16)) \neq 0.$$

Note that for small ϵ, $f(x,\psi)$ is well defined. The free surface is given by

$$\begin{aligned} f(x,1) &= (1 + \epsilon J_1(\mu_1)(S(\epsilon^{1/2}x) + \omega(x)) + \epsilon\theta(x,1))^{1/2} \\ &= 1 + (1/2)\epsilon J_1(\mu_1)S(\epsilon^{1/2}x) + O(\epsilon^2). \end{aligned}$$

Thus $f_1 = 1 + (1/2)\epsilon J_1(\mu_1)S(\epsilon^{1/2}x)$ is the first order approximation of the free surface for a solution of the exact equations (25) to (27) with small $\epsilon > 0$.

In the following, we shall transform the solution (105) of (25) to (27) into a solution of (1) to (5). Note that from (116), the first order derivative of $f(x,\psi)$ with respect to ψ has a singularity at $\psi = 0$. Actually this is caused by the change of variables. From (6), (12) and (24), $\tilde{r} = r^*/R$ and

$$w^* = -c/(2ff_\psi), \quad u^* = -cf_x/(2ff_\psi). \tag{106}$$

By the definition of $f(x,\psi)$ in (116), the relation between $\tilde{r}$ and ψ is

$$\tilde{r} = \psi^{1/2}(1 + \epsilon(J_1(\mu_1\psi^{1/2})/\psi^{1/2})(S(\epsilon^{1/2}x) + \omega(x)) + \epsilon\theta(x,\psi^{1/2}))^{1/2}, \tag{107}$$

which is the formula for changing (x,ψ)-variable to $(x,\tilde{r})$-variable, and

$$w^*(x,\psi) = -c\Big(1+\epsilon(\psi^{1/2}J_1(\mu_1\psi^{1/2}))_\psi(S(\epsilon^{1/2}x)+\omega(x)) \\ + \epsilon\theta(x,\psi^{1/2}) + (\psi^{1/2}/2)\epsilon\theta_r(x,\psi^{1/2})\Big)^{-1}, \tag{108}$$

$$u^*(x,\psi) = -c\epsilon\Big(\psi^{1/2}/4\big(1+\epsilon(J_1(\mu_1\psi^{1/2})/\psi^{1/2})(S(\epsilon^{1/2}x)+\omega(x)) \\ + \epsilon\theta(x,\psi^{1/2})\big)\Big)^{1/2}\Big((J_1(\mu_1\psi^{1/2})/\psi^{1/2})(S(\epsilon^{1/2}x)+\omega(x))_x + \theta_x(x,\psi^{1/2})\Big) \\ \times\Big(1+\epsilon(\psi^{1/2}J_1(\mu_1\psi^{1/2}))_\psi(S(\epsilon^{1/2}x)+\omega(x)) \\ + \epsilon\theta(x,\psi^{1/2}) + (\psi^{1/2}/2)\epsilon\theta_r(x,\psi^{1/2})\Big)^{-1}. \tag{109}$$

We can use implicit function theorem to solve (107) for $\psi^{1/2}$ in terms of $x,\tilde{r}$ with ϵ small enough and write $\psi^{1/2} = \tilde{r} + \epsilon R_1(x,\tilde{r},\epsilon)$ where by Theorem 6, $R_1(x,\tilde{r},\epsilon)$ is twice differentiable for its variables $(x,\tilde{r})$ and bounded for all $-\infty < x < +\infty$, $0 < \tilde{r} < f(x,1)$ and small ϵ. Also $R_1(x,\tilde{r},\epsilon)$ is symmetric in x and decays exponentially as $|x| \to +\infty$. Hence by (108) and (109) $u^*(x,\tilde{r})$ and $w^*(x,\tilde{r})$ are differentiable with respect to the variables x and $\tilde{r}$ and the derivatives are bounded. Finally if we use the same notations $w^*(x^*,r^*), u^*(x^*,r^*)$ when x is replaced by x^*/R and $\tilde{r}$ is replaced by r^*/R in (108) and (109), then $w^*(x^*,r^*)$ and $u^*(x^*,r^*)$ are also differentiable and w^*, u^* defined in (108) and (109) and

$$\eta^*(x^*) = R(1+\epsilon J_1(\mu_1)(S(\epsilon^{1/2}x^*/R)+\omega(x^*/R)) + \epsilon\theta(x^*/R,1))^{1/2} - R$$

satisfy (1) to (5) at $-\infty < x^* < +\infty$, $0 \le r^* < R+\eta^*$ if $p^*(x^*,r^*)$ and $v^*(x^*,r^*)$ are chosen by (15) and (17) accordingly. Therefore we have

THEOREM 7. *Assume the wave speed c satisfies $(2\omega R/c)^2 = \mu_1^2 + \lambda_1\epsilon$ with $\lambda_1 < 0$ and $\epsilon > 0$ is small. Then there exist $u^*(x^*,r^*), v^*(x^*,r^*), w^*(x^*,r^*)$ and $p^*(x^*,r^*)$ differentiable in $0 \le r^* < R+\eta^*, -\infty < x^* < +\infty$ and $\eta^*(x^*)$ twice differentiable, which satisfy (1) to (5) with exponential decay to the uniform rotating flow as $|x^*| \to \infty$. $\eta^*(x^*)$ is given by*

$$\eta^*(x^*) = (R/2)\epsilon J_1(\mu_1)S(\epsilon^{1/2}x^*/R) + O(\epsilon^2)\,.$$

It is obvious that Theorem 7 implies the Main Theorem in the Introduction.

Acknowledgments. The author wishes to thank M. Renardy for valuable suggestions, which have made the proofs much simpler.

References

[1] J. T. Beale, *Exact solitary water waves with capillary ripples at infinity*, Comm. Pure Appl. Math. **44** (1991), 211-257.

[2] T. B. Benjamin, *Theory of the vortex breakdown phenomenon*, J. Fluid Mech. **14** (1962), 593-629.

[3] T. B. Benjamin, *Some developments in the theory of vortex breakdown*, J. Fluid Mech. **28** (1967), 65-84.

[4] M. S. Berger, *Remarks on vortex breakdown*, Mathematical Aspects of Vortex Dynamics, ed. by R. E. Caflisch, SIAM, Philadelphia, PA, 1989, pp. 171-182.

[5] K. Friedrichs and D. Hyers, *The existence of solitary waves*, Comm. Pure Appl. Math. **3** (1954), 517-550.

[6] D. Gilbarg and N. S. Trudinger, *Elliptical partial differential equations of second order*, Springer-Verlag, New York, 1977.

[7] H. P. Greenspan, *The theory of rotating fluids*, Cambridge Univ. Press, New York, 1968.

[8] R. H. J. Grimshaw, *Resonant flow of a rotating fluid past an obstacle: The general case*, Studies in Appl. Math. **83** (1990), 249-269.

[9] J. K. Harvey, *Some observations of the vortex breakdown phenomenon*, J. Fluid Mech. **14** (1962), 585-592.

[10] G. Iooss and K. Kirchgässner, *Water waves for small surface tension: An approach via normal form*, Proc. Royal Soc. Endinburg **122A** (1992), 267-299.

[11] S. Leibovich, *Weakly non-linear waves in rotating fluids*, J. Fluid Mech. **42** (1970), 803-922.

[12] S. Leibovich, *The structure of vortex breakdown*, Ann. Rev. Fluid Mech. **10** (1978), 221-246.

[13] W. M. Ni, *On the existence of global vortex rings*, J. d'Anal. Math. **37** (1980), 208-247.

[14] W. G. Pritchard, *Solitary waves in rotating fluids*, J. Fluid Mech. **42** (1970), 61-83.

[15] S. M. Sun, *Long nonlinear waves in an unbounded rotating jet or rotating two-fluid flow*, Phys. Fluids **6** (1994), 1204-1212.

[16] S. M. Sun, *On the existence of solitary waves in rotating fluids*, Proc. Royal Soc. Endinburg (1995), in press.

[17] S. M. Sun and M. C. Shen, *Exponentially small estimate for the amplitude of capillary ripples of a generalized solitary wave*, J. Math. Anal. Appl. **172** (1993), 533-566.

[18] S. M. Sun and M. C. Shen, *Existence of solitary pressure pulses in a cylindrical fluid-filled elastic tube*, J. Diff. Equations **115** (1995), 224-256.

[19] A. M. Ter-Krikorov, *Théorie exacte des ondes longues stationnaires dans un liquide hétérogéne*, J. Mécanique **2** (1963), 351-375.

[20] E. C. Titchmarsh, *Eigenfunction expansions associated with second-order differential equations I, 2nd ed.*, Oxford Univ. Press, London, 1962.

[21] C.-S. Yih, *Stratified flows, 2nd ed.*, Academic Press, Inc., New York, 1980.

[22] G. N. Watson, *A Treatise on the theory of Bessel functions, 2nd ed.*, The Macmillian Company, New York, 1945.

DEPARTMENT OF MATHEMATICS, VIRGINIA POLYTECHNIC INSTITUTE AND STATE UNIVERSITY, BLACKSBURG, VIRGINIA 24061

E-mail address: sun@math.vt.edu

Contemporary Mathematics
Volume **200**, 1996

On a generalized Kadomtsev-Petviashvili Equation

Michael M. Tom

ABSTRACT. It is shown that in relatively weak function spaces, the initial value problem for a generalized Kadomtsev-Petviashvili equation possesses global solutions. In particular, for the KP I equation, the existence of such solutions is without any restriction on the size of the initial data. Indeed it is shown that for the generalized KP I equation, the critical exponent for the existence of global weak solutions is 4/3.

1. Introduction

We consider the initial-value problem for a generalized Kadomtsev-Petviashvili equation of the form

$$\begin{cases} (u_t + u^p u_x + u_{xxx})_x - u_{yy} = 0, & (x,y) \in \mathbb{R}^2, \qquad t > 0, \\ u(x,y,0) = u_0(x,y), \end{cases} \tag{1.1}$$

When $p = 1$, (1.1) is known as the KP I equation. The KP I equation models the propagation along the x-axis of nonlinear dispersive long waves on the surface of a fluid when the variation along the y-axis proceeds slowly. Thus, to some extent, it can be viewed as a two-dimensional generalization of the Korteweg-de Vries equation. The case $p = 2$ has been found in the modeling of sound waves in antiferromagnetics. As is well known this equation can be written in the form

$$u_t + u_{xxx} + u^p u_x - \partial_x^{-1} u_{yy} = 0, \tag{1.2}$$

where ∂_x^{-1} stands for an anti-derivative in x.

Various authors have studied the initial value problem for the KP I equation. For instance, Fokas and Sung [3] and Zhou [9] have obtained global existence for small data for the KP I equation via the inverse scattering techniques. A general local existence theory has been developed for the generalized KP equations by Saut [4] and Ukai [7] in the classical Sobolev spaces $H^s(\mathbb{R}^2)$ for $s \geq 3$. Saut using a

1991 *Mathematics Subject Classification.* Primary 35Q53; Secondary 35Q25, 35Q10.

The author acknowledges with thanks support from the Council on Research of the Louisiana State University and from the Louisiana Educational Quality Support Fund contract no LEQSF (1995-97)-RD-A-12.

virial computation proved that for $p \geq 4$, certain solutions of the generalized KP I equation cannot remain in the Sobolev space $H^1(\mathbb{R}^2)$ for all time t. This blow-up is essentially due to the transverse effects (i.e $\|u_y(\cdot,t)\|_0$ will blow-up in finite time). Schwarz [6] considered the periodic problem for the KP I equation and showed that global solutions exists provided that the initial data is small in L^2. De Bouard and Saut [2] have shown recently that equation (1.1) does not possess any nontrivial solitary wave solution if $p \geq 4$ and nontrivial solitary waves exist for $1 \leq p < 4$. Wang et al [8] established that the solitary wave solutions for equation (1.1) are unstable for $p > 4/3$ (this instability result however, is not quite correct). The purpose of this paper is to prove the existence of global solutions for the pure initial value problem for the generalized KP I equation in some relatively weak function spaces. In particular, for the KP I equation, the existence of these global weak solutions is without any restriction on the size of the initial data. Indeed, it should be remarked that the exponent for p for which a requirement of some smallness on the initial data is 4/3. As will be apparent from our computations the range of values of p for which our result will be valid is $1 \leq p \leq 4$. The main ingredients used here are some of the conserved quantities satisfied by solutions of equation (1.1). In general, solutions of (1.1) satisfy the following conserved functionals:

$$\Phi_1(u) = \int_{\mathbb{R}^2} u^2 dxdy,$$

$$\Phi_2(u) = \int_{\mathbb{R}^2} [\frac{1}{2}u_x^2 + \frac{1}{2}(\partial_x^{-1}u_y)^2 - \frac{u^{p+2}}{(p+1)(p+2)}]dxdy,$$

and in the case wherein $p = 1$,

$$\Phi_3(u) = \int_{\mathbb{R}^2} [\frac{3}{2}u_{xx}^2 + 5u_y^2 + \frac{5}{6}(\partial_x^{-2}u_{yy})^2 - \frac{5}{2}uu_x^2 - \frac{5}{9}u(\partial_x^{-1}u_y)^2 + \frac{5}{9}u_y\partial_x^{-1}u\partial_x^{-1}u_y - \frac{5}{6}u^2\partial_x^{-2}u_{yy} + \frac{5}{24}u^4]dxdy.$$

We shall employ the following notations. $|\cdot|_p$, $\|\cdot\|_s$ will denote the norms in $L^p(\mathbb{R}^2)$ and $H^s(\mathbb{R}^2)$ respectively, where

$$\|f\|_s^2 = \int_{\mathbb{R}^2} (1+\xi^2+\eta^2)^s |\hat{f}(\xi,\eta)|^2 d\xi d\eta,$$

and $\hat{}$ connotes Fourier transformation. Thus, the norm in $L^2(\mathbb{R}^2)$ will simply be denoted by $\|\cdot\|_0$. Introduce the spaces

$$V_1(\mathbb{R}^2) = \{f | f, f_x, \partial_x^{-1}f_y \in L^2(\mathbb{R}^2)\}, \tag{1.3}$$

and

$$V_2(\mathbb{R}^2) = \{f | f \in V_1(\mathbb{R}^2), f_{xx}, f_y, \partial_x^{-2}f_{yy} \in L^2(\mathbb{R}^2)\} \tag{1.4}$$

By $\partial_x^{-1}f_y \in L^2(\mathbb{R}^2)$, we mean that $\widehat{\partial_x^{-1}f_y} = \frac{\eta}{\xi}\hat{f}(\xi,\eta) \in L^2(\mathbb{R}^2)$. Because of the singularity in the symbol of $\partial_x^{-1}\partial_y$ at $\xi = 0$, we have to make the additional requirement that $\hat{f}(\xi,\eta)$ tend to zero as ξ tends to zero. Thus, $\hat{f}^{(x)}(0,y) = 0$ (the Fourier transform of f in the variable x). This condition is equivalent to

$$\int_{-\infty}^{\infty} f(x,y)dx = 0. \tag{1.5}$$

Similarly by $\partial_x^{-2} f_{yy} \in L^2(\mathbb{R}^2)$, we mean

$$\widehat{\partial_x^{-2} f_{yy}} = \frac{\eta^2}{\xi^2} \hat{f}(\xi, \eta) \in L^2(\mathbb{R}^2),$$

and hence we must require

$$\lim_{\xi \to 0} \frac{\hat{f}^{(x)}(\xi, y)}{\xi} = 0.$$

This condition will be satisfied if (1.5) holds together with

$$\int_{-\infty}^{\infty} \int_{-\infty}^{x} f(x_1, y) dx_1 dx = 0. \tag{1.6}$$

Hence conditions (1.5) and (1.6) will be imposed on any function $f \in V_2(\mathbb{R}^2)$.

Also, observe that by integrating equation (1.1) with respect to x over $\mathbb{R}$, we find that

$$\int_{-\infty}^{\infty} u_{yy} dx = 0. \tag{1.7}$$

Now integrating (1.6) twice with respect to y and noting that u vanishes as $y \to -\infty$, it follows that $\int_{-\infty}^{\infty} u(x, y, t) dx$ depends only on t. In particular,

$$\int_{-\infty}^{\infty} u(x, y, 0) dx = \int_{-\infty}^{\infty} u_0(x, y) dx = \text{constant} = c. \tag{1.8}$$

Because we also that

$$\int_{-\infty}^{\infty} \int_{-\infty}^{\infty} u(x, y, t) dx dy = \int_{-\infty}^{\infty} \int_{-\infty}^{\infty} u_0(x, y) dx dy, \tag{1.9}$$

we must impose the $c = 0$. Hence, if $\int_{-\infty}^{\infty} u_0(x, y) dx = 0$, then $\int_{-\infty}^{\infty} u(x, y, t) dx = 0$ for all subsequent time $t > 0$.

Similarly it can be shown that, if in addition,

$$\int_{-\infty}^{\infty} \int_{-\infty}^{x} u_0(x_1, y) dx_1 dx = 0,$$

then

$$\int_{-\infty}^{\infty} \int_{-\infty}^{x} u(x_1, y, t) dx_1 dx = 0$$

for all $t > 0$. These are the constraints that will be needed to prove the existence of global solutions in $V_1(\mathbb{R}^2)$.

Because

$$(1 + \xi^2 + \eta^2)^{1/2} |\hat{f}(\xi, \eta)|^2 \leq (1 + \xi^2 + \frac{\eta^2}{\xi^2}) |\hat{f}(\xi, \eta)|^2, \tag{1.10}$$

it follows that $V_1(\mathbb{R}^2) \subset H^{1/2}(\mathbb{R}^2)$.

Acknowledgement: The author wishes to acknowledge valuable conversations with J. L. Bona, Gustavo Ponce, and John P. Albert. I am also indebted to Jean-Claude Saut for pointing out to me some nuances about the KP equations. The author also wishes to thank the referee for the comments on this paper.

2. Main Result

The main results of this article are the following existence theorems in the spaces $V_1(\mathbb{R}^2)$ and $V_2(\mathbb{R}^2)$.

THEOREM 2.1. *Let $u_0 \in V_1(\mathbb{R}^2)$ be such that*

$$\text{(C1)} \qquad \int_{-\infty}^{\infty} u_0(x,y)dx = 0 \quad \textit{and} \quad \int_{-\infty}^{\infty}\int_{-\infty}^{x} u_0(x_1,y)dx_1dx = 0.$$

1. *If $1 \le p < \frac{4}{3}$, then the initial value problem (1.1) has a solution u corresponding to the initial data u_0 such that*
$$u \in L^\infty(\mathbb{R}^+; V_1(\mathbb{R}^2)).$$
2. *If $p = \frac{4}{3}$, the same conclusion as in (1) continuous to be valid if $\|u_0\|_0$ is small.*
3. *If $\frac{4}{3} < p \le 4$, the conclusion in (1) holds true if $\|u_0\|_{V_1(\mathbb{R}^2)}$ is sufficiently small.*

It follows therefore that $u \in L^\infty(\mathbb{R}^+; H^{1/2}(\mathbb{R}^2))$.

THEOREM 2.2. *Let $u_0 \in V_2(\mathbb{R}^2)$ with conditions (C1) and*

$$\text{(C2)} \qquad \begin{cases} \displaystyle\int_{-\infty}^{\infty}\int_{-\infty}^{x}\int_{-\infty}^{x_1} u_0(x_2,y)dx_2dx_1dx = 0, \\ \displaystyle\int_{-\infty}^{\infty}\int_{-\infty}^{x}\int_{-\infty}^{x_1}\int_{-\infty}^{x_2} u_0(x_3,y)dx_3dx_2dx_1dx = 0. \end{cases}$$

If $p = 1$, then there exists a solution u to the initial value problem (1.1) corresponding to the initial data u_0 in the class $L^\infty(\mathbb{R}^+; V_2(\mathbb{R}^2))$.

Before proving Thoerems 2.1 and 2.2, we derive some estimates that will come in handy.

LEMMA 2.3. *If $u \in V_1(\mathbb{R}^2)$, then*

$$\|u\|_{L_y^\infty L_x^2} = \sup_{y\in\mathbb{R}} \int_{-\infty}^{\infty} u^2(x,y)dx \le 2\|u_x\|_0\|\partial_x{}^{-1}u_y\|_0. \tag{2.1}$$

PROOF. Note that for each $(x,y) \in \mathbb{R}^2$,

$$u^2(x,y) = \int_{-\infty}^{y} [u^2(x,y)]_y dy,$$

and hence

$$\begin{aligned}\int_{-\infty}^{\infty} u^2(x,y)dx &= 2\int_{-\infty}^{\infty}\int_{-\infty}^{y} u(x,y)u_y(x,y)dydx \\ &= -2\int_{-\infty}^{\infty}\int_{-\infty}^{y} u_x(x,y)\partial_x{}^{-1}u_y(x,y)dydx \\ &\le 2\|u_x\|_0\|\partial_x{}^{-1}u_y\|_0,\end{aligned}$$

which proves the lemma.

□

REMARK 2.4. It is clear from the proof of Lemma 2.3 that the inequality

$$\|u\|_{L_y^\infty L_x^2} = \sup_{y\in\mathbb{R}} \int_{-\infty}^{\infty} u^2(x,y)dx \le 2\|u\|_0\|u_y\|_0 \tag{2.2}$$

is also valid if $u \in H^1(\mathbb{R}^2)$.

LEMMA 2.5. *If* $0 \le p \le 4$, *then* $V_1(\mathbb{R}^2) \subset L^{p+2}(\mathbb{R}^2)$ *in the sense that*

$$\int_{\mathbb{R}^2} |u|^{p+2}dxdy \le 2^p\|u\|_0^{\frac{4-p}{2}}\|\partial_x{}^{-1}u_y\|_0^{p/2}\|u_x\|_0^p. \tag{2.3}$$

Hence

$$\int_{\mathbb{R}^2} |u|^{p+2}dxdy \le 2^p\|u\|_0^{\frac{4-p}{2}}\|u\|_{V_1(\mathbb{R}^2)}^{\frac{3p}{2}}. \tag{2.4}$$

PROOF. The case $p = 0$ is clear. Now suppose $p > 0$, then

$$\int_{-\infty}^{\infty}\int_{-\infty}^{\infty} |u|^{p+2}dxdy \le \int_{-\infty}^{\infty} \sup_x |u|^p \int_{-\infty}^{\infty} u^2dxdy$$

$$\le 2^{\frac{p}{2}} \int_{-\infty}^{\infty}\Big(\int_{-\infty}^{\infty} u^2dx\Big)\Big(\int_{-\infty}^{\infty} u^2dx\Big)^{p/4}\Big(\int_{-\infty}^{\infty} u_x^2dx\Big)^{p/4}dy,$$

where use has been made of the elementary fact that for each $y \in \mathbb{R}$,

$$\sup_{x\in\mathbb{R}} |u(x,y)| \le \sqrt{2}\Big(\int_{-\infty}^{\infty} u^2dx\Big)^{1/4}\Big(\int_{-\infty}^{\infty} u_x^2dx\Big)^{1/4}.$$

Thus

$$\int_{-\infty}^{\infty}\int_{-\infty}^{\infty} |u|^{p+2}dxdy \le 2^{\frac{p}{2}} \int_{-\infty}^{\infty}\Big(\int_{-\infty}^{\infty} u^2dx\Big)^{\frac{p+4}{4}}\Big(\int_{-\infty}^{\infty} u_x^2dx\Big)^{\frac{p}{4}}dy$$

$$\le 2^{\frac{p}{2}}\Big(\int_{-\infty}^{\infty}\int_{-\infty} u_x^2dxdy\Big)^{\frac{p}{4}}\Big(\int_{-\infty}^{\infty}\Big(\int_{-\infty}^{\infty} u^2dx\Big)^{\frac{p+4}{4-p}}dy\Big)^{\frac{4-p}{4}},$$

using Hölder's inequality with conjugate exponents $\frac{4}{p}$ and $\frac{4}{4-p}$. Hence

$$\int_{\mathbb{R}^2} |u|^{p+2}dxdy \le 2^{\frac{p}{2}}\Big(\sup_{y\in\mathbb{R}}\int_{-\infty}^{\infty} u^2dx\Big)^{\frac{p}{2}}\Big(\int_{-\infty}^{\infty}\int_{-\infty}^{\infty} u^2dxdy\Big)^{\frac{4-p}{4}}\Big(\int_{-\infty}^{\infty}\int_{-\infty}^{\infty} u_x^2dxdy\Big)^{\frac{p}{4}}$$

$$\le 2^{\frac{p}{2}}[2\|u_x\|_0\|\partial_x{}^{-1}u_y\|_0]^{p/2}\|u_x\|_0^{p/2}\|u\|_0^{\frac{4-p}{2}},$$

$$\le 2^p\|u\|_0^{\frac{4-p}{2}}\|\partial_x{}^{-1}u_y\|_0^{p/2}\|u_x\|_0^p,$$

where of course Lemma 2.3 has been used. This completes the proof of Lemma 2.5.

□

REMARK 2.6.

1. Observe that in the application of Hölder's inequality, we do need that $0 \le p \le 4$, and hence this argument seems to break down if $p > 4$.
2. The estimate for $\int_{\mathbb{R}^2} |u|^{p+2} dxdy$ offered above is different from but does imply the one obtained in [8] (Lemma 1). Parts (b) and (c) of Theorem 2.1 are direct consequences of this estimate.
3. After this work was completed, the author became aware of [5], where relation (2.3), the key ingredient in proving Theorem 2.1, was mentioned. Lemma 2.5 is related to the imbedding results for anisotropic Sobolev spaces that was studied by Besov, Il'in and Nikolskii [1].
4. Eventhough the solution of the initial-value problem (1.1) makes sense in $V_1(\mathbb{R}^2)$ with $u_0 \in V_1(\mathbb{R}^2)$ together with just the first condition of (C1), the second condition of (C1) is required to justify the proof that Φ_2 is independent of t (cf [4]). Similarly, the solution of the Cauchy problem (1.1) with $p = 1$ makes sense in $V_2(\mathbb{R}^2)$ with $u_0 \in V_2(\mathbb{R}^2)$ together with conditions (C1). However, the additional conditions (C2) are needed to derive the fact that Φ_3 is indeed a conserved functional.

COROLLARY 2.7.

$$|u|_\infty \le 2\|u_x\|_0^{1/4} \|\partial_x{}^{-1} u_y\|_0^{1/4} \|u_{xx}\|_0^{1/4} \|u_y\|_0^{1/4}. \tag{2.5}$$

PROOF.

$$\begin{aligned}
|u|_\infty &= \sup_{y\in\mathbb{R}} \sup_{x\in\mathbb{R}} |u| \\
&\le \sqrt{2} \sup_{y\in\mathbb{R}} \Big(\int_{-\infty}^{\infty} u^2 dx\Big)^{1/4} \Big(\int_{-\infty}^{\infty} u_x^2 dx\Big)^{1/4} \\
&\le \sqrt{2} \Big(\sup_{y\in\mathbb{R}} \int_{-\infty}^{\infty} u^2 dx\Big)^{1/4} \Big(\sup_{y\in\mathbb{R}} \int_{-\infty}^{\infty} u_x^2 dx\Big)^{1/4} \\
&\le \sqrt{2} [2\|u_x\|_0 \|\partial_x{}^{-1} u_y\|_0]^{1/4} [2\|u_{xx}\|_0 \|u_y\|_0]^{1/4} \\
&\le 2\|u_x\|_0^{1/4} \|\partial_x{}^{-1} u_y\|_0^{1/4} \|u_{xx}\|_0^{1/4} \|u_y\|_0^{1/4}.
\end{aligned}$$

Similarly, it is easy to see that

$$|u|_\infty \le 2\|u\|_0^{1/4} \|u_y\|_0^{1/2} \|u_{xx}\|_0^{1/4}. \tag{2.6}$$

□

Proof of Theorem 2.1. Given $\delta > 0$, consider the regularized problem

$$\begin{cases} u_t + u^p u_x + u_{xxx} - \partial_x^{-1} u_{yy} + \delta(u_{xxxx} - u_{yy}) = 0, & (x,y) \in \mathbb{R}^2, \quad t > 0, \\ u(x,y,0) = u_0(x,y), \end{cases} \tag{2.7}$$

Define the space $H^{-2,\infty}(\mathbb{R}^2)$ by

$$H^{-2,\infty}(\mathbb{R}^2) = \{f \in H^{\infty}(\mathbb{R}^2)/ \quad \partial_x^{-2} f \in H^{\infty}(\mathbb{R}^2)\},$$

where $H^{\infty}(\mathbb{R}^2) = \cap_{k\in N} H^k(\mathbb{R}^2)$ and the conditions (C1) imposed on $H^{-2,\infty}(\mathbb{R}^2)$. Then $H^{-2,\infty}(\mathbb{R}^2) \subset V_1(\mathbb{R}^2)$. Let $\{u_{0\delta}\}_{\delta>0}$ be a sequence in $H^{-2,\infty}(\mathbb{R}^2)$ such that $u_{0\delta} \to u_0$ in $V_1(\mathbb{R}^2)$ as $\delta \to 0$, since $H^{-2,\infty}(\mathbb{R}^2)$ is dense in $V_1(\mathbb{R}^2)$. For $\delta > 0$ fixed, the classical theory of parabolic equations asserts that the initial value problem (2.7) possesses a solution u_δ in the class $C^{\infty}(\mathbb{R}^+; H^{-2,\infty}(\mathbb{R}^2))$. Droping the subscript, we now proceed to derive *a priori* bounds on u_δ independent of δ. Any solution in $H^{-2,\infty}(\mathbb{R}^2)$ with initial data satisfying (C1) will inherit the same conditions for all subsequent $t > 0$ as earlier eluded to, hence $\partial_x^{-2} u_{yy}$ makes sense. Multiply equation (2.7) by u and integrate over $\mathbb{R}^2$ to obtain

$$\frac{1}{2}\frac{d}{dt}\int_{\mathbb{R}^2} u^2 + \delta \int_{\mathbb{R}^2} [u_{xx}^2 + u_y^2]\,dxdy = 0,$$

hence by integrating with respect to t from 0 to T there inspires the relation

$$\|u\|_0^2 + 2\delta \int_0^T [\|u_{xx}\|_0^2 + \|u_y\|_0^2]\,dt = \|u_0\|_0^2. \tag{2.8}$$

Therefore

$$\sup_{t\in[0,T]} \|u(t)\|_0^2 \le \|u_0\|_0^2, \tag{2.9}$$

and

$$\delta \int_0^T [\|u_{xx}\|_0^2 + \|u_y\|_0^2]dt \le c\|u_0\|_0^2. \tag{2.10}$$

Next, multiply equation (2.7) by $-u_{xx} + \partial_x^{-2}\partial_y^2 u - \frac{u^{p+1}}{p+1}$ (justified in light of the conditions (C1) on the initial data [4]) and integrate over $\mathbb{R}^2$ to obtain

$$
\begin{aligned}
&\int_{\mathbb{R}^2} \Big[-u_{xx} + \partial_x^{-2} u_{yy} - \frac{1}{p+1} u^{p+1}\Big] u_t \\
&\qquad + \int_{\mathbb{R}^2} u^p u_x \Big[-u_{xx} + \partial_x^{-2} u_{yy} - \frac{1}{p+1} u^{p+1}\Big] \\
&\qquad + \int_{\mathbb{R}^2} u_{xxx} \Big[-u_{xx} + \partial_x^{-2} u_{yy} - \frac{1}{p+1} u^{p+1}\Big] \\
&\qquad - \int_{\mathbb{R}^2} \partial_x^{-1} u_{yy} \Big[-u_{xx} + \partial_x^{-2} u_{yy} - \frac{1}{p+1} u^{p+1}\Big] \\
&\qquad + \delta \int_{\mathbb{R}^2} u_{xxxx} \Big[-u_{xx} + \partial_x^{2} u_{yy} - \frac{1}{p+1} u^{p+1}\Big] \\
&\qquad - \delta \int_{\mathbb{R}^2} u_{yy} \Big[-u_{xx} + \partial_x^{-2} u_{yy} - \frac{1}{p+1} u^{p+1}\Big] \\
&\qquad = 0.
\end{aligned}
\tag{2.11}
$$

After several integrations by parts, equation (2.11) reduces to

$$
\begin{aligned}
&\frac{d}{dt} \int_{\mathbb{R}^2} \Big[\frac{u_x^2}{2} + \frac{(\partial_x^{-1} u_y)^2}{2} - \frac{u^{p+2}}{(p+1)(p+2)}\Big] dxdy \\
&\qquad + \delta \int_{\mathbb{R}^2} \Big[u_{xxx}^2 + 2u_{xy}^2 + \big(\partial_x^{-1} u_{yy}\big)^2\Big] dxdy \\
&\qquad = \delta \int_{\mathbb{R}^2} u^p u_{xx}^2 - \delta \int_{\mathbb{R}^2} u^p u_y^2 + \delta p \int_{\mathbb{R}^2} u^{p-1} u_x^2 u_{xx}.
\end{aligned}
\tag{2.12}
$$

Now we estimate the terms on the right hand side of relations (2.12) as follows. By Lemma 2.5

$$
\begin{aligned}
\int_{\mathbb{R}^2} u^p u_{xx}^2 &\le \Big(\int_{\mathbb{R}^2} |u|^{p+2}\Big)^{p/p+2} \Big(\int_{\mathbb{R}^2} |u_{xx}|^{p+2}\Big)^{\frac{2}{p+2}} \\
&\le C \Big[\|u\|_0^{\frac{4-p}{2}} \|\partial_x{}^{-1} u_y\|_0^{p/2} \|u_x\|_0^{p}\Big]^{\frac{p}{p+2}} \Big[\|u_{xx}\|_0^{\frac{4-p}{2}} \|u_{xy}\|_0^{p/2} \|u_{xxx}\|_0^{p}\Big]^{\frac{2}{p+2}} \\
&\le C \|u\|_0^{\frac{p(4-p)}{p+2}} \|\partial_x^{-1} u_y\|^{\frac{p^2}{2(p+2)}} \|u_x\|_0^{\frac{p^2}{p+2}} \|u_{xx}\|_0^{\frac{4-p}{p+2}} \|u_{xy}\|_0^{\frac{p}{p+2}} \|u_{xxx}\|_0^{\frac{2p}{p+2}}.
\end{aligned}
\tag{2.13}
$$

Similarly

$$\int_{\mathbb{R}^2} u^p u_y^2 \le \left(\int_{\mathbb{R}^2} |u^{p+2}\right)^{\frac{p}{p+2}} \left(\int_{\mathbb{R}^2} |u_y|^{p+2}\right)^{\frac{2}{p+2}}$$

(2.14)

$$\le C\left[\|u\|_0^{\frac{4-p}{2}} \|\partial_x^{-1} u_y\|_0^{\frac{p}{2}} \|u_x\|_0^{p}\right]^{\frac{p}{p+2}} \left[\|u_y\|_0^{\frac{4-p}{2}} \|\partial_x^{-1} u_{yy}\|_0^{\frac{p}{2}} \|u_{xy}\|_0^{p}\right]^{\frac{2}{p+2}}$$

$$= C\|u\|_0^{\frac{p(4-p)}{p+2}} \|\partial_x^{-1} u_y\|_0^{\frac{p^2}{2(p+2)}} \|u_x\|_0^{\frac{p^2}{(p+2)}} \|u_y\|_0^{\frac{4-p}{(p+2)}} \|\partial_x^{-1} u_{yy}\|_0^{\frac{p}{(p+2)}} \|u_{xy}\|_0^{\frac{2p}{(p+2)}}.$$

If $p = 1$, the last term on the right hand side of (2.12) is $\int_{\mathbb{R}^2} u_x^2 u_{xx} = 0$. Thus, we need only estimate this term for $p > 1$. By Hölder's inequality,

$$\int_{\mathbb{R}^2} u^{p-1} u_x^2 u_{xx} \le \left(\int_{\mathbb{R}^2} |u|^{p+2}\right)^{\frac{p-1}{p+2}} \left(\int_{\mathbb{R}^2} |u_x|^{p+2}\right)^{\frac{2}{p+2}} \left(\int_{\mathbb{R}^2} |u_{xx}|^{p+2}\right)^{\frac{2}{p+2}}$$

(2.15)

$$\le C\left[\|u\|_0^{\frac{4-p}{2}} \|\partial_x^{-1} u_y\|_0^{\frac{p}{2}} \|u_x\|_0^{p}\right]^{\frac{p-1}{p+2}} \left[\|u_x\|_0^{\frac{4-p}{2}} \| u_y\|_0^{\frac{p}{2}} \|u_{xx}\|_0^{p}\right]^{\frac{2}{p+2}}$$

$$\cdot \left[\|u_{xx}\|_0^{\frac{4-p}{2}} \|u_{xy}\|_0^{\frac{p}{2}} \|u_{xxx}\|_0^{p}\right]^{\frac{1}{p+2}}$$

$$= \|u\|_0^{\frac{(p-1)(4-p)}{2(p+2)}} \|\partial_x^{-1} u_y\|_0^{\frac{p(p-1)}{2(p+2)}} \|u_x\|_0^{\frac{p(p-1)}{p+2}} \|u_x\|_0^{\frac{4-p}{p+2}} \|u_y\|_0^{\frac{p}{p+2}}$$

$$\cdot \|u_{xx}\|_0^{\frac{2p}{p+2}} \|u_{xx}\|_0^{\frac{4-p}{2(p+2)}} \|u_{xy}\|_0^{\frac{p}{2(p+2)}} \|u_{xxx}\|_0^{\frac{p}{p+2}}.$$

By using the inequalities

(2.16)

$$\|u_x\|_0 \le \|u\|_0^{\frac{1}{2}} \|u_{xx}\|_0^{\frac{1}{2}},$$

$$\|u_{xx}\|_0 \le \|u_x\|_0^{\frac{1}{2}} \|u_{xxx}\|_0^{\frac{1}{2}},$$

$$\|u_y\|_0 \le \|u_{xy}\|_0^{\frac{1}{2}} \|\partial_x^{-1} u_y\|_0^{\frac{1}{2}},$$

in some of the terms on the right hand side of (2.15), we obtain

(2.17)

$$\int_{\mathbb{R}^2} u^{p-1} u_x^2 u_{xx} \le C\|u\|_0^{\frac{p(4-p)}{p+2}} \|\partial_x^{-1} u_y\|_0^{\frac{p^2}{2(p+2)}} \|u_x\|_0^{\frac{p^2}{p+2}} \|u_{xx}\|_0^{\frac{4-p}{p+2}} \|u_{xy}\|_0^{\frac{p}{p+2}} \|u_{xxx}\|_0^{\frac{2p}{p+2}}.$$

Thus relation (2.12) can be rewritten as

$$\frac{d}{dt}\int_{\mathbb{R}^2}\Big[\frac{u_x^2}{2}+\frac{(\partial_x^{-1}u_y)^2}{2}-\frac{u^{p+2}}{(p+1)(p+2)}\Big]dxdy$$

$$+\delta\int_{\mathbb{R}^2}\Big[u_{xxx}^2+2u_{xy}^2+\big(\partial_x^{-1}u_{yy}\big)^2\Big]dxdy \tag{2.18}$$

$$\leq c\delta\|u\|_0^{\frac{p(4-p)}{p+2}}\|\partial_x^{-1}u_y\|_0^{\frac{p^2}{2(p+2)}}\|u_x\|_0^{\frac{p^2}{p+2}}$$

$$\cdot\Big[\|u_{xx}\|_0^{\frac{4-p}{p+2}}\|u_{xy}\|_0^{\frac{p}{p+2}}\|u_{xxx}\|_0^{\frac{2p}{p+2}}$$

$$+\|u_y\|_0^{\frac{4-p}{p+2}}\|\partial_x^{-1}u_{yy}\|_0^{\frac{p}{(p+2)}}\|u_{xy}\|_0^{\frac{2p}{p+2}}\Big].$$

Letting

$$X_1^2=\int_{\mathbb{R}^2}\Big[u_x^2+\big(\partial_x^{-1}u_y\big)^2\Big]\,dxdy,$$

$$X_2^2=\int_{\mathbb{R}^2}\big[u_{xx}^2+u_y^2\big]\,dxdy,$$

$$X_3^2=\int_{\mathbb{R}^2}\Big[u_{xxx}^2+u_{xy}^2+\big(\partial_x^{-1}u_{yy}\big)^2\Big]\,dxdy,$$

and integrating relations (2.18) with respect to t from 0 to T, it follows that

$$X_1^2+\delta\int_0^T X_3^2dt$$

$$\leq C_0\left(\|u_0\|_{V_1(\mathbb{R}^2)}\right)+c\|u\|_0^{\frac{4-p}{2}}\|\partial_x^{-1}u_y\|_0^{\frac{p}{2}}\|u_x\|_0^{p} \tag{2.19}$$

$$+c\delta\int_0^T\|u\|_0^{\frac{p(4-p)}{p+2}}X_1^{\frac{3p^2}{2(p+2)}}X_2^{\frac{4-p}{p+2}}X_3^{\frac{3p}{p+2}}dt$$

$$\leq C_0\left(\|u_0\|_{V_1(\mathbb{R}^2)}\right)+c\|u_0\|_0^{\frac{4-p}{2}}X_1^{\frac{3p}{2}}$$

$$+C\delta\int_0^T\|u_0\|_0^{\frac{p(4-p)}{p+2}}X_1^{\frac{3p^2}{2(p+2)}}X_2^{\frac{4-p}{p+2}}X_3^{\frac{3p}{p+2}}dt.$$

Now observe that an application of Young's inequality to the last term on the right hand side of (2.19) yields the following

$$c\delta\int_0^T \|u_0\|_0^{\frac{p(4-p)}{p+2}} X_1^{\frac{3p^2}{2(p+2)}} X_2^{\frac{4-p}{p+2}} X_3^{\frac{3p}{p+2}}\,dt$$

$$\le \delta\epsilon\int_0^T X_3^2 dt + c(\epsilon)\delta\int_0^T \left[\|u_0\|_0^{\frac{p(4-p)}{p+2}} X_1^{\frac{3p^2}{2(p+2)}} X_2^{\frac{4-p}{p+2}}\right]^{\frac{2(p+2)}{4-p}} dt$$

$$\le \delta\epsilon\int_0^T X_3^2 dt + c(\epsilon)\delta\int_0^T \|u_0\|_0^{2p} X_1^{\frac{3p^2}{4-p}} X_2^2 dt$$

$$\le \delta\epsilon\int_0^T X_3^3 dt + c(\epsilon)\|u_0\|_0^{2p}\sup_t X_1^{\frac{3p^2}{4-p}}\delta\int_0^T X_2^2 dt$$

$$\le \delta\epsilon\int_0^T X_3^2 dt + c(\epsilon)\|u_0\|_0^{2p+2}\sup_t X_1^{\frac{3p^2}{4-p}}$$

by using relation (2.10). It therefore follows that,

$$\sup_t X_1^2 + \delta(1-\epsilon)\int_0^T X_3^2 dt$$

$$\le C_0\|u_0\|_{V_1(\mathbb{R}^2)} + C\|u_0\|_0^{\frac{4-p}{2}}\sup_t X_1^{\frac{3p}{2}} \tag{2.20}$$

$$+ c(\epsilon)\|u_0\|_0^{2p+2}\sup_t X_1^{\frac{3p^2}{4-p}}.$$

By choosing ϵ appropriately, we find that if $\frac{3p^2}{4-p} < 2$ and $\frac{3p}{2} < 2$, which are both true if $p < \frac{4}{3}$, (the former is $(3p-4)(p+2) < 0$) that

$$\sup_t X_1^2 \le C\left(\|u_0\|_{V_1(\mathbb{R}^2)}\right), \tag{2.21}$$

and

$$\delta\int_0^T X_3^2 dt \le C\left(\|u_0\|_{V_1(\mathbb{R}^2)}\right). \tag{2.22}$$

If $p = \frac{4}{3}$ inequalities (2.21) and (2.22) are both valid provided that $\|u_0\|_0$ is sufficiently small. If $\frac{4}{3} < p \le 4$, inequalities (2.21) and (2.22) will hold true if $(\|u_0\|_{V_1(\mathbb{R}^2)})$ is not too large.

By (2.21) and (2.22),

(2.23) $\quad u_\delta$ is bounded independently of δ in $L^\infty(\mathbb{R}^+; V_1(\mathbb{R}^2))$, and

$$\sqrt{\delta}u_\delta \text{ is bounded independently of } \delta \text{ in } L^2(\mathbb{R}^+; V_3(\mathbb{R}^2)), \tag{2.24}$$

where

$$V_3(\mathbb{R}^2) = \left\{ f | f, f_x, \partial_x^{-1} f_y, f_y, f_{xx}, f_{xy}, f_{xxx}, \partial_x^{-1} f_{yy} \in L^2(\mathbb{R}^2) \right\}.$$

With this *a priori* bounds at hand, the passage to the limit as $\delta \to 0$ now follows by using a diagonalization process, the compactness of the injection $V_1(B_R) \subset L^2(B_R)$ for all bounded open balls $B_R \subset \mathbb{R}^2$ (see Lemma 3.3 [2]) and Aubin's lemma. The details are not brief but can be justified in light of the *a priori* bounds at hand and hence is omitted.

□

Proof of Theorem 2.2. To prove Theorem 2.2, we proceed similarly to derive *a priori* bounds that are independent of δ on u_δ. Multiply the regularized equation (2.7) by

$$\begin{aligned} &3u_{xxxx} - 10u_{yy} + \frac{5}{3}\partial_x^{-4}u_{yyyy} + \frac{5}{2}u_x^2 + 5uu_{xx} \\ &- \frac{10}{9}\left(\partial_x^{-1}u_y\right)^2 - \frac{5}{9}\partial_x^{-1}u\partial_x^{-1}u_{yy} + \frac{5}{9}\partial_x^{-1}\left(\partial_x^{-1}uu_{yy}\right) \\ &- \frac{10}{9}\partial_x^{-1}\left(u_y\partial_x^{-1}u_y\right) - \frac{10}{9}\partial_x^{-1}\left(u\partial_x^{-1}u_{yy}\right) - \frac{5}{3}u\partial_x^{-2}u_{yy} \\ &- \frac{5}{6}\partial_x^{-2}\left(u^2\right)_{yy} + \frac{5}{6}u^3 = I(u), \end{aligned} \tag{2.25}$$

(the conditions (C1) and (C2) are needed to justify the definition of $\partial_x^{-4}u_{yyyy}$ as in Theorem 2.1) and integrate the resulting expression over $\mathbb{R}^2$ to obtain

$$\begin{aligned} &\int_{\mathbb{R}^2} I(u)u_t + \int_{\mathbb{R}^2} I(u)u_{xxx} + \int_{\mathbb{R}^2} uu_x I(u) \\ &- \int_{\mathbb{R}^2} I(u)\partial_x^{-1}u_{yy} + \delta\int_{\mathbb{R}^2}(u_{xxxx} - u_{yy})I(u) = 0, \end{aligned} \tag{2.26}$$

Now

$$\int_{\mathbb{R}^2} I(u)u_t = \frac{d}{dt}\int_{\mathbb{R}^2}\left[\frac{3}{2}u_{xx^2} + 5u_y^2 + \frac{5}{6}\left(\partial_x^{-2}u_{yy}\right)^2\right.$$

$$- \frac{5}{2}uu_x^2 - \frac{5}{9}u\left(\partial_x^{-1}u_y\right)^2 + \frac{5}{9}u_y\partial_x^{-1}u\partial_x^{-1}u_y \tag{2.27}$$

$$\left. - \frac{5}{6}u^2\partial_x^{-2}u_{yy} + \frac{5}{24}u^4\right] dxdy.$$

A tedious but straightforward calculation shows that

$$\int_{\mathbb{R}^2} I(u)\left[u_{xxx} + uu_x - \partial_x^{-1}u_{yy}\right] dxdy = 0, \tag{2.28}$$

and

$$\delta\int_{\mathbb{R}^2}(u_{xxxx} - u_{yy})I(u)dxdy$$

$$= \delta\int_{\mathbb{R}^2}\left[3u_{xxxx}^2 + 13u_{xxy}^2 + \frac{35}{3}u_{yy}^2 + \frac{5}{3}\left(\partial_x^{-2}u_{yyy}\right)^2\right]$$

$$+ \delta\left\{5\int_{\mathbb{R}^2} u_{xx}^3 - 5\int_{\mathbb{R}^2} uu_{xxx}^2 + 5\int_{\mathbb{R}^2} u_x u_{xx} u_{yy}\right.$$

$$- 5\int_{\mathbb{R}^2} uu_{xy}^2 - \frac{10}{9}\int_{\mathbb{R}^2} u_{xxxx}\left(\partial_x^{-1}u_y\right)^2$$

$$- \frac{10}{9}\int_{\mathbb{R}^2} u_y\partial_x^{-1}u_y\partial_x^{-1}u_{yy} + \frac{10}{3}\int_{\mathbb{R}^2} uu_{yy}\partial_x^{-2}u_{yy}$$

$$+ \frac{5}{3}\int_{\mathbb{R}^2} u_y^2\partial_x^{-2}u_{yy} + \frac{5}{3}\int_{\mathbb{R}^2} u\partial_x^{-1}u_{yy}u_{xxx}$$

$$+ \frac{10}{9}\int_{\mathbb{R}^2} u_y\partial_x^{-1}u_y u_{xxx} - \frac{5}{3}\int_{\mathbb{R}^2} u\partial_x^{-2}u_{yy}u_{xxx}$$

$$+ \frac{5}{3}\int_{\mathbb{R}^2} uu_y u_{xxy} + \frac{5}{3}\int_{\mathbb{R}^2} u^3 u_{xxxx}$$

$$\left. -\frac{5}{6}\int_{\mathbb{R}^2} u^3 u_{yy}\right\}. \tag{2.29}$$

We now estimate the terms in the braces of (2.29). But first, let

$$X_4^2 = \int_{\mathbb{R}^2} \left[u_{xxxx}^2 + u_{xxy}^2 + u_{yy}^2 + \left(\partial_x^{-2} u_{yyy}\right)^2\right] dxdy.$$

Then

$$\begin{aligned}
\left|\int_{\mathbb{R}^2} u_{xx}^3 dxdy\right| &\le |u_{xx}|_\infty \|u_{xx}\|_0^2 \\
&\le C\|u_{xxx}\|_0^{\frac14}\|u_{xy}\|_0^{\frac14}\|u_{xxxx}\|_0^{\frac14}\|u_{xxy}\|_0^{\frac14}\|u_{xx}\|_0^2 \\
&\le C X_2^2 X_3^{\frac12} X_4^{\frac12},
\end{aligned}$$

$$\begin{aligned}
\left|\int_{\mathbb{R}^2} u u_{xxx}^2 dxdy\right| &\le C|u|_\infty \|u_{xxx}\|_0^2 \\
&\le C\|u_x\|_0^{\frac14}\|\partial_x^{-1}u_y\|_0^{\frac14}\|u_{xx}\|_0^{\frac14}\|u_y\|_0^{\frac14}\|u_{xxx}\|_0^2 \\
&\le C X_1^{\frac12} X_2^{\frac12} X_3^2,
\end{aligned}$$

$$\begin{aligned}
\left|\int_{\mathbb{R}^2} u_x u_{xx} u_{yy} dxdy\right| &\le |u_x|_\infty \|u_{xx}\|_0 \|u_{xy}\|_0 \\
&\le C\|u_{xx}\|_0^{\frac14}\|u_y\|_0^{\frac14}\|u_{xxx}\|_0^{\frac14}\|u_{xy}\|_0^{\frac14}\|u_{xx}\|_0\|u_{xy}\|_0 \\
&\le C X_2^{\frac32} X_3^{\frac32},
\end{aligned}$$

$$\left|\int_{\mathbb{R}^2} u u_{xy}^2 dxdy\right| \le C X_1^{\frac12} X_2^{\frac12} X_3^2,$$

$$\begin{aligned}
\left|\int_{\mathbb{R}^2} \left(\partial_x^{-1}u_y\right)^2 u_{xxxx}\right| &\le |\partial_x^{-1}u_y|_\infty \|\partial_x^{-1}u_y\|_0 \|u_{xxxx}\|_0 \\
&\le C\|u_y\|_0^{\frac14}\|\partial_x^{-2}u_{yy}\|_0^{\frac14}\|u_{xy}\|_0^{\frac14}\|\partial_x^{-1}u_{yy}\|_0^{\frac14}\|\partial_x^{-1}u_y\|_0\|u_{xxxx}\|_0 \\
&\le C X_1 X_2^{\frac14} X_3^{\frac12} X_4 \|\partial_x^{-2}u_{yy}\|_0^{\frac14},
\end{aligned}$$

$$\int_{\mathbb{R}^2} u_y \partial_x^{-1}u_y \partial_x^{-1}u_{yy} \le |\partial_x^{-1}u_y|_\infty \|u_y\|_0 \|\partial_x^{-1}u_{yy}\|_0$$

$$\leq C\|u_y\|_0^{\frac14}\|\partial_x^{-2}u_{yy}\|_0^{\frac14}\|u_{xy}\|_0^{\frac14}\|\partial_x^{-1}u_{yy}\|_0^{\frac14}\|u_y\|_0\|\partial_x^{-1}u_{yy}\|_0$$

$$\leq CX_2^{\frac54}X_3^{\frac32}\|\partial_x^{-2}u_{yy}\|_0^{\frac14},$$

$$\left|\int_{\mathbb{R}^2} u\ u_{yy}\ \partial_x^{-2}u_{yy}\right| \leq |u|_\infty\|u_{yy}\|_0\|\partial_x^{-2}u_{yy}\|_0$$

$$\leq C\|u_y\|_0^{\frac14}\|\partial_x^{-2}u_{yy}\|_0^{\frac14}\|u_{xy}\|_0^{\frac14}\|\partial_x^{-1}u_{yy}\|_0^{\frac14}\|u_{yy}\|_0\|\partial_x^{-2}u_{yy}\|_0$$

$$\leq CX_1^{\frac12}X_2^{\frac12}X_4\|\partial_x^{-2}u_{yy}\|_0,$$

$$\left|\int_{\mathbb{R}^2} u_y^2\partial_x^{-2}u_{yy}\right| \leq |u_y|_\infty\|u_y\|_0\|\partial_x^{-2}u_{yy}\|_0$$

$$\leq c\|u_{xy}\|_0^{\frac14}\|\partial_x^{-1}u_{yy}\|_0^{\frac14}\|u_{xxy}\|_0^{\frac14}\|u_{yy}\|_0^{\frac14}\|u_y\|_0\|\partial_x^{-2}u_{yy}\|_0$$

$$\leq CX_2X_3^{\frac12}X_4^{\frac12}\|\partial_x^{-2}u_{yy}\|_0,$$

$$\left|\int_{\mathbb{R}^2} u\ \partial_x^{-1}\ u_{yy}\ u_{xxx}\right| \leq |u|_\infty\|\partial_x^{-1}u_{yy}\|_0\|u_{xxx}\|_0$$

$$\leq C\|u_x\|_0^{\frac14}\|\partial_x^{-1}u_y\|_0^{\frac14}\|u_{xx}\|_0^{\frac14}\|u_y\|_0^{\frac14}\|\partial_x^{-1}u_{yy}\|_0\|u_{xxx}\|_0$$

$$\leq CX_1^{\frac12}X_2^{\frac12}X_3^2,$$

$$\left|\int_{\mathbb{R}^2} u_y\ \partial_x^{-1}\ u_y\ u_{xxx}\right| \leq |\partial_x^{-1}\ u_y|_\infty\|u_y\|_0\|u_{xxx}\|_0$$

$$\leq C\|u_y\|_0^{\frac14}\|\partial_x^{-2}u_{yy}\|_0^{\frac14}\|u_{xy}\|_0^{\frac14}\|\partial_x^{-1}u_{yy}\|_0^{\frac14}\|u_y\|_0\|u_{xxx}\|_0$$

$$\leq CX_2^{\frac54}X_3^{\frac32}\|\partial_x^{-2}u_{yy}\|_0^{\frac14},$$

$$\left|\int_{\mathbb{R}^2} u\ \partial_x^{-2}\ u_{yy}\ u_{xxx}\right| \leq |u|_\infty\|\partial_x^{-2}u_{yy}\|_0\|u_{xxx}\|_0$$

$$\leq C\|u_x\|_0^{\frac14}\|\partial_x^{-1}u_y\|_0^{\frac14}\|u_{xx}\|_0^{\frac14}\|u_y\|_0^{\frac14}\|\partial_x^{-2}u_{yy}\|_0\|u_{xxx}\|_0$$

$$\leq CX_1^{\frac12}X_2^{\frac12}X_3\|\partial_x^{-2}u_{yy}\|_0,$$

$$\left|\int_{\mathbb{R}^2} u\, u_y\, u_{xxy}\right| \le |u|_\infty \|u_y\|_0 \|u_{xxy}\|_0$$

$$\le C\|u_x\|_0^{\frac14} \|\partial_x^{-1} u_y\|_0^{\frac14} \|u_{xx}\|_0^{\frac14} \|u_y\|_0^{\frac14} \|u_y\|_0 \|u_{xxy}\|_0$$

$$\le X_1^{\frac12} X_2^{\frac32} X_4,$$

$$\left|\int_{\mathbb{R}^2} u^3 u_{xxxx}\right| \le |u|_6^3 \|u_{xxxx}\|_0$$

$$\le C\|\partial_x^{-1} u_y\|_0 \|u_x\|_0^2 \|u_{xxxx}\|_0$$

$$\le C X_1^3 X_4,$$

$$\left|\int_{\mathbb{R}^2} u^3\, u_{yy}\right| \le |u|_6^3 \|u_{yy}\|_0$$

$$\le C X_1^3 X_4.$$

Hence (2.26) becomes

$$\frac{d}{dt}\int_{\mathbb{R}^2}\left[\frac32 u_{xx}^2 + 5u_y^2 + \frac56\left(\partial_x^{-2}u_{yy}\right)^2 - \frac52 u u_x^2 - \frac59 u\left(\partial_x^{-2}u_y\right)^2\right.$$

$$\left. + \frac59 u_y \partial_x^{-1} u \partial_x^{-1} u_y - \frac56 u^2 \partial_x^{-2} u_{yy} + \frac{5}{24} u^4\right] dxdy$$

$$+ \delta \int_{\mathbb{R}^2}\left[3u_{xxxx}^2 + 13u_{xxy}^2 + \frac{35}{3}u_{yy}^2 + \frac53\left(\partial_x^{-2}u_{yyy}\right)^2\right]$$

$$\le C\delta\left\{X_2^2 X_3^{\frac12} X_4^{\frac12} + X_1^{\frac12} X_2^{\frac12} X_3^2 + X_2^{\frac32} X_3^{\frac32}\right.$$

$$+ X_1 X_2^{\frac14} X_3^{\frac12} X_4 \|\partial_x^{-2} u_{yy}\|_0^{\frac14} + X_2^{\frac54} X_3^{\frac32} \|\partial_x^{-2} u_{yy}\|_0^{\frac14}$$

$$+ X_1^{\frac12} X_2^{\frac12} X_4 \|\partial_x^{-2} u_{yy}\|_0 + X_2 X_3^{\frac12} X_4^{\frac12} \|\partial_x^{-2} u_{yy}\|_0$$

$$\left. + X_1^{\frac12} X_2^{\frac12} X_3 \|\partial_x^{-2} u_{yy}\|_0 + X_1^3 X_4\right\}. \tag{2.30}$$

If we now let

$$\tilde{X}_2^2 = \int_{\mathbb{R}^2} \left[u_{xx}^2 + u_y^2 + \left(\partial_x^{-2} u_{yy}\right)^2\right] dxdy,$$

then with applications of Young's inequality, the terms in the braces of (2.30) can be estimated above by

$$C_0(\eta)X_4^2 + C_1(\eta)X_3^2 + C_2(\eta)X_2^2\tilde{X}_2^2 + C_3(\eta) + C_4,$$

where C_4 depends only on η and $\|u_0\|_{V_1(\mathbb{R}^2)}$. Hence if relations (2.30) is integrated with respect to t, and with appropriate choice of η and use of (2.21) and (2.22), we will find that

$$\tilde{X}_2^2 + \delta \int_0^T X_4^2 dt$$

$$\leq C\left(\|u_0\|_{V_2(\mathbb{R}^2)}\right) + C(\eta) \sup_t \tilde{X}_2^2,$$

and hence obtain

$$\sup_t \tilde{X}_2^2 \leq \left(\|u_0\|_{V_2(\mathbb{R}^2)}\right), \tag{2.31}$$

and

$$\delta \int_0^T X_4^2 dt \leq \left(\|u_0\|_{V_2(\mathbb{R}^2)}\right). \tag{2.32}$$

Of course, in obtaining relations (2.31) and (2.32), we had to estimate some of the terms in the first bracket of (2.30), for instance

$$\int_{\mathbb{R}^2} uu_x^2 \leq \|u\|_0 |u_x|_4^2$$

$$\leq C\|u\|_0 \|u_x\|_0^{\frac{1}{2}} \|u_y\|_0^{\frac{1}{2}} \|u_{xx}\|_0$$

$$\leq C\left(\|u_0\|_{V_1(\mathbb{R}^2)}\right) \tilde{X}_2^{\frac{3}{2}},$$

where use has been made of Lemma 2.5, and

$$\int_{\mathbb{R}^2} u_y \partial_x^{-1} u \, \partial_x^{-1} u_y = \frac{1}{2} \int_{\mathbb{R}^2} u \left(\partial_x^{-1} u_y\right)^2$$

$$\leq C \|u\|_0 |\partial_x^{-1} u_y|_4^2$$

$$\leq C \|u\|_0 \|\partial_x^{-1} u_y\|_0^{\frac{1}{2}} \|\partial_x^{-2} u_{yy}\|_0^{\frac{1}{2}} \|u_y\|_0$$

$$\leq C \left(\|u_0\|_{V_1(\mathbb{R}^2)}\right) \tilde{X}_2^{\frac{3}{2}}.$$

This completes the proof of Thoerem 2.2.

□

REMARK 2.8. In a forth coming paper in preparation, this type of theory will be extended to a generalization of the Kadomtsev-Petviashvili equations of the form

$$(u_t + u^p u_x - D_x^\alpha u_x)_x - u_{yy} = 0, \tag{2.33}$$

where $D_x = (-\partial_x^2)^{1/2}$ and $\alpha \geq 1$. Indeed, it will be shown that for $\alpha \geq 2$, global weak solutions exist for large initial data provided that $p < \frac{4\alpha}{4+\alpha}$.

References

1. O. V. Besov, V. P. Il'in, S. M. Nikolskii, Integral Representations of functions Imbedding Theorems vol 1, J. Wiley (1978).
2. A. De Bouard, J. -C. Saut, *Solitary waves of generalized Kadomtsev-Petviashvili equations*, Annales Institut Henri Poincaré, Analyse Linéaire (to appear).
3. A. S. Fokas and L.Y. Sung, *On the solvability of the N-wave, Davey-Stewartson and Kadomtsev-Petviashvili equations*, Inverse Problems **8** (1992), 673-708.
4. J.-C.Saut, *Remarks on the Generalized Kadomtsev-Petviashvili equations*, Ind. U. Math. J. **42** (1993), 1011-1026.
5. J. -C. Saut, *Recent results on the Generalized Kadomtsev-Petviashvili Equations*, Acta Appl. Mat **39** (1995), 477-487.
6. M. Schwarz, *Periodic Solutions of Kadomtsev-Petviashvili*, Adv. Math. **66** (1987), 217-233.
7. S. Ukai, *Local solutions of the Kadomtsev-Petviashvili equations,*, J. Rac. Sci. Univ. Tokyo Sect. 1A Math. **36** (1989), 193-209.
8. X.P. Wang, M.J. Ablowitz, H. Segur, *Wave Collapse and instability of solitary waves of a generalized Kadomtsev-Petvaiashvili equation*, Physics D **78** (1994), 241-265.
9. X. Zhou, *Inverse scattering transform for the time dependent Schrödinger equations with application to the KI P equations*, Comm. Math. Phys. **128** (1990,), 551-564.

DEPARTMENT OF MATHEMATICS, LOUISIANA STATE UNIVERSITY, BATON ROUGE, LA 70803 USA.

E-mail address: tom@math.lsu.edu

Contemporary Mathematics
Volume **200**, 1996

A phenomenological description of soliton splitting during run up

E. van Groesen

Abstract. In this paper a simple model is proposed to describe the splitting of an initial single solitary wave that runs into shallower water into two solitary waves. Different from results in the literature that use inverse scattering theory for the Korteweg - de Vries equation to find the splitting once a single (deformed) wave has arrived at a shallower region of constant depth, in this paper a quasi-static approximation is proposed to capture also the changes during run up. The model is completely based on qualitative properties of the energy and mass of single solitary waves as function of amplitude. With these relations, the splitting process can be described qualitatively in complete agreement with results from numerical calculations.

1. Introduction

In this contribution we investigate one of these fascinating phenomena in wave propagation that appear as a low dimensional dynamical system. Consider a solitary wave like profile, running to the right at constant speed above a flat part of the bottom. When the wave enters a region of variations in the bottom, it will deform since nonlinear and dispersive effects will not balance eachother any longer as they did above the flat bottom. For increasing depth, dispersion will become more dominant, and constituent waves will disperse away. We will consider here the opposite effect: the deformation of a wave during run up. Depending on details of the topography, various deformations can be expected and have partly been described in the literature.
In fact, the experimental and numerical results of Beji & Battjes, [**2**] were a motivation for this investigation. They studied a periodic wave train over a bar. The combination of a decrease and subsequent increase of depth makes it difficult to understand their intriguing results. Therefore we restrict ourselves and study the run-up on its own.

1991 *Mathematics Subject Classification.* Primary 35Q20, 76B15, 76B25, Secondary 35B20, 35A35.

Part of the research is sponsored by the Commission of the European Communities, Directorate General XII-B, Joint Research Project CI1*-CT93-0018 between the Department of Mathematics, Institut Teknologi Bandung, Indonesia, and the Faculty of Applied Mathematics, University of Twente, The Netherlands.

To understand the basic underlying process, we consider the simplest possible topography: a smooth (monotone) transition between two constant depths, decreasing in the direction of propagation of the wave. Although the slope will be assumed to be small, the total variation in depth will be of order one. As a motivation for the mathematical approach of this paper, and to have a partial, practical justification of the results, consider Fig. 1. This shows the wave profiles from numerical results obtained by Van Daalen who carefully calculated the deformation for the full set of nonlinear gravity waves using boundary elements methods, see [**3**]. The same qualitative phenomenon is described by the model that will be presented in Section 3; see Fig. 9.

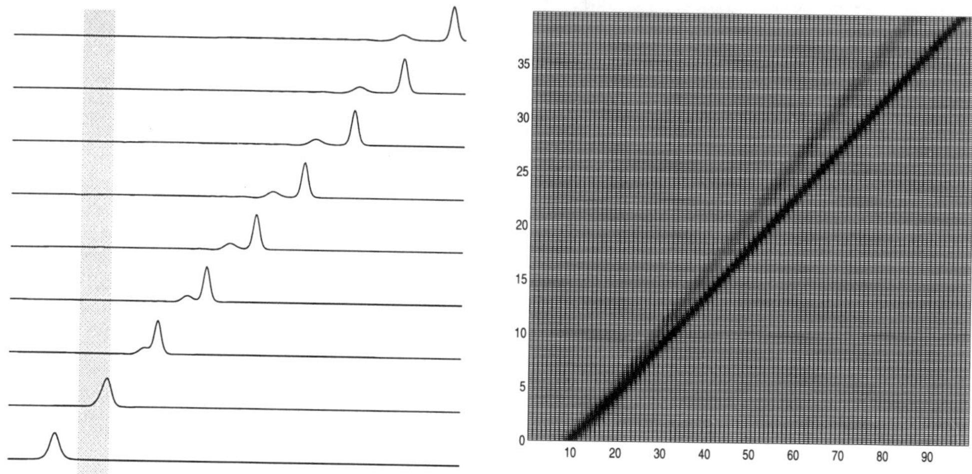

FIGURE 1. Numerical calculations of an initial solitary wave of .1 m during run up from a depth of .5 m to a depth of .35 m. The calculations are based on the complete set of surface wave equations for inviscid fluid without surface tension. At the left the wave profiles are shifted vertically with increasing time and the region of linear decrease in depth is indicated; at the right the splitting is shown in grey tones.

The graphic representation of the calculations shows that for the given topography the initial single solitary wave splits into two (noticible) waves, each of which resembles quite well a solitary wave. Although experimental or numerical results that show the smooth separation process in such a clean way cannot be found in the literature, the deformation of waves by variable bottom has attracted the attention of many scientists, and both numerical and analytical results are available, in particular from the seventies: Madsen & Mei, [**9**], Tappert & Zabusky [**16**], Johnson [**6, 7**], Ono [**13**], Miles [**10, 11**]; see also Newell [**12**]. The description presented in this paper differs from these papers in the following way.

For the distortion when entering shallower water, the general approach in the literature consists of two steps. First it is assumed that during run up, a single initial hump of solitary wave type above a bottom at depth h_0 deformes within a family of *single* hump profiles. Two parameters are used to describe the details of the hump (amplitude and width, for instance), and the change of these parameters are calculated. Then a single hump is obtained at the smaller depth h. Secondly, the distortion of this single hump, that differs from a pure solitary wave profile, is investigated on the new constant depth h. Using the Korteweg-de Vries equation as an analytical model, inverse scattering techniques are applied to predict the number of solitary waves that emerge from this hump ("fission" of solitons). The result of this analysis is that the number of emerging solitons N depends on the depth ration according to the following formula:

$$\frac{2}{N(N+1)} < \left(\frac{h}{h_0}\right)^{9/4} < \frac{2}{N(N-1)}.$$

In particular, two solitons are expected when $(h/h_0)^{9/4} > 1/3$, i.e. when $h/h_0 > .614$.
In this paper we will follow a somewhat different approach. Essentially the idea is to describe the wave running up not as a single hump but as a gradually developing separation process of two solitary waves. (Conditions on the maximal depth ratio for which two waves will appear are not required beforehand, but will follow from the analysis.) The governing mechanism that will determine the separation will be the basic physical laws of energy and mass conservation. The aim is to give a simple description of this phenomenon, using only variables that appear to be necessary from the phenomenon, being the amplitude and position of each of the waves.
In Section 2 the consequence of the conservation properties of energy and mass will be investigated solely from an analysis of the dependence of the energy and mass on the parameters of a one-soliton solution. In fact, only qualitative properties of these functions, like convexity and monotonicity, turn out to be essential to describe the qualitative properties of the splitting process. Explicit analytic expressions for these functions are given in Section 3 based on a model equation that has been proposed by Van Groesen & Pudjaprasetya [**5**] to describe uni-directional waves above a slowly varying bottom. This model equation is a variant of the KdV-equation (Korteweg & De Vries, 1895) [**8**], with coefficients that depend explicitly on the bottom topography and is equivalent with an equation derived by Newell [**12**]. Although the restrictions that underly the validity of this equation may not be fulfilled for steeper bottom variations, the qualitative properties of the energy and mass functions obtained are in agreement with results obtained from calculations for the complete set of surface wave equations. In section 4 we describe the elements of a mathematical justification of the simple approximation presented here. Section 5 contains some conclusions, and suggestions for further research.

2. Qualitative description

Before specifying any evolution equation, we present a simple description of the phenomenon in this section.
First consider the constituent waves appearing in the phenomenon. Denote a single solitary wave above a flat bottom by S_1. Except from the depth of the bottom, other parameters are the "position" of the wave (say the position of its crest) and its amplitude (or a related quantity), to be denoted by γ. In the description to follow we will use a *quasi-homogeneous* evolution. Based on the assumption that the wave is "concentrated" in space (vanishingly small outside a region around the crest), we will take as approximation a solitary wave above a flat bottom of which the depth is determined by the depth below the actual position of the crest. In

that way, γ and the depth h can be taken as independent variables. Instead of the depth h, we will use in the following the quantity c defined by

$$c = \sqrt{gh}$$

where g is the gravitational accelaration, and we will write $S_1(c, \gamma)$ accordingly. This defines a two-dimensional "wave manifold", a collection of waveshapes: at each depth a one-parameter family with parameter indicating the height of the wave.

Each wave carries a certain amount of mass and has certain energy; we will denote these quantities in the following by $\mathcal{M}(c, \gamma)$ and $\mathcal{E}(c, \gamma)$ respectively.

To describe the run up, start with a wave above a flat bottom at depth h_0, that has mass m_0 and energy e_0. For decreasing depth, assuming the wave to remain a single wave with adjusting value of γ, energy conservation alone would determine the governing evolution (parameterized by c). In fact,

$$\mathcal{E}(c, \gamma) = e_0$$

describes a curve in the c, γ-plane, and so implicitly γ as a function of c; see Fig. 2. (The figures presented in this section are based on the model to be described in the next section. However, the arguments given in this section depend only on qualitative properties to be described shortly.)

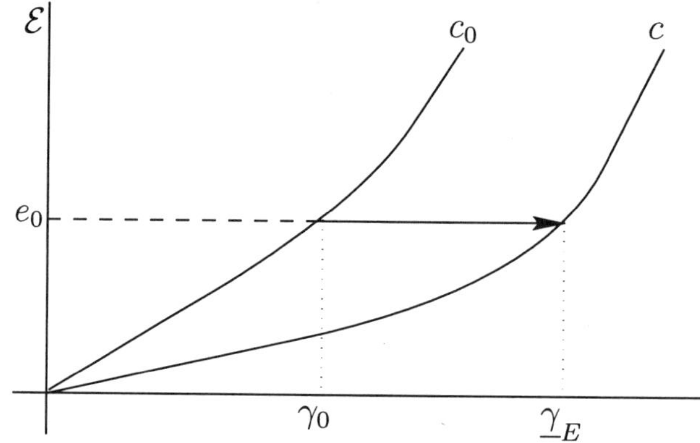

FIGURE 2. Energy vs momentum γ for two different values of c. The arrow would indicate the running up of one solitary wave of constant energy (for which the momentum has to increase), not taken into account mass conservation.

In general it cannot be expected that mass will be conserved in this energy-conserved evolution. It is shown in Fig.3. (depicted for the model of the next section) that the mass decreases and that there is a mass-surplus when modelling with a single solitary wave. One way to overcome the violation in mass conservation for a single wave is to introduce some additional tail at the back of the wave that is designed to account for the mass surplus. In literature various models have been used: shelves, non-flat equilibrium surfaces, etc.; see e.g. Newell, [**12**], Pudjaprasetya & Van Groesen [**15**].

Here we will take another point of view, and assume that instead of a single wave (plus tail), the profile will evolve during run up into the superposition of *two* waves which together satisfy energy and mass conservation. (Assuming that there

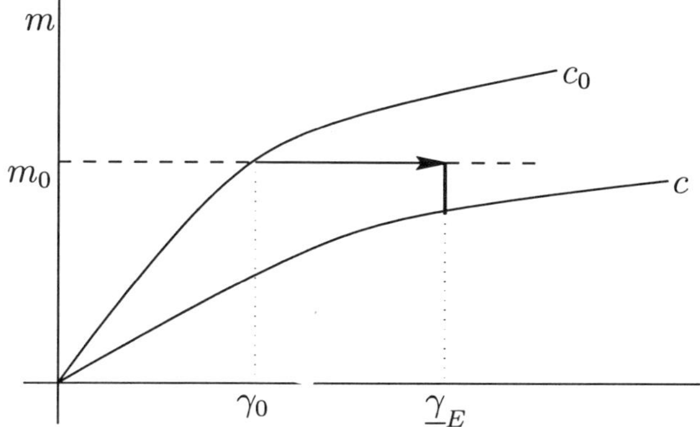

FIGURE 3. Mass vs momentum; the two curves for different value of c indicate that during run up of a single solitary wave that conserves energy, there is mass surplus.

will result only two waves will put some conditions on the topography; in general a breaking up into more than two waves can be expected.)
More precisely, we make the following two simplifying assumptions (which seem to be somewhat contradictory at first sight):

- both constituent waves are taken as if they are at each instant above the same bottom depth,
- the energy and mass of the combined waves are the sum of these quantities of the two constitutent waves.

The last assumption is a consequence of the finite extension of the waves when the constituent waves are sufficiently separated. In that case the energy and mass, which are integrals over local densities, are additive.
Closely related to the additivity argument is the motivation for the choice of the parameter γ that, for a single wave, should be related to its amplitude. However, a quantity like amplitude is not well defined for an arbitrary wave profile, and it cannot be expected to be additive. A parameter that is globally defined, and is additive when waves separate, is preferred. This motivates to use as parameter γ some integral of a local density. As will become clear in the specific model, a natural choice for γ is the value of the horizontal momentum, proportional to the L_2-norm but depending on the topography. Motivating the nomenclature, above a horizontal bottom this quantity is conserved as a consequence of translation invariance.
The first assumption and the choice of γ together imply that the phenomenon of splitting can symbolically be written as

$$S = S_1(c, \gamma_+) + S_1(c, \gamma_-)$$

where γ_-, γ_+ are parameters that characterize the two waves. Note that we do not aim to specify the precise positions of the constituent waves. This will need an expression for the velocity of the wave, which will be found in the model to be presented in Section 3.
The second assumption provides the equations that should determine the values γ_- and γ_+ as a function of c. From energy and mass conservation there result the

conservation conditions:

$$\begin{cases} \mathcal{E}(c,\gamma_+) + \mathcal{E}(c,\gamma_-) & = & e_0 \equiv \mathcal{E}(c_0,\gamma_0) \\ \mathcal{M}(c,\gamma_+) + \mathcal{M}(c,\gamma_-) & = & m_0 \equiv \mathcal{M}(c_0,\gamma_0) \end{cases}$$

where c_0, γ_0 are the values of the initial single wave.

Given the assumptions made above, the whole description will be based on an analysis of these two equations. *The physical properties of the water wave problem are reflected in qualitative information about the functions $\mathcal{E}(c,\gamma)$ and $\mathcal{M}(c,\gamma)$.* We state the properties of these functions as found for the model problem in the next section; these properties are also verified numerically for the complete set of equations.

In investigating the two conditions above, it is simplest to take $\gamma = \gamma_- + \gamma_+, \xi = \gamma_-$ as parameters; the conservation conditions then require to find γ, ξ from

$$\begin{cases} \mathcal{E}(c,\gamma-\xi) + \mathcal{E}(c,\xi) & = & e_0, \\ \mathcal{M}(c,\gamma-\xi) + \mathcal{M}(c,\xi) & = & m_0. \end{cases}$$

We first investigate each condition separately and then combine the results.

Energy-conservation.

Based on the model and numerical calculations we will use the fact that *the function $\mathcal{E}(c,\gamma)$ is (strictly) convex in γ, and, in the relevant parameter interval, an increasing function of c.*

As a consequence, there exists an interval, $[\underline{\gamma}_E, \bar{\gamma}_E]$ and a monotonically increasing function ξ_E such that energy conservation is satisfied. In fact,

$$\xi_E(\underline{\gamma}_E) = 0, \; \xi_E(\bar{\gamma}_E) = \tfrac{1}{2}\bar{\gamma}_E,$$

where $\underline{\gamma}_E, \bar{\gamma}_E$ are defined by

$$\mathcal{E}(c, \underline{\gamma}_E) = e_0, \; \mathcal{E}(c, \tfrac{1}{2}\bar{\gamma}_E) = \tfrac{1}{2}e_0$$

These results are obvious by noting that the function

$$\xi \mapsto \mathcal{E}(c,\gamma-\xi) + \mathcal{E}(c,\xi)$$

is (strictly) convex and symmetric around $\frac{1}{2}\gamma$ at which it attains its least value $2E(c, \frac{1}{2}\gamma)$ (see Fig 4.).

Mass conservation.

Based on the model and numerical calculations we will use the fact that *the function $\mathcal{M}(c,\gamma)$ is strictly concave in γ and increasing in c.*

From this we find that

$$\xi \mapsto \mathcal{M}(c,\gamma-\xi) + \mathcal{M}(c,\xi)$$

is strictly concave. Hence there exists an interval $[\underline{\gamma}_M, \bar{\gamma}_M]$ and a monotonically decreasing function ξ_M such that mass conservation is satisfied. The values are given by

$$\mathcal{M}(c, \tfrac{1}{2}\underline{\gamma}_M) = \tfrac{1}{2}m_0, \; \mathcal{M}(c, \bar{\gamma}_M) = m_0,$$

and the function ξ_m satisfies

$$\xi_M(\underline{\gamma}_M) = \tfrac{1}{2}\underline{\gamma}_M \text{ and } \xi_(\bar{\gamma}_M) = 0;$$

see Fig. 5.

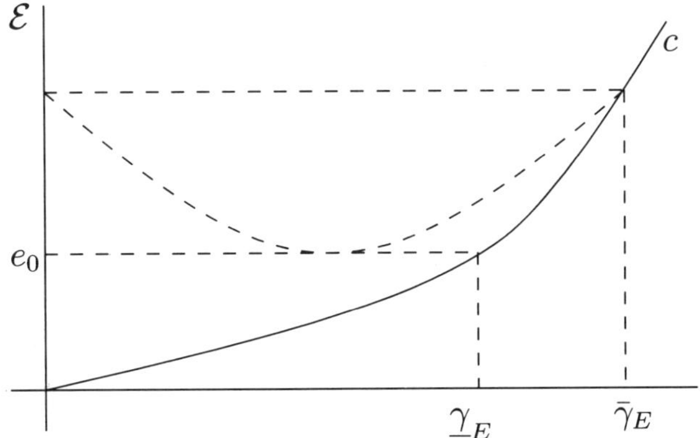

FIGURE 4. Definition of the limiting values of γ for which there exists a superposition of two constituent waves of total energy e_0.

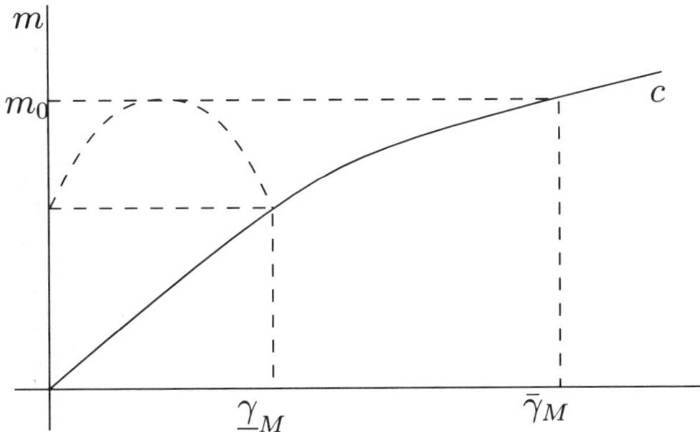

FIGURE 5. Definition of the limiting values of γ for which there exists a superposition of two constituent waves of total mass m_0.

Combined energy and mass conservation

Taken together, the two resulting curves ξ_E, and ξ_M found above (depicted schematically in Fig. 6) imply that a unique common value $\hat{\xi}$ exists, which determines the values γ_- and γ_+ of the constituent waves with total mass and energy as desired, if the following conditions are satisfied

$$\underline{\gamma}_M < \bar{\gamma}_E < \bar{\gamma}_M .$$

For the model to be presented, the last inequality is always satisfied. The first one leads to a lower bound for the depth and corresponding value of c. When an approximation for the energy is taken ($\varepsilon = 0$ in the expression to be given in the next section) there results as condition approximately $h/h_0 > .54$.

These results indicate that in this model, no matter how little the depth decrease, a second wave starts to split off from the initial single wave. From the description here, this process continues until the difference in depth exceeds a certain value, $h/h_0 \approx .54$. In approaching this critical value, the two constituent waves would become almost equal ($\gamma_- \approx \gamma_+ \approx \frac{1}{2}\bar{\gamma}_E$) and a symmetric situation would result.

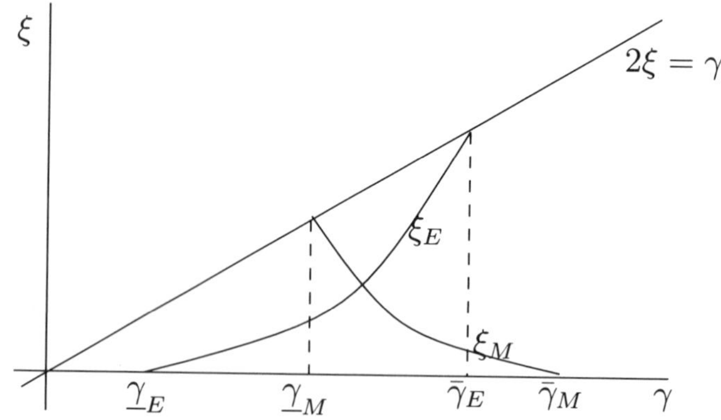

FIGURE 6. ξ vs γ curves for energy and mass conservation.

The geometric picture is depicted in Figs. 7 and 8. Observe that, while γ_- increases monotonically, the change in γ_+ is not monotone: after an initial increase, the value decreases with decreasing depth before the symmetric situation is attained. This is not realistic from a modelling point of view; in another paper ([**14**]) this is investigated in more detail. If we take the approximation to be valid until γ_+ attains its maximum value, this leads to another bounf for the maximal depth ratio, approximately $h/h_0 > .7$. Note that this last critical value is somewhat larger than the result described in the introduction obtained with the model of a changing single-hump with subsequent fission by inverse scattering theory.

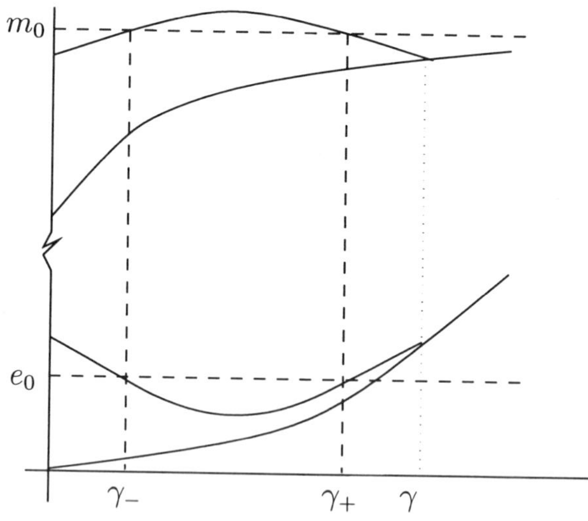

FIGURE 7. Combined conservation of mass and energy.

3. KdV-top model

The properties of the determining functions for energy and mass as stated above will now be obtained for a model equation that describes uni-directional

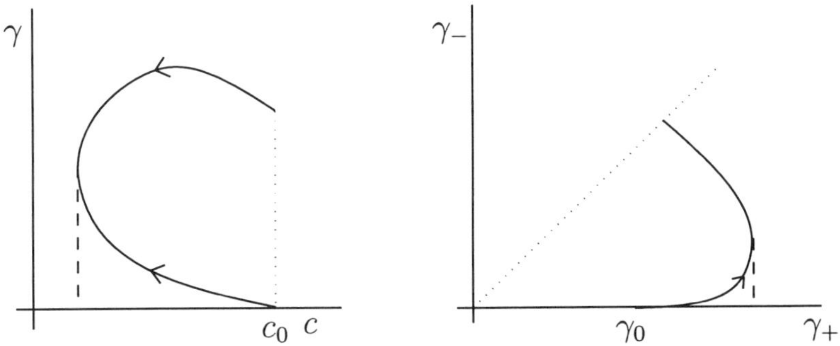

FIGURE 8. Values of γ_+ and γ_-, at the left as function of c and at the right parameterized by c, during evolution for which energy and mass are conserved. Note that during run up (decreasing c), the value of γ_+ attains a maximal value before the symmetric situation (the turning point where $\gamma_- = \gamma_+$) is reached.

waves above a slowly varying bottom. The equation is given here in the form as derived in [**5**]; it is equivalent to an equation derived by Newell [**12**].
The equation for the wave elevation η has been derived under the usual assumptions of Boussinesq properties of the wave shapes (long, low waves) and small bottom variations so that reflections can be neglected; a small parameter ε measures the Boussinesq and bottom effects. The equation is easiest described (and reflects the derivation) in the form of a Hamiltonian system like

$$\partial_t \eta = -\Gamma \delta H(\eta)$$

where the structure map is given by $\Gamma = \sqrt{c(x)}\partial_x \sqrt{c(x)}$, and the Hamiltonian, which is an approximation for the total energy, by

$$H(\eta) = \int [\tfrac{1}{2}\eta^2 + \varepsilon \left(-\frac{1}{12} h^2 \eta_x^2 + \frac{1}{4h}\eta^3 \right)].$$

This equation is called *KdV-top,* since it reduces for waves above constant depth to the familiar Korteweg-de Vries (KdV)-equation but incorporates the effect of topography.
The skew symmetry of the operator Γ implies the Hamiltonian structure of the equation. As one consequence, energy is conserved during the evolution of any solution: $\partial_t H(\eta) = 0$. Also, since Γ is degenerate, $\Gamma 1/\sqrt{c(x)} = 0$, there is a Casimir functional that is the generalization of the total mass functional above a flat bottom:

$$C(\eta) = \int \frac{\eta}{\sqrt{c(x)}}$$

The generalization of the horizontal momentum is given by

$$I(\eta) = \tfrac{1}{2} \int \frac{\eta^2}{c(x)};$$

above a constant bottom this functional is conserved, which property is lost when translation symmetry is destroyed by bottom variations.
These properties of this model equation resemble the corresponding properties for the full wave equations. Therefore we based the previous analysis on the energy and mass conservation, while the horizontal momentum (not constant) was used merely

as a convenient parameter, which has a clear physical interpretation nonetheless. The specific form of the operator Γ is interesting in itself. Even excluding Boussinesq effects, the linear equation

$$\partial_t \eta = -\Gamma \eta$$

describes aready the main properties of a wave (of arbitrary form) running up: decrease of wave length and speed, and increase of amplitude; this follows from the exact solution that is easily written down; see [**5**].
Using the KdV-top model, Van Beckum [**1**] calculated the running up of a single solitary wave using consistent Fourier truncation, see Fig 9, and obtained the same qualitative result for the splitting as in Fig. 1.

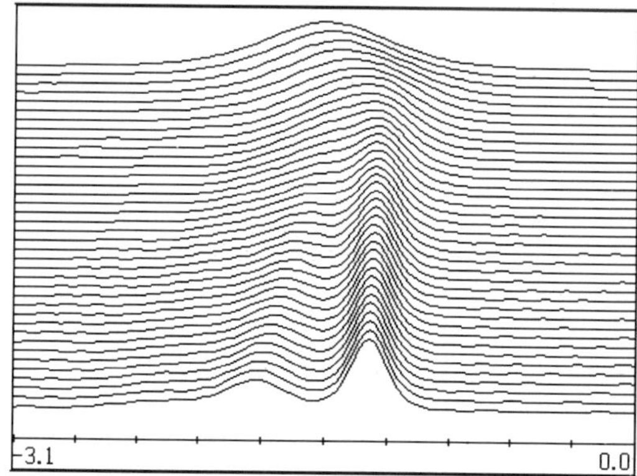

FIGURE 9. Soliton splitting as calculated with the KdV-top equation. For clarity, the process is depicted in a frame moving with the waves, and profiles are translated vertically downwards for increasing time.

As is well known, the KdV-equation (above flat bottom) admits solitary waves. These waves can actually be described as constrained minimizers of the energy at prescribed momentum. Denoting the energy, mass and momentum above a flat bottom by $H(c,\eta), C(c,\eta), I(c,\eta)$ respectively, the *profile* S_1 of a solitary wave is characterized for each $\gamma > 0$ as a solution of

$$S_1(c,\gamma) \in Min\{H(c,\eta) \mid I(c,\eta) = \gamma\}.$$

The equation is found from Lagranges multiplier rule $\delta H = \lambda \delta I$. The explicit expression for the solitary wave profile is given by

$$S_1(c,\gamma)(x) = A \operatorname{sech}(B(x-\varphi)$$

where the dependence on γ and c is explicitly given by

$$A = 3gc^{-4/3}(\frac{\gamma}{4})^{2/3}, \; B = \frac{3}{2}g^2 c^{-11/3}(\frac{\gamma}{4})^{1/3}.$$

Observe that both coefficients depend monotonically on γ and on c. The dynamic solution is a translation of this profile with speed λ. The expression for the speed

$$\lambda(c,\gamma) = c + \frac{3}{2}\varepsilon g^2 c^{-7/3}(\frac{\gamma}{4})^{2/3}$$

shows that it increases monotonically with respect to γ. The variable φ is an arbitrary phase factor that determines the position of (the crest of) the wave. In the quasi-homogeneous description presented above, φ is adapted such that the crest of the wave is above the point at the bottom where the depth is related to c, i.e. $\sqrt{gh(\varphi)} = c$.

From these expressions for the solitary waves, the functions for energy and mass can be found explicitly. They are given by

$$\mathcal{E}(c, \gamma) = c\gamma + ac^{-7/3}\gamma^{5/3}, \ \mathcal{M}(c, \gamma) = bc^{11/2}\gamma^{1/3}$$

with $a = 18\varepsilon g^2/4^{5/3} \approx 35\varepsilon, b = 4^{2/3}/9 \approx .28$.

As stated before, these functions have the qualitative properties of the function $\mathcal{E}$ and $\mathcal{M}$ in the previous section and were used to produce the graphs.

4. Elements of the mathematical justification

At first sight it is rather remarkable that a simple description as presented in Section 2 can describe the actual process correctly. In fact it means that a two-dimensional description suffices to capture the most relevant properties.

However, there is a clear mathematical reason, that can only be described rudimentary here; for a more extensive treatment, see Fledderus & Van Groesen, [**4**].

The main problem is that although the bottom is weakly sloping, the actual deformation from the initial state is of order one since the decrease in depth is of order one. However, the use of the specific functions S_1 in the quasi-homogeneous description guarantees that at each instant the approximation is but a slight deformation form an exact solution: the wave forms used in the description are exact solutions of the corresponding homogenized system. Moreover, as becomes obvious from the constrained minimizing property stated for these waves, these forms are stable in a suitable sense (taking into account the constraint and the degenaration from translation symmetry). Having this property, the next important fact is the correct choice of the parameter dynamics: the projection of the true solution into the two-dimensional manifold with parameteres c, γ. Here the dynamics is simply based on the conservation properties of mass and energy. From a more convincing mathematical point of view, the relevant observation is that it can be shown that these conditions are equivalent to Fredholm necessary solvability conditions that should be satisfied in order that the error is uniformly small during the evolution where changes of the order one appear. This first order condition, combined with the stability property of the waves, suffice to prove the uniform validness of the approximation as constructed for the given KdV-top equation.

The relation with the complete set of surface wave equations should then be justified by referring to the construction of KdV-top as a consistent model for specific wave evolutions. In this process some quantitive differences can be expected since this can already be shown to be the case for soliton solutions from Boussinesq equations compared to solitons from the KdV equation; see also [**1**].

5. Conclusions

Although the simple description predicts the splitting process quite well qualitatively, the underlying assumptions should be refined. In particular, the assumption that the two separated waves are taken at the same depth should be improved. In another paper, [**14**], this will be described in more detail, but the results merely modify the analysis as presented here in an inessential way. Another point for further investigation is the critical value of the depth ratio at which the two-wave approximation breaks down. If it is conjectured that the critical depth is where the largest wave reaches its maximal amplitude, a subsequent splitting of this largest wave, and consequently also of the smaller one, should be expected. Unfortunately, reliable numerical calculations, or experiments, that can show this pairwise splitting are not available yet.

References

[1] F.P.H. van Beckum. *"Hamiltonian consistent discretisation of wave equations"*. Thesis, University of Twente, 1995.

[2] B. Beji and J.A. Battjes. "Numerical simulation of nonlinear wave propagation over a bar". *Coastal Eng.*, 23:1–16, 1994.

[3] E.F.G. Daalen. "Solitary wave split-up due to a sloping bottom: Theory, numerics and experiments". *Comp. Math.*, 1995.

[4] E.R. Fledderus and E. van Groesen. "Deformation of coherent structures". *Rep. Progr. Phys.*, 59:1–89, 1996.

[5] E. van Groesen and S.R. Pudjaprasetya. "Uni-directional waves over slowly varying bottom. Part I: Derivation of a KdV-type of equation". *Wave Motion*, 18:345–370, 1993.

[6] R.S. Johnson. "Some numerical solutions of a variable coefficient Korteweg - de Vries equation". *J. Fluid Mech*, 54:81–91, 1972.

[7] R.S. Johnson. "On the development of a solitary wave moving over an uneven bottom". *Proc. Camb. Phil. Soc.*, 73:183–203, 1973.

[8] D.J. Korteweg and G. de Vries. "On the change of form of long waves advancing in a rectangular canal, and on a new type of long stationary waves". *Phil. Mag.*, 39:422–443, 1895.

[9] O.S. Madsen and C.C. Mei. "The transformation of a solitary wave over an uneven bottom ". *J. Fluid Mech.*, 39:781–791, 1969.

[10] J.W. Miles. "On the Korteweg - de Vries equation for a gradually varying channel". *J. Fluid Mech.*, 91:181–190, 1979.

[11] J.W. Miles. "Solitary waves". *Annual Rev. Fluid Mech.*, 12:11–44, 1980.

[12] A.C. Newell. *Solitons in mathematics and physics.* SIAM, Philadelphia, 1985.

[13] H. Ono. "Wave propagation in an inhomogeneous anharmonic lattice". *J.Phys.Soc. Jap.*, 32:332–336, 1972.

[14] S.R. Pudjaprasetya and E. van Groesen. "Model for the splitting of a solitary wave running into shallower water,". *in preparation*, 1995.

[15] S.R. Pudjaprasetya and E. van Groesen. "Uni-directional waves over slowly varying bottom, part II Deformation of travelling waves". *Wave Motion*, 1995.

[16] F.D. Tappert and N.J. Zabusky. "Gradient-induced fission of solitons". *Phys.Rev.Lett*, 27:1774–1776, 1971.

Applied Mathematics, University of TWENTE, P.O. Box 217, 7500 AE Enschede, The Netherlands

E-mail address: groesen@math.utwente.nl

Contemporary Mathematics
Volume **200**, 1996

Asymptotic stability of nonlinear bound states in conservative dispersive systems

M.I. Weinstein

In memory of Professor T. Brooke Benjamin

Abstract

This is a survey of results and methods of proof for stability and asymptotic stability of solitary waves in nonlinear dispersive systems.

1 Introduction

Nonlinear bound states (e.g. solitary waves, kinks, vortices) have been observed to play an important role in many systems in which the dynamics are dominated by a competition between nonlinearity and dispersion. A very well known example is the Korteweg - de Vries (KdV) equation, (3.17) with $p = 1$ [5],[14], for which the large time evolution of well-localized initial data is a finite train of clearly identifiable solitary waves moving to the right, plus dispersive radiation which trails behind. Although KdV is a completely integrable Hamiltonian system, this type of asymptotic resolution of the nonlinear dynamics into nonlinear bound states plus "radiation" appears to extend to many equations which are not integrable. For nonintegrable equations, the most tractable question of this nature is the local question of stability of nonlinear bound states. We shall be concerned with the local variant of the asymptotic resolution problem, the question of asymptotic stability.

In this paper I will survey results on the asymptotic stability of solitary waves in various nonlinear dispersive systems of Hamiltonian type. The approach to asymptotic stability described in this work

1991 *Mathematics Subject Classification.* Primary 35Q53, 35Q55.
This work is supported, in part, by a grant from the U.S. National Science Foundation.

was initiated in joint work with A. Soffer in [26],[27],[28] for a class of nonlinear Schrödinger equations (NLS) and in the context of model water wave equations of Korteweg - de Vries (KdV) type with R.L. Pego in [19], [20]. It has recently been applied to the regularized long wave (RLW) or BBM equation (see Benjamin, Bona & Mahony [4] and Peregrine [24]) in joint work with J.R. Miller in [16],[17]. Both KdV and RLW arise in the asymptotic description of free surface water waves of small amplitude and long wavelength. NLS is a canonical envelope equation arising in hydrodynamics and nonlinear optics. In this paper, we shall focus on the recent work on the RLW equation

$$(I - \partial_x^2)\partial_t u + \partial_x(u + \frac{1}{2}u^2). \tag{1.1}$$

This paper is structured as follows. In section 2 we review the Lyapunov stability theory of solitary waves. This approach has its origins in the work of Boussinesq [5]. The modern mathematical treatment was initiated in the work of T.B. Benjamin [3]; see also the work of Arnol'd [2] and the survey [11]. In section 3, we present a heuristic discussion of why it is expected that solitary waves of RLW be asymptotically stable, and then state the relevant theorem. We also review the results for KdV type equations and a class of nonlinear Schrödinger equations. In section 4, the general strategy employed to prove asymptotic stability is discussed, and in section 5 its implementation is outlined in the context of RLW.

2 The solitary wave and Benjamin's approach to orbital Lyapunov stability

The RLW equation has a two parameter family of solitary wave solutions:

$$u_c(y) = u_c(x - ct + \gamma) \tag{2.2}$$

for any speed $c > 1$ and spatial shift $\gamma \in \mathbb{R}$. These solutions correspond to the homoclinic orbit of the second order ordinary differential equation:

$$cu_c'' - (c-1)u_c + \frac{1}{2}u_c^2 = 0. \tag{2.3}$$

Explicitly, we have

$$u_c(x) = 3(c-1)\text{sech}^2\left(\frac{1}{2}\sqrt{\frac{c-1}{c}}x\right). \tag{2.4}$$

An alternative, and quite fruitful, point of view is that the solitary, u_c, wave is a *critical point* of an energy functional:

$$\mathcal{E}[u;c] = \mathcal{H}[u] - c\mathcal{N}[u],$$

i.e. $\delta\mathcal{E}[u_c] = 0$. Here

$$\mathcal{H}[u] = \int \frac{1}{2}u^2 + \frac{1}{6}u^3 \, dx \tag{2.5}$$

$$\mathcal{N}[u] = \int \frac{1}{2}u^2 + \frac{1}{2}(\partial_x u)^2 \, dx \tag{2.6}$$

are respectively, the conserved Hamiltonian and momentum functionals of RLW.

We now briefly outline the H^1 orbital Lyapunov stability theory. This theory states that if RLW is solved with initial data u_0 which is close to a solitary wave $u_c(x)$, then at all times, $t \neq 0$, the solution $u(x,t)$ will remain close to the family of spatial translates (the group orbit of the solitary wave)

$$\{u_c(x+\gamma) \mid \gamma \in \mathbb{R}\}. \tag{2.7}$$

More precisely, let $H^1(\mathbb{R})$ denote the Sobolev space of functions f which together with their first derivatives are square-integrable. For any $\epsilon > 0$, there exists $\delta > 0$ such that if

$$\inf_s \|u_0(\cdot + s) - u_c(\cdot)\|_{H^1} < \delta, \tag{2.8}$$

then for all $t \neq 0$

$$\inf_s \|u(\cdot + s, t) - u_c(\cdot)\|_{H^1} < \epsilon. \tag{2.9}$$

In his pioneering work on the stability of solitary waves, T.B. Benjamin [3] gave a proof of this result (see also [6]). This work has been very influential and Benjamin's approach has been refined and applied to a large class of Hamiltonian systems with symmetry; see for example, [15], [7], [10], [30], [31]. See also the related work [8]. I would like to present a short exposition of the main idea.

Let ϵ be a positive number. We have, for any $t \neq 0$

$$\begin{aligned}
\epsilon &\sim \mathcal{E}[u(\cdot,0)] - \mathcal{E}[u_c], \quad \text{by (2.8) for } \delta \text{ sufficiently small} \\
&= \mathcal{E}[u(\cdot,t)] - \mathcal{E}[u_c], \quad \text{by conservation laws} \\
&= \mathcal{E}[u(\cdot+\gamma,t)] - \mathcal{E}[u_c], \quad \text{by translation invariance} \\
&\equiv \mathcal{E}[u_c + v(\cdot,t)] - \mathcal{E}[u_c], \quad \text{definition of the perturbation}, v \\
&\sim (Lv(t), v(t)), \quad \text{by Taylor expansion,}
\end{aligned} \tag{2.10}$$

for $||v(t)||_{H^1}$ small. The operator L is given by

$$L = -c\partial_x^2 + (c-1) - u_c(x),$$

and is the second variation of the functional $\mathcal{E}$ at u_c.

If L were positive definite, that is

$$(Lv(t), v(t)) \geq C||v(t)||_{H^1}^2 \tag{2.11}$$

for some $C > 0$, it would follow from (2.10) that the size of perturbation would remain small in norm for all time $t \neq 0$. In fact, L is not positive definite. The relevant facts to note about the spectrum of the self-adjoint operator L are: (i) L has exactly one strictly negative (simple) eigenvalue, (ii) zero is a simple eigenvalue, with corresponding zero mode associated with translation invariance of RLW, and (iii) the continuous spectrum of L runs from c to $+\infty$.

The two *bad* directions associated with the two dimensional non-positive subspace of L are treated as follows. First, we note that the function $v(\cdot, t)$ appearing in the quadratic form in (2.11) is not arbitrary but is constrained by the dynamics of RLW. In particular, since $\mathcal{N}$ is conserved, we have that infinitesimally $v(\cdot, t)$ lies in the tangent space of the manifold of constant $\mathcal{N}$. The implied codimension one constraint, $v(\cdot, t) \perp u_c$, can be shown by constrained variational analysis to imply that the quadratic form in (2.11) is, in fact, non-negative on the relevant class of v's, *i.e.* $C \geq 0$. Finally, at each $t \neq 0$ we choose $\gamma = \gamma(t)$, so as to minimize the distance of the solution to the solitary wave orbit, as in (2.9). The additional implied codimension one constraint on $v(\cdot, t)$, $v(\cdot, t) \perp \partial_y u_c$ ensures that (2.11) holds with C strictly positive. The neglected superquadratic terms in (2.10) are easily controlled, if $v(t = 0)$ is sufficiently small in H^1, and only lead to a small modification in the constant C. Thus (2.11) applies and we have $||v(t)||_{H^1}$ remains small for all time.

3 Asymptotic stability - Heuristic discussion and Theorems

The above Lyapunov stability result states essentially that the *shape* of the solitary wave is stable under sufficiently small H^1 perturbations. The result, however, does not address the asymptotic behavior of such solutions as $t \to +\infty$, *i.e.* does the solution converge, in some sense, as $t \to +\infty$ to an asymptotic solitary wave of asymptotic speed c_+ and phase shift γ_+ or does it wander without end about the group orbit of the solitary wave? Note that in a finite dimensional Hamiltonian system, the fact that the flow preserves volume in phase space precludes asymptotic stability, but in an infinite dimensional system, different (non-equivalent) norms can be used to measure different phenomena.

Here is a heuristic explanation of solitary wave asymptotic stability of RLW waves. Think of the solution as being composed of nonlinear bound states (solitary waves of various amplitudes and speeds) and dispersive radiation. Suppose we perturb an exact solitary wave of speed $c_0 > 1$ slightly. We expect then, for a while, that the perturbation will be composed of two parts (i) small amplitude solitary waves ($c \sim 1$) and (ii) small amplitude dispersive waves. By (2.4), small amplitude solitary waves travel slowly, relative to the dominant wave. Regarding small amplitude dispersive waves, in a frame of reference moving with speed c_0, these are seen to satisfy the constant coefficient equation

$$(I - \partial_y^2)(\partial_t - c_0\partial_y)u + \partial_y u = 0. \tag{3.12}$$

Solutions of (3.12) are superpositions of plane waves, $e^{i(kx-\omega(k)t)}$, with group velocity $\omega'(k) < 0$, implying that dispersive waves travel to the left. Therefore, the dominant solitary wave appears to outrun the kinds of distortions generated by a small local perturbation. Hence, if we *ride with* the solitary wave, then locally we will see asymptotic convergence to a solitary wave, of possibly different speed, c_+ (not necessarily equal to c_0) and phase shift, γ_+. Note that structures which escape behind the dominant solitary wave (e.g. small solitary waves) will not necessarily decay to zero, but a measure of the perturbation, localized say about the peak of the dominant solitary wave, should decay to zero.

The mathematical tool which captures this effect is an appropriate *local energy norm.* We introduce, for any $a \geq 0$, the function space:

$$H_a^1 = \{v \mid e^{ax}v(x) \in H^1(\mathbb{R})\}, \quad \text{with} \quad \|v\|_{H_a^1} = \|e^{ax}v\|_{H^1}. \tag{3.13}$$

The theorem on asymptotic stability of solitary waves of RLW is as follows.

Theorem 3.1 *[16],[17]: Let $u_c(x - ct + \gamma)$ be an RLW solitary wave, where $c > 1$ and $\gamma \in \mathbb{R}$. Consider the initial value problem for RLW with data*

$$u_0(x) = u(x,0) = u_{c_0}(x + \gamma_0) + v_0(x) \in H^2 \cap H_a^1, \tag{3.14}$$

where

$$0 < a < a_*(c) = \sqrt{\frac{2}{2+\sqrt{8c+1}}}\sqrt{\frac{c-1}{c}}. \tag{3.15}$$

(a) There exist $C > 0$, $\epsilon > 0$, $c_* > 1$, and some b with $-ac_0 + a/(1-a^2) < -b < 0$, so that if $c_0 \in (1, c_*)$ and $\|v_0\|_{H^1} + \|v_0\|_{H_a^1} < \epsilon$, then for all $t \geq 0$ we have

$$\begin{aligned} \|u(\cdot,t) - u_{c_+}(\cdot - c_+t + \gamma_+)\|_{H^1} &\leq C\epsilon, \\ \|u(\cdot + c_+t - \gamma_+, t) - u_{c_+}(\cdot)\|_{H_a^1} &\leq C\epsilon e^{-bt} \end{aligned} \tag{3.16}$$

for some $c_+ > 1$, $\gamma_+ \in \mathbb{R}$ *with* $|c_0 - c_+| < C\epsilon$, $|\gamma_0 - \gamma_+| < C\epsilon$.
(b) The conclusion of (a) holds for all speeds $c \geq c_*$, *except possibly for an exceptional set of values which has no finite accumulation point.*

Finally, we state the asymptotic stability results for the KdV- type and NLS equations.

Theorem 3.2 *[20] Consider the generalized Korteweg - de Vries equation:*

$$\partial_t u + u^p \partial_x u + \partial_x^3 u = 0. \tag{3.17}$$

Let $u_c(x - ct + \gamma)$, $c > 0$, $\gamma \in \mathbb{R}$, *be a solitary wave solution. Suppose* $0 < a < \sqrt{c/3}$.
(a) Let $0 < b < a(c - a^2)$. *Then there exists* $C > 0$ *such that if* ϵ *is sufficiently small, we have the following. Consider the initial value problem for KdV* ($p = 1$) *and mKdV* ($p = 2$) *with data*

$$u(x, 0) = u_c(x + \gamma) + v_0(x). \tag{3.18}$$

Assume that $v_0 \in H^2 \cap H^1_a$, *with* $||v_0||_{H^1} + ||v_0||_{H^1_a} < \epsilon$. *Then there exist* $c_+ > 0$ *and* $\gamma_+ \in \mathbb{R}$, *such that the estimates (3.17) hold.*
(b) Suppose $3 \leq p < 4$ *and that* p *does not lie in a certain finite exceptional set* $\mathbf{E}$ *(see [20]). Then, there exists a number,* $-\lambda_\#(p) \to 0$ *as* $p \to 4^-$, *such that for any* b *satisfying* $-\lambda_\#(p) < -b < 0$, *the conclusion of part (a) holds. Therefore, the rate of decay of perturbations degrades as* $p \to 4^-$.

Remark: It is known that for $p > 4$ the solitary wave of gKdV is unstable [7], [23]. The number, $-\lambda_\#(p)$, is a *resonance pole*; it is an isolated singularity of the resolvent (**not** an eigenvalue) of the linearized operator, $\partial_y \left(-\partial_y^2 + c - u_c^p(y)\right)$ about the solitary wave. It is found by analytic continuation of the resolvent across the essential spectrum; see [20], [23].

Nonlinear Schrödinger equations (NLS): We only sketch a statement of the results and refer to [26],[27] for a technically precise formulation. In this work we consider a class of nonlinear Schrödinger equations

$$\begin{aligned} i\partial_t \Phi &= (-\Delta + V)\,\Phi + \lambda|\Phi|^{m-1}\Phi, \quad 1 < m < \frac{n+2}{n-2} \qquad (3.19) \\ \Phi(x, 0) &= \Phi_0(x) \in H^1(\mathbb{R}^n). \end{aligned}$$

The nonlinear bound states in this problem are solitary standing waves of the form: $\psi_E(x)e^{-iEt}$, where ψ_E satisfies the equation

$$\left(-\Delta + V + \lambda|\psi_E|^{m-1}\right)\psi_E = E\psi_E. \tag{3.20}$$

The basic hypotheses on the potential, $V(x)$, are:
(i) V is sufficiently smooth and decaying at infinity
(ii) V is "nonresonant", a condition which holds generically, and
(iii) the linear Schrödinger operator, $H = -\Delta + V(x)$, acting on L^2 has exactly one negative eigenvalue $E_* < 0$, with corresponding L^2 normalized eigenfunction ψ_*.

Hypothesis (iii) implies the existence of a family of nonlinear bound states, solutions $\{\psi_E e^{i\gamma}\}$, of (3.20), which bifurcate from the zero solution at $(E, \psi) = (E_*, = 0)$, where E lies between E_* and $E_* + sgn(\lambda)\eta$ for η sufficiently small and $\gamma \in [0, 2\pi)$.

The asymptotic stability theorem states that if we consider NLS with the initial data

$$\Phi(0) = (\psi_{E_0} + \phi_0)\, e^{i\gamma_0} \tag{3.21}$$

for which ϕ_0 is sufficiently small in H^1, then for $t \neq 0$ the solution has the structure:

$$\Phi(t) = e^{-i\int^t E(s)ds + i\gamma(t)} \left(\psi_{E(t)} + \phi(t)\right), \tag{3.22}$$

where $E(t) \to E^{\pm}$ and $\gamma(t) \to \gamma^{\pm}$ as $t \to \pm\infty$, and where the dispersive part of the solution, $\phi(t)$ decays to zero in L^{m+1} and in a local energy sense:

$$|| < x >^{-\sigma} \phi(t) ||_{L^2} = \mathcal{O}\left(< t >^{-1-\delta}\right), \ \delta > 0$$

Here, $< x > \equiv \sqrt{1 + |x|^2}$.

4 Strategy for proving asymptotic stability

In this section I sketch the strategy of proof used for the above theorems. Suppose we have a nonlinear evolution equation

$$\partial_t u = Au + N(u), \tag{4.23}$$

for which the initial value problem is well posed in an appropriate function space. Here, A denotes a linear operator and $N(\cdot)$ is a nonlinear operator. Assume also that the equation (4.23) has a family of equilibria, indexed by a parameter, $\vec{p}$,

$$\mathcal{M} \equiv \{u_*(x; \vec{p}) : \ \vec{p} \in \mathcal{P}\},$$

and that $u(x; \vec{p})$ varies smoothly as $\vec{p}$ varies continuously over a set $\mathcal{P}$. (Note that by a linear, time-dependent change of spatial variable or phase, solitary traveling waves of RLW and KdV, and standing waves of NLS can be viewed as equilibria.) We seek to understand the dynamical system (4.23) in a neighborhood of $\mathcal{M}$. If the initial data lies close to $\mathcal{M}$, then it is natural to attempt to represent the solution in the form

$$u(x,t) = u_*(x, \vec{p}(t)) + v(x,t). \tag{4.24}$$

The change of variables $u(x,t) \longrightarrow (\vec{p}(t), v(x,t))$ can be thought of as an introduction of coordinates in a neighborhood of the "center" manifold of equilibria. Thus, we view the solution as being composed of a part which is an evolution along the manifold of equilibria, $\mathcal{M}$, and a part which is "normal" to $\mathcal{M}$. The evolution of $\vec{p}(t)$, which specifies the dynamics along $\mathcal{M}$, is to be determined so that $v(\cdot, t)$ decays in a suitable sense. The time-dependent parameters $\vec{p}(t)$ are sometimes referred to as *collective coordinates*; see for example [12], [13], [18],[29].

The equation governing v is:

$$\begin{aligned} \partial_t v &= \mathcal{A}v + \sum_j \partial_{p_j} u_*(x, \vec{p}(t)) \partial_t p_j(t) + \mathcal{F}(v, \vec{p}), \ \ \textit{where} \qquad (4.25) \\ \mathcal{A} &= A + N'\left(u_*(\cdot, \vec{p}_r)\right). \end{aligned}$$

Here we have linearized about some *reference* equilibrium, $u_*(\cdot, \vec{p}_r)$.

The key steps in the proof of asymptotic stability are:

(I) *Linear spectral theory of* $\mathcal{A}$: Since the systems under consideration are Hamiltonian, Lyapunov stability implies that the L^2 spectrum lies on the imaginary axis. More refined spectral analysis is needed, however. To obtain local decay estimates, properties of the operator in weighted spaces are needed. The spectral theory in weighted spaces, reveals the sense in which we can expect the perturbation of the solitary wave to decay to zero. For KdV type equations and RLW equations the operator is consider in the space $L^2(e^{ay}\ dy)$. For NLS, the appropriate space is $L^2(< x >^{-\sigma}\ dx)$.

Symmetries of the equation (e.g. translation invariance in space or phase) and infinitesimal variations in the speed or frequency give rise to a generalized eigenspace at zero energy, spanned by so-called *zero modes*.

(II) *Derivation of coupled equations governing the nondecaying bound state and the decaying "dispersive" parts of the solution*: Introduce a projection operator, P, associated with the zero modes. We require that $Pv(\cdot, t) = 0$ for all $t \geq 0$, and derive from (4.25) the dynamical system:

$$\begin{aligned} \partial_t v &= \mathcal{A}v + (I - P)\,\tilde{\mathcal{F}}(v, \vec{p}), & (4.26) \\ \partial_t \vec{p}(t) &= \mathcal{S}(v, \vec{p}(t)). & (4.27) \end{aligned}$$

This system governs the interaction of the "radiative" part of the solution, v, with the, finite dimensional, " bound state part of the solution".

(III) *Analysis of the linear evolution operator* e^{At}, *in particular, decay estimates in local energy norms*
(IV) *Convergence to an asymptotic nonlinear bound state via á priori estimates of (4.26-4.27)*

5 Asymptotic stability for RLW

In this section an outline of the proof of Theorem 3.1 for RLW is given, with an emphasis on parts (I)-(III) of §4.

The manifold of equilibrium, $\mathcal{M}$, is the family of solitary waves

$$\mathcal{M} = \{u(x - ct + \gamma) : \ c > 1, \ \gamma \in \mathbb{R}\}.$$

Thus, in a manner similar to [20], we seek a solution of RLW in the form of a dominant modulated solitary wave plus a remainder term:

$$u(x,t) = u_{c(t)}\left(x - \int_0^t c(s)\ ds + \gamma(t)\right) + v\left(x - \int_0^t c(s)\ ds + \gamma(t), t\right). \tag{5.28}$$

Here, the speed $c(t)$ and phase $\gamma(t)$ vary with time, and are unknown functions to be determined by our scheme. Note that if $c(t) = c$ and $\gamma(t) = \gamma$ are constant in time, then the solitary wave component $u_{c(t)}(x - \int_0^t c(s)\ ds + \gamma(t))$ reduces to $u_c(x - ct + \gamma)$, an exact solitary wave solution of RLW. At a later stage in the analysis, it is also necessary to make the change of time variable:

$$\tau(t) = \int_0^t c(s)\ ds - \gamma(t). \tag{5.29}$$

This was not required in [20].

Substitution of the Ansatz (5.28) into RLW yields an evolution equation for the perturbation v in the moving frame, $y(x,t) = x - \int_0^t c(s)\ ds + \gamma(t)$:

$$\partial_t v = Av + \mathcal{F}(v; u_{c(t)}, \dot{c}(t), \dot{\gamma}(t)). \tag{5.30}$$

The operator, A, is the linearized operator about the solitary wave, u_{c_0}, in a frame moving at speed $c_0 > 1$ and is given by:

$$\begin{aligned} A &= (I - \partial_y^2)^{-1}\partial_y L_{c_0} \\ L_{c_0} &= -c_0\partial_y^2 + c_0 - 1 - u_{c_0}. \end{aligned} \tag{5.31}$$

Based on the heuristic discussion of §3, we do not expect any L^p norms of v to decay to zero with increasing time. Indeed, we expect small nondecaying solitary waves to be left behind a rightward advancing, "large" solitary wave. For this reason, we study an appropriate spatial localization of the perturbation in the frame of reference moving with the dominant wave. In particular we study the evolution of the *weighted perturbation* $w = e^{ay}v$, which satisfies the equation:

$$\partial_t w = A_a w + \mathcal{G}, \tag{5.32}$$

where $\mathcal{G}$ is nonlinear and $A_a = e^{ay}Ae^{-ay}$. The goal is now to show that $\|v(\cdot,t)\|_{H^1}$ remains bounded and $\|w(\cdot,t)\|_{H^1}$ decays to zero as $t \to \infty$.
(I) *Linear spectral theory*: A key ingredient in the proof is a detailed spectral analysis of the operators A and A_a in the space $L^2(\mathbb{R})$. Under the transformation $A \mapsto A_a$, where a is as in Theorem 3.1, the essential spectrum of A is mapped to a curve in open left half plane with a vertical asymptote. To the right of this curve, the only point in the spectrum is $\lambda = 0$; discrete eigenvalues remain eigenvalues in the weighted space.

In particular, it must be shown that the linearized eigenvalue problem $AY = \lambda Y$ (as well as $A_aY = \lambda Y$) has no nonzero eigenvalues in the *closed* right half plane. Thus, the only neutral eigenvalue allowable is $\lambda = 0$ with eigenmodes derived from symmetries of the equation. This *strong spectral stability property* [21],[22] plays a central role in proofs of asymptotic stability for other nonlinear dispersive systems [26],[27],[28], [20]. The key tool in this analysis is the Evans function, $D_{RLW}(\lambda,\gamma)$, an analytic function of the eigenvalue parameter, λ, and the parameter $\gamma^2 \equiv (c-1)/c$, whose zeros, λ, in the closed right half plane are L^2 eigenvalues of the RLW linearized eigenvalue problem [20],[23],[1],[9]. $D(\lambda)$ has an interpretation as a transmission coefficient at energy λ, when viewing the eigenvalue equation as a scattering problem.

We consider the cases of small and large $|\lambda|$ separately. For small values of $|\lambda|$, we exploit the convergence of RLW to KdV under a certain scaling, an approach used recently by Pego and Weinstein to study Boussinesq type equations [22], [21]. Using this approach we find that the scaled Evans function D_{RLW} converges to the Evans function associated with the KdV equation. The latter can be explicitly computed [20], and is seen to have zero as its only root, a generalized eigenvalue of multiplicity two. The continuity of $D_{RLW}(\lambda;\gamma)$ and the argument principle for analytic functions enable us to conclude that in a neighborhood of the origin, zero is the only eigenvalue of the RLW stability problem, for $0 < (c-1)/c \equiv \gamma^2$ sufficiently small.

For large values of $|\lambda|$, we first relate the resolvent of A_a to the resolvent of its constant coefficient part A_a^∞ by compact operator $C(\lambda,\gamma)$:

$$(\lambda - A_a)^{-1} = (I - C(\lambda,\gamma))^{-1} (\lambda - A_a^\infty)^{-1}. \tag{5.33}$$

We then use norm estimates on $C(\lambda,\gamma)$ for γ small and λ large to show that the norm of $C(\lambda,\gamma)$ is smaller than one and therefore that A_a has no eigenvalues for λ sufficiently large. This latter argument is related

to an alternative WKB - type analysis in [21], where it is shown for large λ that $D(\lambda) \sim 1$.
(II) *Coupled evolution equations for the bound state and decaying "dispersive" parts of the solution*: To avoid secular growth of w and v as $t \to \infty$, we require that $Pw \equiv 0$, where P denotes the projection onto the (two-dimensional) generalized nullspace of A_a. This requirement gives rise to a system of ordinary differential equations for $c(t)$ and $\gamma(t)$, called *modulation equations*, which are coupled to the infinite-dimensional evolution equation for w. In this way, the original RLW is decomposed into a system of three equations, governing the interaction of the dispersive and bound components of the solution:

$$\begin{aligned} \partial_t w &= A_a w + (I - P)\,\mathcal{G}\,(v, w, c(t), \gamma(t)) && (5.34)\\ c'(t) &= \mathcal{S}_c(v, c, \gamma) && (5.35)\\ \gamma'(t) &= \mathcal{S}_\gamma(v, c, \gamma). && (5.36) \end{aligned}$$

(III) *Decay estimates for the evolution operator* e^{At}:
The next step is to show that the operator A_a generates a semigroup with exponentially decaying H^1 norm on the spectral complement of its zero eigenspace. That is, solutions of the linear equation $\partial_t w = A_a w$ with initial data in the range of $Q \equiv I - P$ decay exponentially to zero in H^1 as $t \to \infty$. To prove this decay estimate, we use a criterion due to Prüss [25] which requires bounds on the operator norm of the resolvent $(\lambda I - A_a)^{-1}$. In particular, we show

$$\Re(\lambda) > -b \quad \Longrightarrow \quad \left\|(\lambda I - A_a Q)^{-1}\right\| \le M \tag{5.37}$$

for some $M < \infty$, $-b < 0$, and this implies the estimate:

$$||e^{A_a t} Q f||_{H^1} \le C e^{-bt} ||f||_{H^1}. \tag{5.38}$$

The proof of (5.37) requires operator estimates like those used to rule out large eigenvalues.

References

[1] J. Alexander, R. Gardner, and C.K.R.T. Jones, *A topological invariant arising in the stability analysis of traveling waves*, J. Reine Angew. Math. **410** 167-212 (1990)

[2] V.I. Arnold, *Conditions for nonlinear stabiliity of stationary curvilinear flows of an ideal fluid*, Dokl. Mat. Nauk. **162**, 773-777 (1965)

[3] T.B. Benjamin, *The stability of solitary waves*, Proc. Roy. Soc. Lond. A **328** 153-183 (1972)

[4] T.B. Benjamin, J.L. Bona and J.J Mahony, *Model equations for long waves in nonlinear dispersive systems*, Phil. Trans. Roy. Soc. Lond. **A272** 47-78 (1972)

[5] M.J. Boussinesq, *Essai sur la théorie des eaux courantes, Mémoire présentés par divers savants á l'Académie des Sciences Inst. France (séries 2)* **23**, 1-680 (1877)

[6] J.L. Bona, *On the stability theory of solitary waves*, Proc. Roy. Soc. Lond. **A344** 363-374 (1975)

[7] J.L. Bona, P.E. Souganidis, and W.A. Strauss, *Stability and instability of solitary waves of Korteweg - de Vries type,* Proc. Roy. Soc. Lond. **A411** 395-412 (1987)

[8] T. Cazenave, P.-L. Lions, *Orbital stability of ground states of nonlinear Schrödinger equations*, Commun. Math. Phys. **85**, 549-561 (1982)

[9] J.W. Evans, *Nerve axon equations, IV: The stable and unstable impulse*, Indiana J. Math. **24**, 1169-1190 (1975)

[10] M. Grillakis, J. Shatah, W. Strauss, *Stability of solitary waves in the presence of symmetry, I.* J. Func. Anal. **74**, 160-197 (1988)

[11] D.D. Holm, J.E. Marsden, T. Ratiu, and A. Weinstein, *Nonlinear stability of fluid and plasma equilibria*, Physics Reports **123**, 1-116 (1985)

[12] J.P. Keener and D.W. McLaughlin, *Solitons under perturbation*, Phys. Rev. **A16**, 777-790 (1977)

[13] Y. Kodama, and M.J. Ablowitz, *Perturbations of solitons and solitary waves,* Stud. in Appl. Math. **64**, 225-245 (1981)

[14] D.J. Korteweg and G. de Vries, *On the change of form of long waves advancing in a rectangular channel, and on a new type of long stationary waves*, Phil. Mag. **39**, 422-443 (1895)

[15] E.W. Laedke and K.H. Spatschek, *Stability theorem for KdV type equations*, J. Plasma Phys. **32** 263-272 (1984)

[16] J.R. Miller, *Asymptotic stability for the regularized long wave equation*, Ph.D. thesis, University of Michigan, 1994.

[17] J.R. Miller and M.I. Weinstein, *Asymptotic stability of solitary waves of the regularized long wave equation*, Comm. Pure Appl. Math. **49** (1996), no. 4, 399–441.

[18] A.C. Newell, *Near-integrable systems, nonlinear tunneling and solitons in slowly changing media,* pp. 127–129 in: *Nonlinear Evolution Equations Solvable by the Inverse Spectral Transform*, F. Calogero, ed., Pitman, London, 1978.

[19] R.L. Pego and M.I Weinstein, *On asymptotic stability of solitary waves*, Phys. Lett. A **162**, 263–268 (1992)

[20] R.L. Pego and M.I Weinstein, *Asymptotic stability of solitary waves*, Commun. Math. Phys. **164**, 305-349 (1994)

[21] R.L. Pego and M.I Weinstein, *Convective linear stability of solitary waves for Boussinesq equations*, submitted to Studies in Applied Mathematics.

[22] R.L. Pego and M.I Weinstein, *On strong spectral stability of solitary waves of Boussinesq equations*, Proceedings of IUTAM Symposium on Nonlinear Waves in Fluid Systems, Hannover, 1994.

[23] R.L. Pego and M.I Weinstein, *Eigenvalues and solitary wave instability*, Phil. Trans. Roy. Soc. Lond. **A340**, 47-94 (1992)

[24] D.H. Peregrine, *Calculations of the development of an undular bore*, J. Fluid Mech. **25**, 321 (1966)

[25] J. Prüss, *On the spectrum of* C_0 *semigroups*, Trans. Amer. Math. Soc. **284**, 847-857 (1984)

[26] A. Soffer and M.I. Weinstein, *Multichannel nonlinear scattering for nonintegrable equations*, in Integrable Systems and Applications, Balabane, M., Lochak, P. and Sulem, C., eds., Springer Lec. Notes in Physics No. 34 (1988)

[27] A. Soffer and M.I. Weinstein, *Multichannel nonlinear scattering for nonintegrable equations*, Comm. Math. Phys. **133**, 119-146 (1990)

[28] A. Soffer and M.I. Weinstein, *Multichannel nonlinear scattering for nonintegrable equations, II: The case of anisotropic potentials and data*, J. Diff. Eq. **98**, 376-390 (1992)

[29] M.I. Weinstein, *Modulational stability of ground states of nonlinear Schrödinger equations,* SIAM J. Math. Anal. **16**, 472-491 (1985)

[30] M.I. Weinstein, *Lyapunov stability of ground states of nonlinear dispersive evolution equations*, Commun. Pure Appl. Math. **39**, 51-68 (1986)

[31] M.I. Weinstein, *Existence and dynamic stability of solitary wave solutions of equations arising in long wave propagation,* Comm. Partial Diff. Eq. **12**, 1133-1173 (1987)

Department of Mathematics, University of Michigan, Ann Arbor, MI 48109

Current address: Program in Applied and Computational Mathematics, Princeton University, Princeton, NJ 08544

Selected Titles in This Series

(Continued from the front of this publication)

171 **Rüdiger Göbel, Paul Hill, and Wolfgang Liebert, Editors,** Abelian group theory and related topics, 1994

170 **John K. Beem and Krishan L. Duggal, Editors,** Differential geometry and mathematical physics, 1994

169 **William Abikoff, Joan S. Birman, and Kathryn Kuiken, Editors,** The mathematical legacy of Wilhelm Magnus, 1994

168 **Gary L. Mullen and Peter Jau-Shyong Shiue, Editors,** Finite fields: Theory, applications, and algorithms, 1994

167 **Robert S. Doran, Editor,** C^*-algebras: 1943–1993, 1994

166 **George E. Andrews, David M. Bressoud, and L. Alayne Parson, Editors,** The Rademacher legacy to mathematics, 1994

165 **Barry Mazur and Glenn Stevens, Editors,** p-adic monodromy and the Birch and Swinnerton-Dyer conjecture, 1994

164 **Cameron Gordon, Yoav Moriah, and Bronislaw Wajnryb, Editors,** Geometric topology, 1994

163 **Zhong-Ci Shi and Chung-Chun Yang, Editors,** Computational mathematics in China, 1994

162 **Ciro Ciliberto, E. Laura Livorni, and Andrew J. Sommese, Editors,** Classification of algebraic varieties, 1994

161 **Paul A. Schweitzer, S. J., Steven Hurder, Nathan Moreira dos Santos, and José Luis Arraut, Editors,** Differential topology, foliations, and group actions, 1994

160 **Niky Kamran and Peter J. Olver, Editors,** Lie algebras, cohomology, and new applications to quantum mechanics, 1994

159 **William J. Heinzer, Craig L. Huneke, and Judith D. Sally, Editors,** Commutative algebra: Syzygies, multiplicities, and birational algebra, 1994

158 **Eric M. Friedlander and Mark E. Mahowald, Editors,** Topology and representation theory, 1994

157 **Alfio Quarteroni, Jacques Periaux, Yuri A. Kuznetsov, and Olof B. Widlund, Editors,** Domain decomposition methods in science and engineering, 1994

156 **Steven R. Givant,** The structure of relation algebras generated by relativizations, 1994

155 **William B. Jacob, Tsit-Yuen Lam, and Robert O. Robson, Editors,** Recent advances in real algebraic geometry and quadratic forms, 1994

154 **Michael Eastwood, Joseph Wolf, and Roger Zierau, Editors,** The Penrose transform and analytic cohomology in representation theory, 1993

153 **Richard S. Elman, Murray M. Schacher, and V. S. Varadarajan, Editors,** Linear algebraic groups and their representations, 1993

152 **Christopher K. McCord, Editor,** Nielsen theory and dynamical systems, 1993

151 **Matatyahu Rubin,** The reconstruction of trees from their automorphism groups, 1993

150 **Carl-Friedrich Bödigheimer and Richard M. Hain, Editors,** Mapping class groups and moduli spaces of Riemann surfaces, 1993

149 **Harry Cohn, Editor,** Doeblin and modern probability, 1993

148 **Jeffrey Fox and Peter Haskell, Editors,** Index theory and operator algebras, 1993

147 **Neil Robertson and Paul Seymour, Editors,** Graph structure theory, 1993

146 **Martin C. Tangora, Editor,** Algebraic topology, 1993

145 **Jeffrey Adams, Rebecca Herb, Stephen Kudla, Jian-Shu Li, Ron Lipsman, and Jonathan Rosenberg, Editors,** Representation theory of groups and algebras, 1993

144 **Bor-Luh Lin and William B. Johnson, Editors,** Banach spaces, 1993

(See the AMS catalog for earlier titles)